T0320647

Plasmonic Nanoelectronics and Sensing

Plasmonic nanostructures provide new ways of manipulating the flow of light, with nanostructures and nanoparticles exhibiting optical properties never before seen in the macro-world. Covering plasmonic technology from fundamental theory to real-world applications, this work provides a comprehensive overview of the field.

- Discusses the fundamental theory of plasmonics, enabling a deeper understanding of plasmonic technology
- Details numerical methods for modeling, design, and optimization of plasmonic nanostructures
- Includes step-by-step design guidelines for active and passive plasmonic devices, demonstrating the implementation of real devices in the standard CMOS nanoscale electronic–photonic integrated circuit to help cut design, fabrication, and characterization time and cost
- Includes real-world case studies of plasmonic devices and sensors, explaining the benefits and downsides of different nanophotonic integrated circuits and sensing platforms.

Ideal for researchers, engineers, and graduate students in the fields of nanophotonics and nanoelectronics as well as optical biosensing.

Er-Ping Li is a Principal Scientist and Director of Nanophotonics and Electronics at the Institute of High Performance Computing, A*STAR, Singapore. He is a Fellow of the IEEE and of the Electromagnetics Academy, USA.

Hong-Son Chu is a Scientist at the Nanophotonics and Electronics Department of the Institute of High Performance Computing, A*STAR, Singapore. He is a member of the Optical Society of America, the IEEE, and the Materials Research Society.

EuMA High Frequency Technologies Series

Series Editor
Peter Russer, Technical University of Munich

Homayoun Nikookar, *Wavelet Radio*
Thomas Zwick, Werner Wiesbeck, Jens Timmermann, and Grzegorz Adamiuk (Eds), *Ultra-wideband RF System Engineering*
Er-Ping Li and Hong-Son Chu, *Plasmonic Nanoelectronics and Sensing*

Forthcoming

Peter Russer, Johannes Russer, Uwe Siart, and Andreas Cangellaris, *Interference and Noise in Electromagnetics*
Maurizio Bozzi, Apostolos Georgiadis, and Ke Wu, *Substrate Integrated Waveguides*
Luca Roselli (Ed.), *Green RFID Systems*
George Deligeorgis, *Graphene Device Engineering*
Luca Pierantoni and Fabio Coccetti, *Radiofrequency Nanoelectronics Engineering*
Alexander Yarovoy, *Introduction to UWB Wireless Technology and Applications*

Plasmonic Nanoelectronics and Sensing

ER-PING LI and HONG-SON CHU
A*STAR Institute of High Performance Computing, Singapore

Shaftesbury Road, Cambridge CB2 8EA, United Kingdom

One Liberty Plaza, 20th Floor, New York, NY 10006, USA

477 Williamstown Road, Port Melbourne, VIC 3207, Australia

314–321, 3rd Floor, Plot 3, Splendor Forum, Jasola District Centre, New Delhi – 110025, India

103 Penang Road, #05–06/07, Visioncrest Commercial, Singapore 238467

Cambridge University Press is part of Cambridge University Press & Assessment, a department of the University of Cambridge.

We share the University's mission to contribute to society through the pursuit of education, learning and research at the highest international levels of excellence.

www.cambridge.org
Information on this title: www.cambridge.org/9781107027022

First published 2014

A catalogue record for this publication is available from the British Library

ISBN 978-1-107-02702-2 Hardback

Contents

Contributors

1 Fundamentals of plasmonics and
2 Plasmonic properties of metal nanostructures
Yuriy A. Akimov
*A*STAR Institute of High Performance Computing, Singapore*

3 Frequency-domain methods for modeling plasmonics
Zhengtong Liu
*A*STAR Institute of High Performance Computing, Singapore*

4 Time-domain simulation for plasmonic devices
Iftikhar Ahmed and Eng Huat Khoo
*A*STAR Institute of High Performance Computing, Singapore*

6 Silicon-based active plasmonic devices for on-chip integration
Dim-Lee Kwong, Guo-Qiang Lo, and Shiyang Zhu
*A*STAR Institute of Microelectronics, Singapore*

7 Plasmonic biosensing devices and systems
Lin Wu and Ping Bai
*A*STAR Institute of High Performance Computing, Singapore*

Preface

Data communication and information processing are driving the rapid development of ultra-high speed and ultra-compactness in nano-photo-electronic integration. Plasmonics technology has in recent years demonstrated the promise to overcome the size mismatch between microscale photonic and nanoscale electronic integration, and it likely will be crucial for the next generation of on-chip optical nano-interconnects, enabling the deployment of small-footprint and low-energy integrated circuitry.

The phenomenon of surface plasmons was first observed in the Lycurgus cup, which is a Roman glass cage cup in the British Museum, London, UK. This special cup is made of a dichroic glass that shows a different color depending on the condition of illumination. Specifically, in daylight, the cup appears to have a green color, which means that light is being reflected from the cup; however, when a light is shone into the cup and transmitted through the glass, it appears to have a red color. Today, we know that this fascinating behavior is due to nanoscopic-scale gold and silver particles embedded in the glass. However, it took 1500 years and doubtless countless fantastic interpretations for a plausible explanation to emerge. In the last few decades, the phenomenon of surface plasmons has been extensively studied both theoretically and experimentally, and there have been attempts to use it for various applications ranging from solar-cell energy and sensing to nanophotonic devices.

This book presents the results from many years of our collective research in the fields of nanoplasmonics and its applications. It presents state-of-the-art plasmonics device modeling and design techniques, with novel developments in particular in CMOS-compatible integrated circuits and sensing technologies. We hope this book can serve as a good basis for further progress in this field, both in academic research and for industrial applications. The book consists of seven chapters, contributed by Yuriy Akimov, Zhengtong Liu, Iftikhar Ahmed, Eng Huat Khoo, Er-Ping Li, Hong-Son Chu, Wu Lin, and Bai Ping, from the Department of Electronics and Photonics, Institute of High Performance Computing, Singapore, and Shiyang Zhu, Patrick Guo-Qiang Lo, and Dim-Lee Kwong from the Institute of Microelectronics, Agency for Science Technology and Research, Singapore.

Chapter 1 introduces the fundamentals of plasmonics associated with Maxwell's theory and applications in plasmonics. Chapter 2 provides an introduction to the plasmonic properties of metal nanostructures. Chapter 3 presents the modeling and simulation of plasmonics associated with plasmonic devices by implementation of frequency-domain numerical methods. In Chapter 4, time-domain simulation methods, in particular the finite-difference time-domain method, are introduced for passive and active plasmonic device design. Chapter 5 describes the development of various passive plasmonic waveguides, in particular CMOS-compatible devices for on-chip nanoelectronic integration, and Chapter 6 presents CMOS-compatible active plasmonic devices for on-chip nanoelectronic integration. Both theoretical studies and experimental results are presented in these two chapters. The recent development of plasmonics for biosensing applications is presented in Chapter 7.

We gratefully acknowledge the research support from the Agency for Science Technology and Research, Singapore. Also acknowledged are the contributors to the book, Drs. Yuriy Akmov, Zhengtong Liu, Iftikhar Ahmed, Eng Huat Khoo, Wu Lin, Bai Ping, Shiyang Zhu, and Patrick Guo-Qiang Lo and Professor Dim-Lee Kwong, who did the really hard work. We also wish to express our gratitude to Mia Balashova and Julie Lancashire from Cambridge University Press for their great assistance in keeping us on schedule. Finally, we are grateful to all the contributors' families, without whose continuing support and understanding this book would never have been published.

We hope that this book will serve as a valuable reference for engineers, researchers, and post-graduate students in the fields of nanophotonics and nanoelectronics as well as optical biosensing. Even though much has been accomplished in these fields, we predict that many more exciting challenges will arise in these areas.

Er-Ping LI and Hong-Son CHU

1 Fundamentals of plasmonics

In this chapter, we give a brief introduction to the classical electrodynamics of metals that constitutes the basis of modern plasmonics. We review the Maxwell equations for electromagnetic fields and consider the main optical properties of metals within the local-response approximation. In conclusion, we give a general classification of plasmons that appear in metal structures.

1.1 Electromagnetic field equations

1.1.1 Maxwell's equations in a medium

Most of the electromagnetic phenomena occurring in metals are well described within the classical electrodynamics based on the *macroscopic Maxwell equations*. These equations assume the use of the statistically averaged (over an ensemble of the equivalent systems) electric and magnetic fields. Practically, the averaging is performed in space over "physically small" volumes, which are much smaller than the wavelength, but much longer than the mean interatomic distance. Within this approach, we neglect all field fluctuations that occur at atomic scales and consider only the macroscopic response of the medium.

In the absence of external charges and currents, the macroscopic Maxwell equations for electromagnetic fields in a medium can be written as follows:[1]

$$\nabla \times \boldsymbol{E} = -\frac{1}{c}\frac{\partial \boldsymbol{B}}{\partial t}, \qquad \nabla \cdot \boldsymbol{E} = 4\pi\rho, \tag{1.1}$$

$$\nabla \times \boldsymbol{B} = \frac{1}{c}\frac{\partial \boldsymbol{E}}{\partial t} + \frac{4\pi}{c}\boldsymbol{j}, \qquad \nabla \cdot \boldsymbol{B} = 0, \tag{1.2}$$

where $\boldsymbol{E}$ is the electric field, $\boldsymbol{B}$ is the magnetic induction, ρ is the induced internal charge density, $\boldsymbol{j}$ is the induced electric current density, and c is the speed of light in vacuum. The induced charges and currents comprise the medium's response to the electromagnetic field, as a result of its polarization and magnetization.

[1] Throughout this chapter, all quantities and equations are written in the Gaussian unit system for a more natural description of electromagnetic fields. For conversion to the SI unit system, we refer the reader to the textbook by Jackson [1].

In general, the induced charges and currents are given with the *polarization* $\boldsymbol{P}$ and *magnetization* $\boldsymbol{M}$ fields,

$$\rho = -\nabla \cdot \boldsymbol{P}, \tag{1.3}$$

$$\boldsymbol{j} = \frac{\partial \boldsymbol{P}}{\partial t} + c\nabla \times \boldsymbol{M}, \tag{1.4}$$

that allow us to rewrite the macroscopic Maxwell equations in a simpler form,

$$\nabla \times \boldsymbol{E} = -\frac{1}{c}\frac{\partial \boldsymbol{B}}{\partial t}, \quad \nabla \cdot \boldsymbol{D} = 0, \tag{1.5}$$

$$\nabla \times \boldsymbol{H} = \frac{1}{c}\frac{\partial \boldsymbol{D}}{\partial t}, \quad \nabla \cdot \boldsymbol{B} = 0, \tag{1.6}$$

where $\boldsymbol{D}$ and $\boldsymbol{H}$ are the auxiliary fields called the electric displacement and magnetic field, which are introduced to account for the polarization and magnetization of the medium,

$$\boldsymbol{D} = \boldsymbol{E} + 4\pi \boldsymbol{P}, \tag{1.7}$$

$$\boldsymbol{H} = \boldsymbol{B} - 4\pi \boldsymbol{M}. \tag{1.8}$$

Thus, the Maxwell equations in a medium give us the relation between two pairs of electric $\{\boldsymbol{E}, \boldsymbol{D}\}$ and magnetic $\{\boldsymbol{B}, \boldsymbol{H}\}$ fields. In this sense, Eqs. (1.5) and (1.6) do not form a closed set of equations until we provide *material relations* for the medium's response to electric and magnetic fields. In general, these relations are given by field-dependent functions for polarization $\boldsymbol{P} = \boldsymbol{P}(\boldsymbol{E})$ and magnetization $\boldsymbol{M} = \boldsymbol{M}(\boldsymbol{B})$ vectors that eventually result in material relations for the auxiliary fields $\boldsymbol{D} = \boldsymbol{D}(\boldsymbol{E})$ and $\boldsymbol{H} = \boldsymbol{H}(\boldsymbol{B})$.

1.1.2 Material equations

Establishing the relations for $\boldsymbol{D}(\boldsymbol{E})$ and $\boldsymbol{H}(\boldsymbol{B})$ is the key issue, since it describes how the medium responds to electromagnetic fields. In general, these relations are nonlinear. However, for $\boldsymbol{E}$ and $\boldsymbol{B}$ fields that are not too high, the auxiliary fields $\boldsymbol{D}(\boldsymbol{E})$ and $\boldsymbol{H}(\boldsymbol{B})$ can be approximated with linear functions. This is the so-called *linear-electrodynamics* approach. In this approximation, the response of the medium at a given point $\boldsymbol{r}$ and moment t is assumed to be a linear function of electromagnetic fields at any point $\boldsymbol{r}'$ taken at all preceding moments $t' < t$ in accordance with the causality principle,

$$D_i(t, \boldsymbol{r}) = \int_{-\infty}^{t} \mathrm{d}t' \int \mathrm{d}\boldsymbol{r}'\, \varepsilon_{ij}(t, t', \boldsymbol{r}, \boldsymbol{r}') E_j(t', \boldsymbol{r}'), \tag{1.9}$$

$$B_i(t, \boldsymbol{r}) = \int_{-\infty}^{t} \mathrm{d}t' \int \mathrm{d}\boldsymbol{r}'\, \mu_{ij}(t, t', \boldsymbol{r}, \boldsymbol{r}') H_j(t', \boldsymbol{r}'). \tag{1.10}$$

Note that here we write the dependence $\boldsymbol{B}(\boldsymbol{H})$ instead of $\boldsymbol{H}(\boldsymbol{B})$. It is caused by the symmetry of Maxwell's equations observed with respect to the field pairs $\{\boldsymbol{E}, \boldsymbol{D}\}$ and

$\{\boldsymbol{H}, \boldsymbol{B}\}$. Following this symmetry, it is more natural to consider the dependence $\boldsymbol{B}(\boldsymbol{H})$, rather than $\boldsymbol{H}(\boldsymbol{B})$. For this reason, the field $\boldsymbol{H}$ is commonly called the magnetic field, by analogy with the electric field $\boldsymbol{E}$, although it is actually an auxiliary quantity.

The functions $\varepsilon_{ij}(t, t', \boldsymbol{r}, \boldsymbol{r}')$ and $\mu_{ij}(t, t', \boldsymbol{r}, \boldsymbol{r}')$ in Eqs. (1.9) and (1.10) characterize the efficiency of the material response transfer from one point of space and time to another. For a medium that is homogeneous in space and time, the functions ε_{ij} and μ_{ij} depend on the differences $t - t'$ and $\boldsymbol{r} - \boldsymbol{r}'$. In this case,

$$D_i(t, \boldsymbol{r}) = \int_{-\infty}^{t} \mathrm{d}t' \int \mathrm{d}\boldsymbol{r}' \, \varepsilon_{ij}(t - t', \boldsymbol{r} - \boldsymbol{r}') E_j(t', \boldsymbol{r}'), \tag{1.11}$$

$$B_i(t, \boldsymbol{r}) = \int_{-\infty}^{t} \mathrm{d}t' \int \mathrm{d}\boldsymbol{r}' \, \mu_{ij}(t - t', \boldsymbol{r} - \boldsymbol{r}') H_j(t', \boldsymbol{r}'). \tag{1.12}$$

By performing the Fourier transform,

$$G(t, \boldsymbol{r}) = \frac{1}{(2\pi)^2} \iint G(\omega, \boldsymbol{k}) \mathrm{e}^{\mathrm{i}(\boldsymbol{k}\cdot\boldsymbol{r} - \omega t)} \, \mathrm{d}\omega \, \mathrm{d}\boldsymbol{k},$$

of D_i, E_j, B_i, and H_j in the $(t, \boldsymbol{r})$ space, we get the material relations for the fields in the frequency–wavevector space $(\omega, \boldsymbol{k})$,

$$D_i(\omega, \boldsymbol{k}) = \varepsilon_{ij}(\omega, \boldsymbol{k}) E_j(\omega, \boldsymbol{k}), \tag{1.13}$$

$$B_i(\omega, \boldsymbol{k}) = \mu_{ij}(\omega, \boldsymbol{k}) H_j(\omega, \boldsymbol{k}). \tag{1.14}$$

Here, $\varepsilon_{ij}(\omega, \boldsymbol{k})$ and $\mu_{ij}(\omega, \boldsymbol{k})$ are the tensors of complex permittivity and permeability given by

$$\varepsilon_{ij}(\omega, \boldsymbol{k}) = \int_{0}^{\infty} \mathrm{d}t_1 \int \mathrm{d}\boldsymbol{r}_1 \, \varepsilon_{ij}(t_1, \boldsymbol{r}_1) \mathrm{e}^{-\mathrm{i}(\boldsymbol{k}\cdot\boldsymbol{r}_1 - \omega t_1)}, \tag{1.15}$$

$$\mu_{ij}(\omega, \boldsymbol{k}) = \int_{0}^{\infty} \mathrm{d}t_1 \int \mathrm{d}\boldsymbol{r}_1 \, \mu_{ij}(t_1, \boldsymbol{r}_1) \mathrm{e}^{-\mathrm{i}(\boldsymbol{k}\cdot\boldsymbol{r}_1 - \omega t_1)}, \tag{1.16}$$

where $t_1 = t - t'$ and $\boldsymbol{r}_1 = \boldsymbol{r} - \boldsymbol{r}'$.

For an isotropic medium, the properties of which are identical in any direction, $\varepsilon_{ij}(\omega, \boldsymbol{k})$ and $\mu_{ij}(\omega, \boldsymbol{k})$ can be composed of the unit tensor δ_{ij} and the tensor $k_i k_j$, since they are the only two tensors of second rank formed from the vector $\boldsymbol{k}$. In this case, we have

$$\varepsilon_{ij}(\omega, \boldsymbol{k}) = \left(\delta_{ij} - \frac{k_i k_j}{k^2}\right) \varepsilon_{\mathrm{t}}(\omega, \boldsymbol{k}) + \frac{k_i k_j}{k^2} \varepsilon_{\mathrm{l}}(\omega, \boldsymbol{k}), \tag{1.17}$$

$$\mu_{ij}(\omega, \boldsymbol{k}) = \left(\delta_{ij} - \frac{k_i k_j}{k^2}\right) \mu_{\mathrm{t}}(\omega, \boldsymbol{k}) + \frac{k_i k_j}{k^2} \mu_{\mathrm{l}}(\omega, \boldsymbol{k}). \tag{1.18}$$

Thus, among the nine components of each tensor ε_{ij} and μ_{ij}, only two components are independent, namely $\varepsilon_{\mathrm{t}}(\omega, \boldsymbol{k})$ and $\varepsilon_{\mathrm{l}}(\omega, \boldsymbol{k})$ for ε_{ij}, and $\mu_{\mathrm{t}}(\omega, \boldsymbol{k})$ and $\mu_{\mathrm{l}}(\omega, \boldsymbol{k})$ for μ_{ij}. The meaning of those components becomes clear if we write $\boldsymbol{D}$ and $\boldsymbol{B}$ in vector form,

$$\boldsymbol{D}(\omega,\boldsymbol{k}) = \varepsilon_{\rm t}(\omega,\boldsymbol{k})\frac{\boldsymbol{k}\times(\boldsymbol{E}\times\boldsymbol{k})}{k^2} + \varepsilon_{\rm l}(\omega,\boldsymbol{k})\frac{\boldsymbol{k}(\boldsymbol{E}\cdot\boldsymbol{k})}{k^2},$$

$$\boldsymbol{B}(\omega,\boldsymbol{k}) = \mu_{\rm t}(\omega,\boldsymbol{k})\frac{\boldsymbol{k}\times(\boldsymbol{H}\times\boldsymbol{k})}{k^2} + \mu_{\rm l}(\omega,\boldsymbol{k})\frac{\boldsymbol{k}(\boldsymbol{H}\cdot\boldsymbol{k})}{k^2}.$$

According to these expressions, $\varepsilon_{\rm l}(\omega,\boldsymbol{k})$ and $\mu_{\rm l}(\omega,\boldsymbol{k})$ give the medium response to longitudinal electric ($\boldsymbol{E}\times\boldsymbol{k}=0$) and magnetic ($\boldsymbol{H}\times\boldsymbol{k}=0$) fields, while $\varepsilon_{\rm t}(\omega,\boldsymbol{k})$ and $\mu_{\rm t}(\omega,\boldsymbol{k})$ describe the response to transverse electric ($\boldsymbol{E}\cdot\boldsymbol{k}=0$) and magnetic ($\boldsymbol{H}\cdot\boldsymbol{k}=0$) fields.

1.1.3 Temporal and spatial dispersion in metals

In the general case, both tensors ε_{ij} and μ_{ij} depend on the frequency ω and the wavevector $\boldsymbol{k}$. Eventually, any electromagnetic pulse disperses by propagating in the medium, as the Fourier components

$$G(\omega,\boldsymbol{k})e^{i(\boldsymbol{k}\cdot\boldsymbol{r}-\omega t)}$$

with different ω and $\boldsymbol{k}$ (that comprise the pulse in accordance with the Fourier transform) propagate with different phase velocities ω/k. Thus, materials with properties that exhibit frequency and wavevector dependence are *dispersive*. The frequency dependence of the tensors ε_{ij} and μ_{ij} describes the *temporal dispersion* of electromagnetic fields, while the wavevector dependence gives the *spatial dispersion*.

In the optical range, metals feature very strong temporal dispersion. It arises due to the inertia and friction of electrons in metals that make the polarization and magnetization inertial with respect to electric and magnetic fields. Thus, the metal's response at a given moment t is dependent on the values of the electric and magnetic fields at all preceding moments $t' \le t$.

The time interval $\tau = t - t'$ for which the previous history still has a significant effect is defined by the metal's characteristic frequencies $\omega_{\rm s}$. It is obvious that, for electromagnetic fields oscillating at a very high frequency $\omega \gg \omega_{\rm s}$, the electrons do not have enough time to form any significant polarization and magnetization. Eventually, this results in very weak temporal dispersion with

$$\varepsilon_{ij}(\omega\to\infty) = \delta_{ij}, \quad \mu_{ij}(\omega\to\infty) = \delta_{ij}.$$

However, at frequencies ω below or close to the characteristic frequencies $\omega_{\rm s}$, the temporal dispersion increases and becomes significant.

In general, the characteristic frequencies $\omega_{\rm s}$ are different for electric ($\omega_{\boldsymbol{E}}$) and magnetic ($\omega_{\boldsymbol{H}}$) properties of metals. For diamagnetic and paramagnetic metals, the magnetic characteristic frequencies $\omega_{\boldsymbol{H}}$ usually lie far below the optical range, while the electric frequencies $\omega_{\boldsymbol{E}}$ vary from the near infrared to the ultraviolet [2]. Therefore, most diamagnetic and paramagnetic metals lose their magnetic properties early, before reaching the optical range. Thus, starting from the optical frequencies, they feature

$$\mu_{\rm t} = \mu_{\rm l} = 1, \tag{1.19}$$

with the temporal dispersion given by $\varepsilon_{ij}(\omega)$.

For ferromagnetic metals, the situation is different, since they are magnetically anisotropic. Their μ_{ij} is different from (1.18) and contains additional (gyrotropic) components that slowly decrease with frequency [2]. Owing to these gyrotropic components, ferromagnetics can remain anisotropic even at relatively high frequencies. For this reason, we will not consider ferromagnetic metals, and study nonmagnetic materials only.

For magnetically inactive metals that satisfy condition (1.19), the dispersion is given by the dependence $\varepsilon_{ij}(\omega, \boldsymbol{k})$. In the optical range, it is dominated by the frequency dependence, with a weak wavevector dependence. In general, the spatial dispersion of $\varepsilon_{ij}(\omega, \boldsymbol{k})$ comes from the nonlocality of the metal's response to electric fields caused by the transport processes. Physically, it is given by the polarization vector $\boldsymbol{P}(t, \boldsymbol{r})$ defined by the electric fields $\boldsymbol{E}(t, \boldsymbol{r}')$ taken in the vicinity of the point $\boldsymbol{r}$. The region over which the nonlocality takes place is defined by the metal's characteristic length a_{s}. The smaller a_{s}, compared with the wavelength $2\pi/k$, the weaker the spatial dispersion. For metals, the characteristic length is given by the Debye screening length

$$a_{\mathrm{s}} = V_{\mathrm{F}}/\omega_{\mathrm{pf}},$$

where $V_{\mathrm{F}} = (3\pi^2)^{1/3}\hbar n_{\mathrm{f}}^{1/3}/m$ and $\omega_{\mathrm{pf}} = (4\pi e^2 n_{\mathrm{f}}/m)^{1/2}$ are the velocity at the Fermi level and the plasma frequency of the free electrons, with n_{f}, m, and $-e$ being the density, mass, and charge of the free electrons, respectively. In the optical range, this gives us $ka_{\mathrm{s}} \ll 1$. Thus, the spatial dispersion of metals is weak in this frequency range and contributes negligibly small amounts to ε_{l} and ε_{t}.

Following the strict consideration of the electron dynamics in the optical regime [3, 4], the transverse and longitudinal components of the complex permittivity tensor can be written as

$$\varepsilon_{\mathrm{t}}(\omega, \boldsymbol{k}) = \varepsilon(\omega) - \frac{1}{5}\frac{\omega_{\mathrm{pf}}^2}{\omega(\omega + \mathrm{i}\nu)}\frac{k^2 V_{\mathrm{F}}^2}{\omega^2}, \tag{1.20}$$

$$\varepsilon_{\mathrm{l}}(\omega, \boldsymbol{k}) = \varepsilon(\omega) - \frac{3}{5}\frac{\omega_{\mathrm{pf}}^2}{\omega(\omega + \mathrm{i}\nu)}\frac{k^2 V_{\mathrm{F}}^2}{\omega^2}, \tag{1.21}$$

where the first term on the right-hand side in each equation, $\varepsilon(\omega)$, is related to the pure temporal dispersion in the absence of the spatial one, while the second terms give the spatial dispersion (with ν being the electron collision frequency). As one can see, the second terms are proportional to $k^2 a_{\mathrm{s}}^2$ and, hence, contribute negligibly small amounts in the optical range. Therefore, we can consider $\varepsilon_{ij}(\omega, \boldsymbol{k})$ as dependent on the frequency only, when

$$\varepsilon_{ij}(\omega, \boldsymbol{k}) = \varepsilon(\omega)\delta_{ij}, \tag{1.22}$$

with

$$\varepsilon_{\mathrm{t}}(\omega, \boldsymbol{k}) = \varepsilon_{\mathrm{l}}(\omega, \boldsymbol{k}) = \varepsilon(\omega).$$

Thus, magnetically inactive metals are well described within the *local-response* approximation, where the material relations

$$\boldsymbol{D}(\omega) = \varepsilon(\omega)\boldsymbol{E}(\omega), \quad \boldsymbol{B}(\omega) = \boldsymbol{H}(\omega) \tag{1.23}$$

are completely given by the scalar complex dielectric permittivity $\varepsilon(\omega)$.

1.2 The local-response approximation

1.2.1 The energy of an electromagnetic field in metals

According to Maxwell's equations, the distribution of an electromagnetic field in a medium obeys the energy-conservation law. For magnetically inactive metals, it can be written as follows:

$$-\frac{\partial}{\partial t}\frac{E^2+H^2}{8\pi}=\frac{c}{4\pi}\nabla\cdot(\boldsymbol{E}\times\boldsymbol{H})+\boldsymbol{j}\cdot\boldsymbol{E}. \tag{1.24}$$

The left-hand side of this equation gives the decrease of the energy density

$$U_0=\frac{E^2+H^2}{8\pi} \tag{1.25}$$

of the electric $\boldsymbol{E}$ and magnetic $\boldsymbol{H}$ fields in the volume element $\mathrm{d}V$ over the time $\mathrm{d}t$, and the right-hand side represents two mechanisms for that energy change. The first mechanism is the radiation of the energy out of the volume $\mathrm{d}V$. Its contribution is given by the divergence of the Poynting vector,

$$\boldsymbol{S}=\frac{c}{4\pi}\boldsymbol{E}\times\boldsymbol{H}, \tag{1.26}$$

which is considered as the energy flux density of the electromagnetic field. The second mechanism is the work done by the fields on the medium's internal charges, described with the power density

$$Q_{\text{med}}=\boldsymbol{j}\cdot\boldsymbol{E}. \tag{1.27}$$

In general, Q_{med} is comprised of the dissipation power density Q_{dis} and the change of the energy density U_{pol} stored in the medium in the form of the polarization field $\boldsymbol{P}$,

$$Q_{\text{med}}=Q_{\text{dis}}+\frac{\partial U_{\text{pol}}}{\partial t}. \tag{1.28}$$

Thus, the conservation law for the overall electromagnetic field in the medium can be written as

$$-\frac{\partial U}{\partial t}=\nabla\cdot\boldsymbol{S}+Q_{\text{dis}}, \tag{1.29}$$

where $U=U_0+U_{\text{pol}}$ is the total energy density of the three fields $\boldsymbol{E}$, $\boldsymbol{H}$, and $\boldsymbol{P}$.

To derive the energy characteristics in the local-response approximation, we calculate the time-averaged value of Q_{med},

$$\overline{Q_{\text{med}}}=\frac{1}{2}\operatorname{Re}(\boldsymbol{j}\cdot\boldsymbol{E}^*)=\frac{1}{2}\operatorname{Re}\left(\frac{\partial \boldsymbol{P}}{\partial t}\cdot\boldsymbol{E}^*\right), \tag{1.30}$$

assuming that the electromagnetic field is not purely monochromatic, but has a small frequency width,

$$\boldsymbol{E}=\boldsymbol{E}_0(t)\mathrm{e}^{-\mathrm{i}\omega t},\quad \boldsymbol{H}=\boldsymbol{H}_0(t)\mathrm{e}^{-\mathrm{i}\omega t},$$

where $\boldsymbol{E}_0(t)$ and $\boldsymbol{H}_0(t)$ are slowly varying functions, as compared with $\mathrm{e}^{-\mathrm{i}\omega t}$. By doing that, we can capture the effect of the strong temporal dispersion that makes a significant contribution to the energy stored in a metal [2],

$$\overline{Q_{\mathrm{med}}} = \frac{\partial}{\partial t}\frac{|E|^2}{16\pi}\left[\frac{\partial(\omega\varepsilon')}{\partial\omega} - 1\right] + \frac{\omega}{8\pi}\varepsilon''|E|^2. \tag{1.31}$$

By comparing with Eq. (1.28), we find that the time-averaged energy density of the overall electromagnetic field is

$$\overline{U} = \frac{1}{16\pi}\left[\frac{\partial(\omega\varepsilon')}{\partial\omega}|E|^2 + |H|^2\right], \tag{1.32}$$

with the dispersion effect given by the function $\partial(\omega\varepsilon')/\partial\omega$, where ε' is the real part of the complex dielectric permittivity $\varepsilon(\omega)$. For the time-averaged dissipation rate, we obtain

$$\overline{Q_{\mathrm{dis}}} = \frac{\omega}{8\pi}\varepsilon''|E|^2, \tag{1.33}$$

where ε'' is the imaginary part of $\varepsilon(\omega)$. In this way, we derive that the real part of the complex dielectric permittivity defines the energy of the electromagnetic field, while its imaginary part describes the dissipation in the metal.

1.2.2 Properties of the complex dielectric permittivity

As we have already seen, the scalar dielectric permittivity plays an essential role in the description of the interaction of an electrodynamic field with a medium. Therefore, it is of vital importance to understand the main properties of $\varepsilon(\omega)$.

Some common qualities of the scalar dielectric permittivity can be obtained directly from the general definition of $\varepsilon_{ij}(\omega, \boldsymbol{k})$ given by Eq. (1.15). Note the difference between the tensors $\varepsilon_{ij}(t, \boldsymbol{r})$ and $\varepsilon_{ij}(\omega, \boldsymbol{k})$ in that equation – the former is a purely real function, insofar as it relates the real fields $\boldsymbol{D}(t, \boldsymbol{r})$ and $\boldsymbol{E}(t, \boldsymbol{r})$, while the latter is generally complex. Given this peculiarity, we can write the symmetry relation for $\varepsilon(\omega)$ with respect to the frequency ω,

$$\varepsilon(-\omega) = \varepsilon^*(\omega).$$

In terms of the real, $\varepsilon'(\omega)$, and imaginary, $\varepsilon''(\omega)$, parts of $\varepsilon(\omega)$, it can be rewritten as follows:

$$\varepsilon'(-\omega) = \varepsilon'(\omega), \quad \varepsilon''(-\omega) = -\varepsilon''(\omega). \tag{1.34}$$

Also, Eq. (1.15) gives us the value of $\varepsilon(\omega)$ at $\omega \to \infty$,

$$\varepsilon(\omega \to \infty) = 1,$$

or, alternatively,

$$\varepsilon'(\omega \to \infty) = 1, \quad \varepsilon''(\omega \to \infty) = 0. \tag{1.35}$$

The thermodynamic properties of the scalar dielectric permittivity can be obtained if we consider the energy density of the overall electromagnetic field, $\overline{U}$, and the energy dissipation rate, $\overline{Q_{\rm dis}}$, given by Eqs. (1.32) and (1.33), respectively. Following them, under thermodynamic equilibrium, $\varepsilon'(\omega)$ and $\varepsilon''(\omega)$ should feature

$$\frac{\partial}{\partial\omega}\left[\omega\varepsilon'(\omega)\right] \geq 0, \quad \omega\varepsilon''(\omega) \geq 0. \tag{1.36}$$

In addition, the scalar dielectric permittivity, as a function of frequency, is analytic in the upper half-plane [1, 5], and, thus, fulfills the *Kramers–Kronig relations*

$$\varepsilon'(\omega) - 1 = \frac{1}{\pi}\mathcal{P}\int_{-\infty}^{\infty} \frac{\varepsilon''(x)}{x-\omega}\,\mathrm{d}x, \tag{1.37}$$

$$\varepsilon''(\omega) = -\frac{1}{\pi}\mathcal{P}\int_{-\infty}^{\infty} \frac{\varepsilon'(x)-1}{x-\omega}\,\mathrm{d}x, \tag{1.38}$$

where $\mathcal{P}$ denotes the Cauchy principal value. These relations show that, due to the causality principle, the real and imaginary parts of the scalar dielectric permittivity are not independent. In modeling the medium's response, it is convenient to consider $\varepsilon''(\omega)$ as the independent function. Then, the function $\varepsilon'(\omega)$ can be derived from Eq. (1.37) as

$$\varepsilon'(\omega) - 1 = \frac{2}{\pi}\mathcal{P}\int_{0}^{\infty} \frac{x\varepsilon''(x)}{x^2-\omega^2}\,\mathrm{d}x, \tag{1.39}$$

where we have taken into account the symmetry condition for ε'' with respect to ω. Thus, $\varepsilon'(\omega)$ can be completely restored, without any extra assumptions, if we know the dependence $\varepsilon''(\omega)$ in the whole spectral range.

1.2.3 The conduction-electron contribution

Within the local-response approximation, the polarization vector $\boldsymbol{P}(\omega)$ in metals can be written as

$$\boldsymbol{P}(\omega) = \frac{\chi(\omega)}{4\pi}\boldsymbol{E}(\omega), \tag{1.40}$$

where the linear coefficient $\chi(\omega)$ is the scalar susceptibility of the metal. Following the definition of the electric displacement (1.7),

$$\varepsilon(\omega) = 1 + \chi(\omega). \tag{1.41}$$

Thus, the susceptibility gives us the contribution of the polarization to the dielectric permittivity.

Note that any metal contains two types of intrinsic charges – quasi-free *conduction electrons* and the subsystem of *bound charges* comprised of bound electrons and nuclei. Owing to their different mobilities, these two subsystems behave and react to electromagnetic fields differently. Therefore, it is convenient to separate their

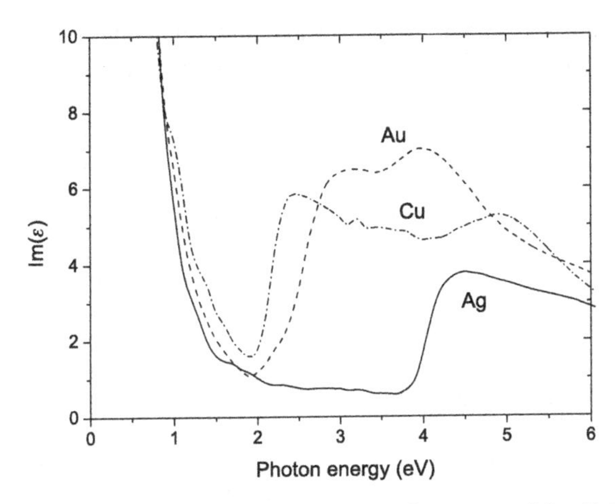

Figure 1.1 The dependence of the imaginary part of the dielectric permittivity ε on the photon energy $\hbar\omega$ for the noble metals. The peaks observed are attributed to the enhanced photon absorption caused by electron interband transitions. The optical data are taken from Ref. [6].

contributions as

$$\chi(\omega) = \chi_{\mathrm{f}}(\omega) + \chi_{\mathrm{b}}(\omega), \tag{1.42}$$

$$\varepsilon(\omega) = 1 + \chi_{\mathrm{f}}(\omega) + \chi_{\mathrm{b}}(\omega), \tag{1.43}$$

where $\chi_{\mathrm{f}}(\omega)$ and $\chi_{\mathrm{b}}(\omega)$ are the susceptibilities of the free-electron subsystem and the subsystem of bound electrons and nuclei.

The contribution of the bound charges to the overall polarization in the optical range is mainly given by the electron *interband transitions* accompanied by photon absorption. For many metals, such transitions appear at energies $\hbar\omega > 1$ eV. As an example, let us consider the noble metals. They have similar electronic configurations with completely filled d-shells: $[\mathrm{Ar}]3\mathrm{d}^{10}4\mathrm{s}^1$ for copper, $[\mathrm{Kr}]4\mathrm{d}^{10}5\mathrm{s}^1$ for silver, or $[\mathrm{Xe}]5\mathrm{d}^{10}6\mathrm{s}^1$ for gold. Therefore, their first interband transitions appear when d-electrons jump to the empty states above the Fermi level. Such transitions are accompanied by the absorption of photons around $\hbar\omega = E_{\mathrm{F}} - E_{\mathrm{d}}$, which is about 2 eV for copper and gold, or 4 eV for silver (Fig. 1.1). Below these values, the susceptibility $\chi_{\mathrm{b}}(\omega)$ goes to zero, so the main contribution to $\chi(\omega)$ is given by the free electrons,

$$|\chi_{\mathrm{b}}(\omega)| \ll |\chi_{\mathrm{f}}(\omega)|, \quad \chi(\omega) \approx \chi_{\mathrm{f}}(\omega).$$

Thus, any metal at energies below the interband transitions can be treated as a free-electron gas.

The Drude model of free electrons

In the local-response approximation, the dynamics of a free-electron gas is well described with the linearized electron-motion equation,

$$\frac{\partial \boldsymbol{v}(t, \boldsymbol{r})}{\partial t} = -\frac{e}{m}\boldsymbol{E}(t, \boldsymbol{r}) - \nu \boldsymbol{v}(t, \boldsymbol{r}), \tag{1.44}$$

where $\boldsymbol{v}$, $-e$, m, and ν are the velocity, charge, mass, and collisional frequency of the electrons. Oscillating in the electric field $\boldsymbol{E}$, electrons create the internal current $\boldsymbol{j}$ given by

$$\boldsymbol{j}(t, \boldsymbol{r}) = -en_{\mathrm{f}}\boldsymbol{v}(t, \boldsymbol{r}), \tag{1.45}$$

where n_{f} is the density of the free electrons. By performing the Fourier transform for $\boldsymbol{v}$, $\boldsymbol{E}$, and $\boldsymbol{j}$, we derive

$$\boldsymbol{j}(\omega) = \mathrm{i}\frac{e^2 n_{\mathrm{f}}}{m(\omega + \mathrm{i}\nu)}\boldsymbol{E}(\omega). \tag{1.46}$$

At the same time, according to the definition of the polarization vector $\boldsymbol{P}_{\mathrm{f}}$, the internal current is given by

$$\boldsymbol{j}(\omega) = -\mathrm{i}\omega\boldsymbol{P}_{\mathrm{f}}(\omega) = -\mathrm{i}\frac{\omega}{4\pi}\chi_{\mathrm{f}}(\omega)\boldsymbol{E}(\omega). \tag{1.47}$$

Thus, we get

$$\chi_{\mathrm{f}}(\omega) = -\frac{\omega_{\mathrm{pf}}^2}{\omega(\omega + \mathrm{i}\nu)}, \tag{1.48}$$

$$\varepsilon(\omega) = 1 - \frac{\omega_{\mathrm{pf}}^2}{\omega(\omega + \mathrm{i}\nu)}, \tag{1.49}$$

where $\omega_{\mathrm{pf}} = (4\pi e^2 n_{\mathrm{f}}/m)^{1/2}$ is the plasma frequency of the free (conduction) electrons in the metal.

The expression derived for $\varepsilon(\omega)$ was proposed first by Paul Drude in 1900 to explain the electrical conduction of metals. In his theory, the response of the electron gas, $\boldsymbol{P}_{\mathrm{f}}(\boldsymbol{E})$, is modeled with the differential equation

$$\frac{\partial^2 \boldsymbol{P}_{\mathrm{f}}}{\partial t^2} + \nu\frac{\partial \boldsymbol{P}_{\mathrm{f}}}{\partial t} = \frac{\omega_{\mathrm{pf}}^2}{4\pi}\boldsymbol{E}. \tag{1.50}$$

It provides a good description of the polarization and conduction of a metal at frequencies below the plasma frequency of the free electrons, ω_{pf}. A comparison of this model with the susceptibility measured for gold is shown in Fig. 1.2. This demonstrates that the Drude model provides an adequate fitting below the electron interband transitions and breaks down in the visible.

1.2.4 The bound-charge contribution

At frequencies closer to the electron interband transitions, the contribution of the bound electrons increases, so one should consider both χ_{f} and χ_{b},

$$\chi(\omega) = \chi_{\mathrm{f}}(\omega) + \chi_{\mathrm{b}}(\omega).$$

The derivation of $\chi_{\mathrm{b}}(\omega)$ is very complicated. It requires a full quantum-mechanical consideration of all possible electron transitions that are particularly sensitive to the metal's electronic configuration. All this makes the problem very tough. Therefore, the

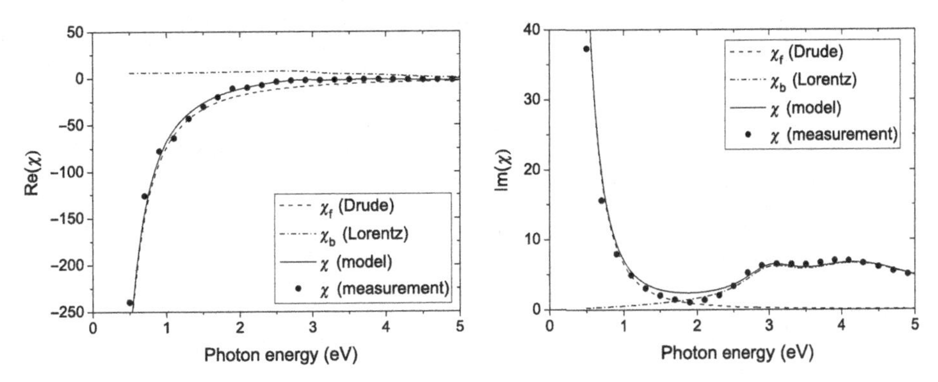

Figure 1.2 Fitting of the susceptibility $\chi(\omega)$ for gold. The experimental data are taken from Ref. [6]. The optimized values of the Drude model for conduction electrons are $\hbar\omega_{\text{pf}} = 8.566$ eV and $\hbar\nu = 0.087$ eV, and those of the Lorentz model for bound charges are $\hbar\omega_{01} = 3.034$ eV, $\hbar\Omega_1 = 2.924$ eV, $\hbar\gamma_1 = 0.848$ eV, $\hbar\omega_{02} = 4.335$ eV, $\hbar\Omega_2 = 7.386$ eV, $\hbar\gamma_2 = 2.217$ eV, $\hbar\omega_{03} = 7.263$ eV, $\hbar\Omega_3 = 10.009$ eV, and $\hbar\gamma_3 = 1.672$ eV.

methods of classical mechanics are commonly adopted to model (or fit) the quantum response of real metals.

The Lorentz oscillator model

The Lorentz oscillator model is a purely classical model that gives a simple picture and very basic insight into the light–matter interaction [7]. It assumes that a bound electron–nucleus system can be treated as an oscillator, where the electron periodically moves around the nucleus with the characteristic frequency given by the interband transition. Following this model, the overall system of all bound electrons and nuclei is comprised of a number N of such oscillators that oscillate with their own frequencies $\omega_{0i} \neq 0$, where $i = 1, 2, \ldots, N$ is the oscillator index. When the medium is placed in a time-varying electric field $\boldsymbol{E}$, the oscillators start to interact with the field. Eventually, they result in the polarization $\boldsymbol{P}_{\text{b}}$ given by

$$\boldsymbol{P}_{\text{b}} = \sum_{i=1}^{N} \boldsymbol{P}_i, \tag{1.51}$$

where the polarization vector of the ith oscillator $\boldsymbol{P}_i(\boldsymbol{E})$ is defined according to the following differential equation:

$$\frac{\partial^2 \boldsymbol{P}_i}{\partial t^2} + \gamma_i \frac{\partial \boldsymbol{P}_i}{\partial t} + \omega_{0i}^2 \boldsymbol{P}_i = \frac{\Omega_i^2}{4\pi} \boldsymbol{E}. \tag{1.52}$$

This is the equation of forced oscillations caused by the applied electric field $\boldsymbol{E}$. Note the similarity between this equation and Eq. (1.50) given by the Drude model. The coefficient γ_i is analogous to the collision frequency ν, and Ω_i is similar to the plasma frequency of the free electrons, ω_{pf}. Thus, γ_i gives the characteristic frequency of the oscillator losses, while Ω_i characterizes the density (or strength) of the oscillators of the ith type.

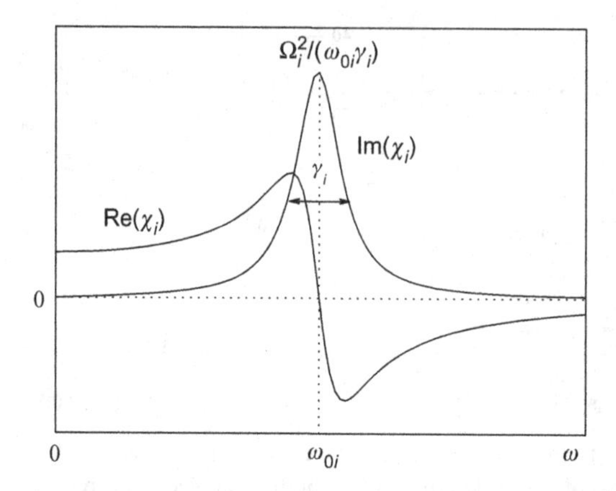

Figure 1.3 Characteristics of the oscillator susceptibility $\chi_i(\omega)$ in the Lorentz model.

Following Eq. (1.51), the susceptibility of bound charges in the Lorentz model is given by the sum over the oscillator index i,

$$\chi_{\mathrm{b}}(\omega) = \sum_{i=1}^{N} \chi_i(\omega), \tag{1.53}$$

where $\chi_i(\omega)$ is defined in accordance with Eq. (1.52),

$$\chi_i(\omega) = \frac{\Omega_i^2}{\omega_{0i}^2 - \omega^2 - \mathrm{i}\gamma_i\omega}. \tag{1.54}$$

Although this expression was derived from the principles of classical mechanics, it is widely used for fitting experimentally measured dielectric permittivities, especially in the region of the interband transitions (Fig. 1.2). It arises from the oscillator susceptibility $\chi_i(\omega)$ that satisfies all of the requirements listed in Section 1.2.2 for the scalar dielectric permittivity. Therefore, Exp. (1.54) is considered as an elementary function that can be used for any type of dielectric permittivity. In addition, the parameters in this model bear very clear spectroscopic meanings – Ω_i, ω_{0i}, and γ_i are the strength, central frequency, and width of the spectral line, as shown in Fig. 1.3.

Note that the oscillator susceptibility (1.54) works very well for the fitting of single symmetric spectral lines in $\chi_{\mathrm{b}}''(\omega)$. However, for peaks with strong asymmetry or for quasi-continuous electron transitions that appear in metals, fitting with the Lorentz model becomes more difficult, insofar as it requires the inclusion of a greater number N of harmonic oscillator terms $\chi_i(\omega)$ with closely spaced central frequencies ω_{0i}.

The quantum-mechanical oscillator model

In a more accurate quantum-mechanical consideration, the susceptibility of the bound electrons and nuclei is given as a sum over all bound states of the metal,

$$\chi_{\mathrm{b}}(\omega) = \sum_{i} \chi_i(\omega), \tag{1.55}$$

where χ_i is the susceptibility of the bound electrons in the ith energy state. Interestingly, χ_i can be written in a similar way to the classical Lorentz model [8],

$$\chi_i(\omega) = \frac{4\pi e^2}{m} \sum_j \frac{f_{ij}}{\omega_{ij}^2 - \omega^2 - \mathrm{i}\gamma_{ij}\omega}, \tag{1.56}$$

where $\omega_{ij} = (E_j - E_i)/\hbar$ is the characteristic frequency of the transition from the initial state i to an excited state j; f_{ij} is the quantum oscillator strength given by the probability of the excitation of an electron from the state i to the state j; and γ_{ij} is the frequency defined by the probability of the inverse transition from the state j to all other states different from i.

The parameters of this model are calculated with the dipole moment operator and require consideration of all electronic states. Therefore, it works well for atomic structures and becomes cumbersome as the number of atoms included increases. Eventually, for large metal objects with quasi-continuous electronic bands, the realization of this model becomes very complicated. For this reason, the classical Lorentz model is more attractive, especially in view of its similarities to the quantum-mechanical model.

The frequency-distributed oscillator model

As has been mentioned, the electron transitions in metal structures are often broad due to the quasi-continuous distribution of electronic states inside the energy bands. This makes the realization of both classical and quantum-mechanical models difficult, since they should involve a huge number of terms defined by the frequency region which is being simulated. The broader the frequency range considered, the greater the number of $\chi_i(\omega)$ which should be included in $\chi_\mathrm{b}(\omega)$. In this case, the model of frequency-distributed oscillators becomes advantageous.

The frequency-distributed oscillator model is a fitting technique that was developed within the framework of the theory of dispersion based on the Kramers–Kronig relations. If we consider the Kramers–Kronig relations given by Eqs. (1.37) and (1.38), we find that they are actually written for susceptibility $\chi(\omega)$. In fact, these relations are applied separately to every susceptibility component [5]. In particular, they must be satisfied for $\chi_\mathrm{b}(\omega)$. Therefore, the real and imaginary parts of $\chi_\mathrm{b}(\omega)$ are related to each other by Eq. (1.39),

$$\chi_\mathrm{b}'(\omega) = \frac{2}{\pi}\mathcal{P}\int_0^\infty \frac{\omega_0 \chi_\mathrm{b}''(\omega_0)}{\omega_0^2 - \omega^2}\,\mathrm{d}\omega_0. \tag{1.57}$$

The written integral can be considered as a contribution of the oscillators continuously distributed over the frequency ω_0. Thus, one can introduce the frequency distribution function of the oscillators

$$f(\omega_0) = \frac{m}{2\pi^2 e^2}\omega_0 \chi_\mathrm{b}''(\omega_0)$$

and write $\chi_\mathrm{b}'(\omega)$ in a similar way to Exp. (1.56),

$$\chi_\mathrm{b}'(\omega) = \frac{4\pi e^2}{m}\mathcal{P}\int_0^\infty \frac{f(\omega_0)\mathrm{d}\omega_0}{\omega_0^2 - \omega^2}, \tag{1.58}$$

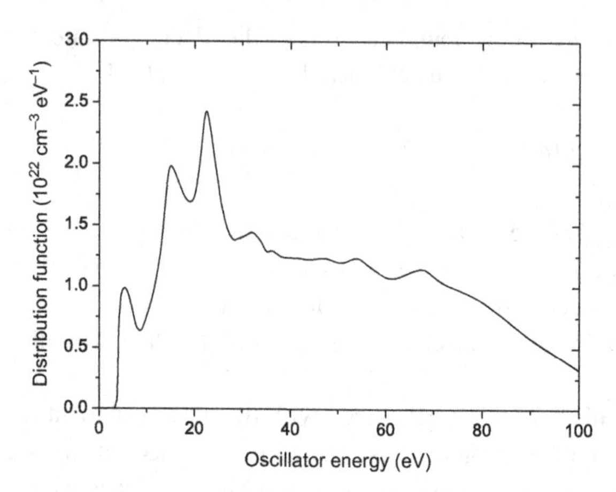

Figure 1.4 The distribution function f of the oscillator energy $\hbar\omega_0$ in silver. The optical data are taken from Ref. [9].

where $f(\omega_0)\mathrm{d}\omega_0$ is the strength of the oscillators in the frequency range $\mathrm{d}\omega_0$ about ω_0. Thus, by specifying the oscillator distribution function $f(\omega_0)$, we can calculate the real part of the susceptibility, $\chi_\mathrm{b}'(\omega)$, as well as the imaginary part,

$$\chi_\mathrm{b}''(\omega) = \frac{2\pi^2 e^2}{m\omega} f(\omega). \tag{1.59}$$

In other words, the response of bound charges in the frequency-distributed oscillator model is completely given by the distribution function $f(\omega_0)$ unique for every metal (Fig. 1.4). This model also can be considered as a generalization of the Lorentz model, where the medium response is given by an infinite number of closely spaced and extremely narrow spectral lines (1.54) with

$$\omega_{0(i+1)} - \omega_{0i} \to 0, \quad \gamma_i \to 0.$$

The advantage of this method is that it does not require any fitting; the distribution function can be obtained directly from the measured values of $\chi_\mathrm{b}''(\omega)$. As a result, one can get a broad-band analytical solution for $\chi_\mathrm{b}(\omega)$ that can easily be continued to complex ω (as is required for some eigenvalue problems). As a drawback of this method, the calculation procedure for $\chi_\mathrm{b}'(\omega)$ given by Eq. (1.58) requires f to be determined over quite a broad range of ω_0.

1.3 Electromagnetic fields in metals

1.3.1 Plasmon classification

In general, the spatial distribution of electromagnetic fields in magnetically inactive metals is given by Maxwell's equations governed by the tensor of complex dielectric permittivity $\varepsilon_{ij}(\omega, \boldsymbol{k})$. In the $(\omega, \boldsymbol{k})$ space, these equations can be written as follows:

$$(\boldsymbol{k} \times \boldsymbol{H})_i = -k_0 \varepsilon_{ij}(\omega, \boldsymbol{k}) E_j, \quad \boldsymbol{k} \cdot \boldsymbol{H} = 0, \tag{1.60}$$

$$\boldsymbol{k} \times \boldsymbol{E} = k_0 \boldsymbol{H}, \quad k_i \varepsilon_{ij}(\omega, \boldsymbol{k}) E_j = 0, \tag{1.61}$$

where $k_0 = \omega/c$ is the free-space wavenumber. From this set of equations, we can derive three linear equations for the electric field components E_j,

$$\left[k^2 \delta_{ij} - k_i k_j - k_0^2 \varepsilon_{ij}(\omega, \boldsymbol{k})\right] E_j = 0. \tag{1.62}$$

In the case of an isotropic medium, the tensor $\varepsilon_{ij}(\omega, \boldsymbol{k})$ can be written as Eq. (1.17) with two independent components, $\varepsilon_{\mathrm{l}}(\omega, \boldsymbol{k})$ and $\varepsilon_{\mathrm{t}}(\omega, \boldsymbol{k})$. For such a medium, equation set (1.62) gives us

$$\boldsymbol{k} \times (\boldsymbol{k} \times \boldsymbol{E}) \left[k^2 - k_0^2 \varepsilon_{\mathrm{t}}(\omega, \boldsymbol{k})\right] + \boldsymbol{k}(\boldsymbol{k} \cdot \boldsymbol{E}) \varepsilon_{\mathrm{l}}(\omega, \boldsymbol{k}) = 0. \tag{1.63}$$

Following this condition, the existence of nontrivial solutions of the field equations in an isotropic medium requires

$$k^2 - k_0^2 \varepsilon_{\mathrm{t}}(\omega, \boldsymbol{k}) = 0 \tag{1.64}$$

for transverse fields and

$$\varepsilon_{\mathrm{l}}(\omega, \boldsymbol{k}) = 0 \tag{1.65}$$

for longitudinal ones.

Conditions (1.64) and (1.65) give us the general "bulk" relations between ω and $\boldsymbol{k}$ supported by an isotropic metal. These are the necessary conditions that must be met by electromagnetic fields in a bulk metal. Additional requirements that electromagnetic fields must meet are the boundary conditions defined by the shape of the metal object, as well as by the adjacent material. The boundary conditions impose constraints on the field wavenumber $\boldsymbol{k}$ and, thus, restrict the frequency ω (in accordance with "bulk" conditions (1.64) and (1.65)) to a discrete set of values ω_{nml}, where the three integer indexes n, m, and l characterize the three-dimensional quantization of the wavevector $\boldsymbol{k}$.

In other words, the ω_{nml} represent the eigenfrequencies of the electromagnetic fields supported by the metal object. They are the solutions of the eigenvalue problem comprised by the Maxwell equations and the boundary conditions. The normal modes corresponding to ω_{nml} have unique distributions of electromagnetic fields that are strongly coupled to the metal intrinsic charges. These modes are commonly considered as quasiparticles called *plasmons*. They appear as a result of the quantization of intrinsic charge oscillations, similar to *photons* and *phonons*, which are the quanta of electromagnetic and mechanical vibrations, respectively.

As quanta of the charge oscillations, plasmons are spatially confined to the metal volume, with the field distribution defined by the object geometry. This is a result of the quantization of the wavevector $\boldsymbol{k}$, which is strictly given by the shape and dimensions of the metal object. As an example, let us consider a metal bar with edge sizes a_x, a_y, and a_z. In this case, the quantization of $\boldsymbol{k}$ appears with the steps of $\Delta k_x \sim 1/a_x$, $\Delta k_y \sim 1/a_y$, and $\Delta k_z \sim 1/a_z$, and results in a discrete set of energy values $E_{nml} = \hbar\omega_{nml}$. It gives us plasmons that are bound in all three dimensions. For this reason, such plasmons are also frequently called *localized plasmons*.

Now, let us fix two dimensions, a_x and a_y, of the bar and start to gradually increase the third one, a_z. In this case, Δk_z decreases until it reaches zero for $a_z \to \infty$. Mathematically, such an increase of a_z weakens the boundary conditions in the z direction. Finally, when the bar becomes infinitely long, the boundary conditions in the z direction lose their strength and the quantization of k_z disappears. This gives us plasmons that are bound in two dimensions (along the x and y axes), but can freely move in the third one (along the z axis). Such plasmons represent propagating electromagnetic waves, and form a new class of quasi-particles called *plasmon polaritons*.

The main difference between plasmon polaritons and localized plasmons is that the former can freely move at least in one direction, while the latter are bound in all three dimensions. For the case being considered here, the plasmon polaritons propagate in the z direction with momentum $p_z = \hbar k_z$ that can take any arbitrary value. As a result, the plasmon polariton energy becomes a function of k_z, when $E_{nm} = \hbar\omega_{nm}(k_z)$, with two indexes n and m, as the quantization of the wavenumber k_z disappears.

In a similar way, we can extend one more dimension of the bar a_y and obtain an infinite metal slab of thickness a_x that can support two-dimensional plasmon polaritons propagating both in the y direction and in the z direction. Such plasmon polaritons are characterized by two independent momenta, $p_y = \hbar k_y$ and $p_z = \hbar k_z$, and their energy is given by $E_n = \hbar\omega_n(k_y, k_z)$, where the index n denotes the quantization of the polariton momentum $p_x = \hbar k_x$.

Regardless of the plasmon type, the general relations between the frequency ω and the wavevector $\boldsymbol{k}$ of the charge oscillations are given by "bulk" equations, (1.64) or (1.65). Following them, plasmons with longitudinal and transverse electric fields obey different laws. For transverse plasmons in isotropic and magnetically inactive metals with $\varepsilon_t(\omega, \boldsymbol{k}) \approx \varepsilon(\omega)$, the relation between ω and $\boldsymbol{k}$ can be written as

$$k^2 = k_0^2 \varepsilon(\omega), \tag{1.66}$$

where the coupling with the charge carriers is described by the local dielectric permittivity $\varepsilon(\omega)$. It is noteworthy that this expression is similar to the photon relation given by $k^2 = k_0^2$. It reveals the analogy between the photons and transverse plasmons caused by their belonging to the same class of transverse electromagnetic fields. In addition, since photons feature only transverse fields, plasmons are considered as a more general class of quasi-particles that enable the full description of electromagnetic interaction in a medium.

To derive the relation between ω and $\boldsymbol{k}$ for longitudinal plasmons, we should consider another "bulk" equation, (1.65). Following this equation, the spatial dispersion plays a crucial role for longitudinal plasmons, since it defines the dependence on the wavevector $\boldsymbol{k}$. By using Eq. (1.21) for $\varepsilon_l(\omega, \boldsymbol{k})$, we get the following law:

$$k^2 = \frac{5}{3}\frac{\omega(\omega + i\nu)}{\omega_{pf}^2}\frac{\omega^2}{V_F^2}\varepsilon(\omega). \tag{1.67}$$

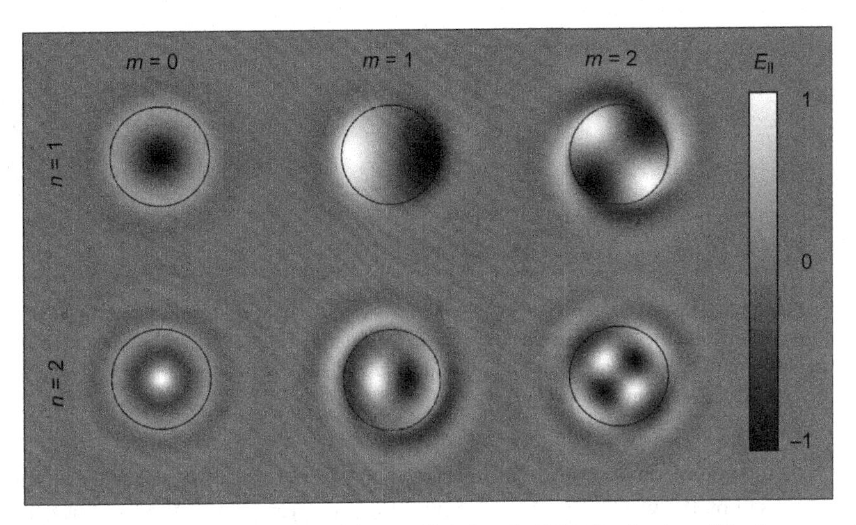

Figure 1.5 Cross-sectional views of the parallel electric field distribution of the transverse bulk plasmon polaritons in a silver nanowire of radius $R = 100$ nm. The field distributions are plotted for the plasmon polaritons with azimuthal numbers $m = 0, 1, 2$ and order indexes $n = 1, 2$ at the energy of $\hbar\omega = 6$ eV. The optical data are taken from Ref. [9].

Thus, the longitudinal plasmons feature a more complex relation for their momentum $\boldsymbol{p} = \hbar\boldsymbol{k}$ and energy $E = \hbar\omega$, given by the transport of conduction electrons in a metal.

1.3.2 Bulk plasmon modes

Recall that relations (1.66) and (1.67) derived for ω and $\boldsymbol{k}$ of the transverse and longitudinal plasmons were obtained from the general properties of an isotropic medium. Therefore, they give fundamental properties of the bulk material, regardless of its shape. After applying the boundary conditions, these relations result in an infinite number of transverse and longitudinal plasmons with specifically patterned electromagnetic fields. They form spatial patterns similar to standing waves with the fields oscillating according to the object's geometry, as shown in Fig. 1.5. Such plasmons feature electromagnetic fields spread out over the whole metal volume, and, therefore, are called *bulk* or *volume plasmons*.

The spatially oscillating fields of bulk plasmons require that the real part of the wavevector $\boldsymbol{k}$ should be greater than its imaginary part (otherwise the fields exponentially decay rather than oscillate). This is the condition of the so-called *transparency regime* of metals, where $\mathrm{Re}(\varepsilon) \geq 0$. Usually it occurs at higher frequencies, starting from the visible or ultraviolet region. The lower boundary of this regime, $\omega = \omega_{\mathrm{p}}$, corresponds to the condition $\mathrm{Re}(\varepsilon) = 0$ and is called the *plasma frequency* of the metal (not to be confused with the free-electron plasma frequency ω_{pf}, which is commonly higher than ω_{p}). Thus, the bulk plasmon energies always lie above the plasma level, $E \geq \hbar\omega_{\mathrm{p}}$.

Another feature of bulk plasmons can be obtained if we compare Eqs. (1.66) and (1.67). According to these equations, the energies of the transverse plasmons are higher

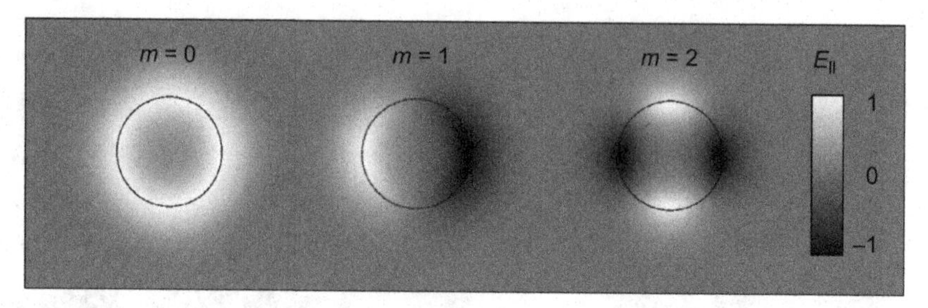

Figure 1.6 Cross-sectional views of the parallel electric field distribution of the surface plasmon polaritons in a silver nanowire of radius $R = 100$ nm. The field distributions are plotted for plasmon polaritons with azimuthal numbers $m = 0, 1, 2$ at the energy $\hbar\omega = 1$ eV. The optical data are taken from Ref. [9].

than those of the longitudinal ones for comparable values of the wavevector $\boldsymbol{k}$. This is caused by the weak spatial dispersion of metals provided by the smallness of the Fermi velocity of the conduction electrons compared with the speed of light, $V_{\mathrm{F}}/c \sim 10^{-3}$. Therefore, the energies of longitudinal plasmons for many optical applications can be treated just as $E = \hbar\omega_{\mathrm{p}}$, ignoring the negligibly small contribution of the spatial dispersion.

1.3.3 Surface plasmon modes

Below the plasma frequency ω_{p}, metals feature negative dielectric permittivity, $\mathrm{Re}(\varepsilon) < 0$. This is the so-called *opacity regime* of metals. According to Eqs. (1.66) and (1.67), the wavenumber k in this frequency range exhibits a large imaginary part, $|\mathrm{Im}(k)| > |\mathrm{Re}(k)|$. Moreover, the dielectric permittivity of metals at lower frequencies very quickly becomes highly negative with $|\mathrm{Re}(\varepsilon)| \gg 1$, resulting in $|\mathrm{Im}(k)| \gg |\mathrm{Re}(k)|$. Therefore, the propagation of any transverse or longitudinal fields inside the bulk metal is forbidden at $\omega < \omega_{\mathrm{p}}$.

In general, the opacity regime allows and supports only *evanescent* fields that exponentially decay into the bulk metal. Such oscillations of the charge carriers form the class of *surface plasmons*. They create spatial field patterns (Fig. 1.6) in which most of the electromagnetic energy is concentrated along the metal boundaries in accordance with the skin effect. In other words, the surface plasmon fields oscillate along the metal boundary, in contrast to the bulk plasmons which propagate in the bulk material.

The necessary condition for the surface plasmons' existence is a substantially negative dielectric permittivity, $\mathrm{Re}(\varepsilon) < 0$. As a result, the energies of surface plasmons always lie below the plasma level, $E < \hbar\omega_{\mathrm{p}}$. Thus, surface and bulk plasmons belong to different frequency/energy ranges. In addition, surface plasmons feature *only* transverse fields, in contrast to bulk ones that can be *both* transverse and longitudinal.

Owing to the evanescent fields, surface plasmon modes exhibit a number of unique properties. In particular, they can concentrate energy into subwavelength regions as small as a few nanometers. This enables them to overcome the diffraction limit for ordinary light and to create extremely intense electromagnetic fields at optical frequencies. For these properties, localized surface plasmons and surface plasmon polaritons of metal nanostructures are commonly considered as promising tools for diverse optical applications ranging from subwavelength lithography and super-lensing to nanoelectronics and biosensing.

References

[1] J. D. Jackson, *Classical Electrodynamics*, 2nd ed. New York: Wiley, 1975.

[2] L. D. Landau, E. M. Lifshitz, and L. P. Pitaevskii, *Electrodynamics of Continuous Media*, 2nd ed. Oxford: Pergamon Press, 1984.

[3] E. M. Lifshitz and L. P. Pitaevskii, *Physical Kinetics*. Oxford: Pergamon Press, 1981.

[4] A. F. Aleksandrov, L. S. Bogdankevich, and A. A. Rukhadze, *Principles of Plasma Electrodynamics*. Heidelberg: Springer-Verlag, 1984.

[5] L. D. Landau and E. M. Lifshitz, *Statistical Physics*. Oxford: Pergamon Press, 1969.

[6] SOPRA database, http://www.sspectra.com/sopra.html.

[7] C. F. Bohren and D. R. Huffman, *Absorption and Scattering of Light by Small Particles*, 2nd ed. New York: Wiley-Interscience, 1998.

[8] J. M. Ziman, *Principles of the Theory of Solids*. Cambridge: Cambridge University Press, 1972.

[9] H. J. Hagemann, W. Gudat, and C. Kunz, "Optical constants from the far infrared to the X-ray region: Mg, Al, Cu, Ag, Au, Bi, C, and Al_2O_3," *J. Opt. Soc. Am.*, vol. 65, pp. 742–744, 1975.

2 Plasmonic properties of metal nanostructures

Plasmons, being the electromagnetic eigenoscillations of intrinsic charges, play an important role in the electrodynamics of metals and determine the main optical properties of metal structures. Following the classification given in Section 1.3.1, there are two types of plasmons – longitudinal and transverse. Both types are inherent to any metal and appear equally in metal structures. Longitudinal plasmons define the optical response of metals to conservative fields, while transverse plasmons define the response to solenoidal fields. Therefore, transverse plasmons are more attractive for optical applications, since they provide resonant interaction with photons, in contrast to longitudinal plasmons, which require very specific conditions, such as certain electron density profiles or applied external magnetic/electric fields, in order for them to interact with photons. In this chapter, we consider transverse plasmons supported by different metal nanostructures in spherical, cylindrical, and planar geometries. By solving eigenvalue and scattering problems, we discuss the properties of these plasmons and study their coupling with incident photons.[1]

2.1 Plasmonic modes in spherical geometry

In this section, we consider the transverse eigenmodes supported in structures with spherical geometry. Within the vector spherical-harmonics formalism, we study the plasmonic modes of a metal sphere and a spherical cavity in a bulk metal. Also, we investigate the scattering of plane waves by metal nanoparticles and make a generalization for the case of a multilayer sphere.

2.1.1 Spherical harmonics

To describe the normal modes in spherical geometry, we introduce the *scalar spherical harmonics* $Y_{lm}(\theta, \phi)$, which are the two-dimensional eigenfunctions of the Poisson equation

$$\nabla^2 Y_{lm}(\theta, \phi) = -\frac{l(l+1)}{r^2} Y_{lm}(\theta, \phi) \tag{2.1}$$

[1] All quantities and equations of this chapter are written in the Gaussian unit system. For conversion to the SI unit system, we refer the reader to the textbook by Jackson [1].

in the spherical coordinates (r, θ, ϕ). We adopt the following form of the scalar spherical harmonics:

$$Y_{lm}(\theta, \phi) = \sqrt{\frac{2l+1}{4\pi}\frac{(l-m)!}{(l+m)!}}\, P_l^m(\cos\theta)\mathrm{e}^{\mathrm{i}m\phi}, \tag{2.2}$$

where $P_l^m(\cos\theta)$ are the associated Legendre polynomials, with the integers m and l being the azimuthal and orbital indexes, respectively. The harmonics defined in such a way represent a complete set of orthonormal functions,

$$\int_0^{4\pi} Y_{lm}(\theta, \phi)Y^*_{l'm'}(\theta, \phi)\mathrm{d}\Omega = \delta_{ll'}\delta_{mm'}, \tag{2.3}$$

where $\delta_{mm'}$ is the Kronecker delta and $\mathrm{d}\Omega = \sin\theta\, \mathrm{d}\theta\, \mathrm{d}\phi$ is the differential solid angle. Owing to the completeness of $Y_{lm}(\theta, \phi)$, any arbitrary scalar function $g(\theta, \phi)$ can be expanded into the spherical harmonics as

$$g(\theta, \phi) = \sum_{l=0}^{\infty}\sum_{m=-l}^{l} A_{lm}Y_{lm}(\theta, \phi), \tag{2.4}$$

where the expansion coefficients are given by

$$A_{lm} = \int_0^{4\pi} g(\theta, \phi)Y^*_{lm}(\theta, \phi)\mathrm{d}\Omega. \tag{2.5}$$

Using the scalar spherical harmonics, we can perform a vector extension to define *vector spherical harmonics* that possess a similar completeness to that of the functions $Y_{lm}(\theta, \phi)$. In general, this extension can be done in different ways. Here, we follow the definition of the vector spherical harmonics given in Ref. [2],

$$\begin{aligned} \boldsymbol{Y}_{lm}^{(1)} &= \frac{\boldsymbol{r}}{r}Y_{lm}(\theta, \phi), \\ \boldsymbol{Y}_{lm}^{(2)} &= \frac{r\,\nabla Y_{lm}(\theta, \phi)}{\sqrt{l(l+1)}}, \\ \boldsymbol{Y}_{lm}^{(3)} &= \frac{\boldsymbol{r}\times\nabla Y_{lm}(\theta, \phi)}{\sqrt{l(l+1)}}. \end{aligned} \tag{2.6}$$

The vectors obtained are orthogonal in real space,

$$\boldsymbol{Y}_{lm}^{(1)}\cdot\boldsymbol{Y}_{lm}^{(2)} = 0, \quad \boldsymbol{Y}_{lm}^{(1)}\cdot\boldsymbol{Y}_{lm}^{(3)} = 0, \quad \boldsymbol{Y}_{lm}^{(2)}\cdot\boldsymbol{Y}_{lm}^{(3)} = 0, \tag{2.7}$$

as well as possessing orthogonality in Hilbert space,

$$\int_0^{4\pi} \boldsymbol{Y}_{lm}^{(j)}\cdot\boldsymbol{Y}_{l'm'}^{(j)*}\,\mathrm{d}\Omega = \delta_{ll'}\delta_{mm'}, \quad j = 1, 2, 3. \tag{2.8}$$

Thus, any vector field $\boldsymbol{G}(\theta, \phi)$ can be expanded into the vector spherical harmonics as follows:

$$\boldsymbol{G} = \sum_{l=0}^{\infty} \sum_{m=-l}^{l} \left[G_{lm}^{(1)} \boldsymbol{Y}_{lm}^{(1)} + G_{lm}^{(2)} \boldsymbol{Y}_{lm}^{(2)} + G_{lm}^{(3)} \boldsymbol{Y}_{lm}^{(3)} \right] \tag{2.9}$$

with the expansion coefficients

$$G_{lm}^{(j)} = \int_0^{4\pi} \boldsymbol{G} \cdot \boldsymbol{Y}_{lm}^{(j)*} \, \mathrm{d}\Omega, \quad j = 1, 2, 3. \tag{2.10}$$

Note that according to the definition of vector spherical harmonics (2.6), the coefficient $G_{lm}^{(1)}$ represents the normal (radial) component of the field $\boldsymbol{G}$, while $G_{lm}^{(2)}$ and $G_{lm}^{(3)}$ give the tangential (angular) field components.

2.1.2 Electromagnetic fields in vector spherical harmonics

The vector spherical harmonics are particularly useful, since they provide the complete angular dependence for any vector field. Thus, by decomposing time-harmonic electric and magnetic fields

$$\boldsymbol{E}(t, \boldsymbol{r}) = \boldsymbol{E}_{\omega}(\boldsymbol{r}) \mathrm{e}^{-\mathrm{i}\omega t}, \quad \boldsymbol{H}(t, \boldsymbol{r}) = \boldsymbol{H}_{\omega}(\boldsymbol{r}) \mathrm{e}^{-\mathrm{i}\omega t} \tag{2.11}$$

into the vector spherical harmonics, we can easily separate their radial and angular dependences,

$$\boldsymbol{E}_{\omega} = \sum_{l=0}^{\infty} \sum_{m=-l}^{l} \left[E_{lm}^{(1)}(r) \boldsymbol{Y}_{lm}^{(1)} + E_{lm}^{(2)}(r) \boldsymbol{Y}_{lm}^{(2)} + E_{lm}^{(3)}(r) \boldsymbol{Y}_{lm}^{(3)} \right], \tag{2.12}$$

$$\boldsymbol{H}_{\omega} = \sum_{l=0}^{\infty} \sum_{m=-l}^{l} \left[H_{lm}^{(1)}(r) \boldsymbol{Y}_{lm}^{(1)} + H_{lm}^{(2)}(r) \boldsymbol{Y}_{lm}^{(2)} + H_{lm}^{(3)}(r) \boldsymbol{Y}_{lm}^{(3)} \right]. \tag{2.13}$$

Using the vector spherical harmonics, we can also decompose Maxwell's equations for transverse electromagnetic fields, if we take into account that, for any arbitrary vector $\boldsymbol{G}$,

$$\nabla \times \boldsymbol{G} = \sum_{l=0}^{\infty} \sum_{m=-l}^{l} \left[-\frac{l(l+1)}{r} G_{lm}^{(3)} \boldsymbol{Y}_{lm}^{(1)} - \left(\frac{\mathrm{d}G_{lm}^{(3)}}{\mathrm{d}r} + \frac{G_{lm}^{(3)}}{r} \right) \boldsymbol{Y}_{lm}^{(2)} \right.$$
$$\left. + \left(-\frac{G_{lm}^{(1)}}{r} + \frac{\mathrm{d}G_{lm}^{(2)}}{\mathrm{d}r} + \frac{G_{lm}^{(2)}}{r} \right) \boldsymbol{Y}_{lm}^{(3)} \right].$$

After this decomposition, the Maxwell equations split into two independent sets of equations that represent the *transverse magnetic* (TM) and *transverse electric* (TE) polarizations. The TM modes feature two nonzero electric components,

$$E_{lm}^{(1)} = -\frac{\mathrm{i}}{k_0 \varepsilon} \frac{l(l+1)}{r} H_{lm}^{(3)}, \quad E_{lm}^{(2)} = -\frac{\mathrm{i}}{k_0 \varepsilon} \left(\frac{\mathrm{d}H_{lm}^{(3)}}{\mathrm{d}r} + \frac{H_{lm}^{(3)}}{r} \right), \tag{2.14}$$

with the governing transverse magnetic field $H_{lm}^{(3)}$ which obeys the equation

$$\frac{\mathrm{d}^2 H_{lm}^{(3)}}{\mathrm{d}r^2} + \frac{2}{r}\frac{\mathrm{d}H_{lm}^{(3)}}{\mathrm{d}r} - \left[\frac{l(l+1)}{r^2} - k_0^2\varepsilon\right] H_{lm}^{(3)} = 0. \tag{2.15}$$

At the same time, the TE modes exhibit the other two magnetic field components,

$$H_{lm}^{(1)} = \frac{\mathrm{i}}{k_0}\frac{l(l+1)}{r}E_{lm}^{(3)}, \quad H_{lm}^{(2)} = \frac{\mathrm{i}}{k_0}\left(\frac{\mathrm{d}E_{lm}^{(3)}}{\mathrm{d}r} + \frac{E_{lm}^{(3)}}{r}\right), \tag{2.16}$$

governed by the transverse electric field $E_{lm}^{(3)}$ which satisfies the equation

$$\frac{\mathrm{d}^2 E_{lm}^{(3)}}{\mathrm{d}r^2} + \frac{2}{r}\frac{\mathrm{d}E_{lm}^{(3)}}{\mathrm{d}r} - \left[\frac{l(l+1)}{r^2} - k_0^2\varepsilon\right] E_{lm}^{(3)} = 0. \tag{2.17}$$

Depending on the region of consideration, the solutions of Eqs. (2.15) and (2.17) are given by two different spherical functions. For any internal region containing the origin $r = 0$, the finite solution is given by the spherical Bessel functions j_l,

$$\left[H_{lm}^{(3)}\right]_{\mathrm{int}}, \left[E_{lm}^{(3)}\right]_{\mathrm{int}} \propto j_l(\sqrt{\varepsilon}k_0 r),$$

while for an external part including infinity, it is given by the spherical Hankel functions h_l,

$$\left[H_{lm}^{(3)}\right]_{\mathrm{ext}}, \left[E_{lm}^{(3)}\right]_{\mathrm{ext}} \propto h_l(\sqrt{\varepsilon}k_0 r),$$

the type of which is defined by the complex value of $\sqrt{\varepsilon}k_0$ in accordance with the requirement for finiteness of the solution at $r \to \infty$,

$$h_l(\sqrt{\varepsilon}k_0 r) = \begin{cases} h_l^{(1)}(\sqrt{\varepsilon}k_0 r), & \text{if } \mathrm{Im}(\sqrt{\varepsilon}k_0) \geq 0, \\ h_l^{(2)}(\sqrt{\varepsilon}k_0 r), & \text{if } \mathrm{Im}(\sqrt{\varepsilon}k_0) < 0. \end{cases}$$

For intermediate regions, one should use a linear combination of the spherical functions $j_l(\sqrt{\varepsilon}k_0 r)$ and $h_l(\sqrt{\varepsilon}k_0 r)$.

2.1.3 Spherical plasmons

Now, let us consider a uniform sphere of radius R and dielectric permittivity $\varepsilon = \varepsilon_{\mathrm{i}}$ embedded in a material with $\varepsilon = \varepsilon_{\mathrm{e}}$. The governing fields of the TM and TE modes inside the sphere, $r \leq R$, can be written as

$$\left[H_{lm}^{(3)}\right]_{\mathrm{int}} = -\mathrm{i}\frac{k_{\mathrm{i}}}{k_0}d_{lm}j_l(k_{\mathrm{i}}r), \tag{2.18}$$

$$\left[E_{lm}^{(3)}\right]_{\mathrm{int}} = c_{lm}j_l(k_{\mathrm{i}}r), \tag{2.19}$$

while for the surrounding medium, $r \geq R$, they are given by

$$\left[H_{lm}^{(3)}\right]_{\mathrm{ext}} = \mathrm{i}\frac{k_{\mathrm{e}}}{k_0}a_{lm}h_l(k_{\mathrm{e}}r), \tag{2.20}$$

$$\left[E_{lm}^{(3)}\right]_{\mathrm{ext}} = -b_{lm}h_l(k_{\mathrm{e}}r), \tag{2.21}$$

where a_{lm}, b_{lm}, c_{lm}, and d_{lm} are unknown constants; and $k_{\rm i} = \sqrt{\varepsilon_{\rm i}}k_0$ and $k_{\rm e} = \sqrt{\varepsilon_{\rm e}}k_0$ are the wavenumbers in the internal and external media, respectively.

Note that, for the eigenmodes, the unknown constants a_{lm}, b_{lm}, c_{lm}, and d_{lm} in Eqs. (2.18)–(2.21) must satisfy the boundary conditions which imply the continuity of the tangential components $H_{lm}^{(3)}$, $E_{lm}^{(2)}$ (for the TM modes) and $E_{lm}^{(3)}$, $H_{lm}^{(2)}$ (for the TE modes) on the sphere boundary, $r = R$. These conditions give us the dispersion relations for the eigenfrequencies of the plasmons supported in a spherical structure. For the TM modes, this relation can be written as

$$q_{\rm i}\psi_l(q_{\rm i})\xi_l'(q_{\rm e}) - q_{\rm e}\xi_l(q_{\rm e})\psi_l'(q_{\rm i}) = 0, \tag{2.22}$$

while for the TE modes, it is given by the following transcendental equation:

$$q_{\rm e}\psi_l(q_{\rm i})\xi_l'(q_{\rm e}) - q_{\rm i}\xi_l(q_{\rm e})\psi_l'(q_{\rm i}) = 0, \tag{2.23}$$

where $\psi_l(q_{\rm i}) = q_{\rm i}j_l(q_{\rm i})$ and $\xi_l(q_{\rm e}) = q_{\rm e}h_l(q_{\rm e})$ are the Riccati–Bessel functions, with $q_{\rm i} = k_{\rm i}R$ and $q_{\rm e} = k_{\rm e}R$.

Note that the two dispersion relations (2.22) and (2.23) do not depend on the azimuthal index m, so the eigenfrequencies of the normal modes are determined by the orbital index l only, $\omega = \omega_l$. Owing to the dissipation losses, these frequencies are complex, $\omega_l = \omega_l' + {\rm i}\omega_l''$, where the real part ω_l' gives the plasmon energy, $E_l = \hbar\omega_l'$, and the *negative* imaginary part ω_l'' characterizes the lifetime of the plasmons, $\tau_l = -1/(2\omega_l'')$.

In general, there are infinitely many eigenmodes of both polarizations for every orbital number l, $\omega_l = \omega_{ln}$ (n is the mode order number). This is a consequence of the oscillating behavior of the functions $\psi_l(q_{\rm i})$ and $\xi_l(q_{\rm e})$ for real $q_{\rm i}$ and $q_{\rm e}$. It forms an infinite pattern of *bulk plasmons*, the electromagnetic energy of which is spread out over the volume of the structure.

The existence of the bulk TM and TE eigenmodes is a common feature of media with a positive dielectric permittivity. For metals, it corresponds to energies above the plasma level. Thus, the bulk plasmons exist at $E_{ln} \geq \hbar\omega_{\rm p}$. Below the plasma frequency $\omega_{\rm p}$, the dielectric permittivity of any metal is negative. This gives us one more TM eigenmode for every orbital index l. These are the so-called *surface plasmons*. Their wavenumbers $k_{\rm e}$ and $k_{\rm i}$ exhibit very large imaginary parts. As a result, surface plasmons concentrate their energy near the metal interface according to the distribution of the induced surface charge.

For illustration, we consider two complementary structures – a silver particle and a cavity made in bulk silver. Note the difference between the dispersion equations (2.22) and (2.23) for these two structures – for the particle case $h_l = h_l^{(2)}$, since ${\rm Im}(k_{\rm e})$ is always negative, whereas for the cavity case h_l is given by $h_l^{(1)}$, since ${\rm Im}(k_{\rm e}) > 0$. Figures 2.1 and 2.2 show the energy and lifetime of the plasmons calculated according to the dispersion equations of the TM and TE eigenmodes for the first five orbital indexes l. They demonstrate that both structures support bulk TM and TE plasmons above the plasma frequency of silver, and that only surface TM modes are maintained below $\hbar\omega_{\rm p} \approx 3.7$ eV.

We point out that the right choice of the spherical Hankel function type is the crucial point in the eigenvalue problem, because $h_l^{(1)}$ and $h_l^{(2)}$ provide physically different

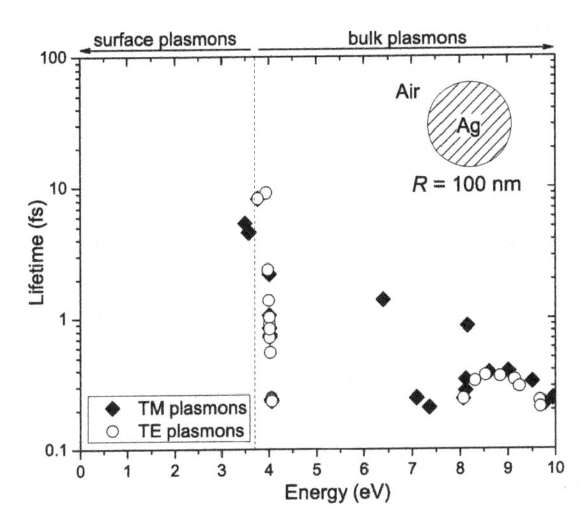

Figure 2.1 Surface and bulk plasmons of a spherical silver nanoparticle with $R = 100$ nm. The plasmons plotted have been calculated for the orbital indexes $l \leq 5$. The dashed line shows the level of the plasma energy $\hbar\omega_p$ in silver. The optical data are from Ref. [3].

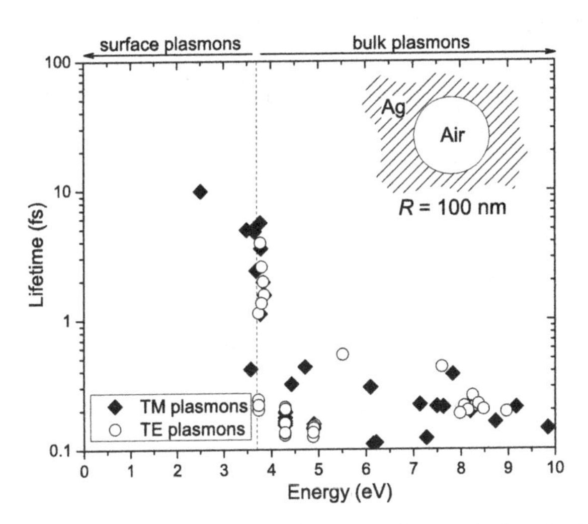

Figure 2.2 Surface and bulk plasmons of a spherical cavity in bulk silver. The radius of the cavity is $R = 100$ nm. The plasmons plotted have been calculated for the orbital indexes $l \leq 5$. The dashed line shows the level of the plasma energy $\hbar\omega_p$ in silver. The optical data are from Ref. [3].

solutions for the power flux distribution in the structure. Often, using the wrong type of Hankel function results in a misleading conclusion that the plasmon fields are radiative into the surrounding medium by nature (see, for example, the classical books by Stratton [4, Chapter IX] and by Bohren and Huffman [5, Chapter 4], or the original paper by Fuchs and Kliewer [6]). In fact, the functions $h_l^{(1)}$ give *diverging* spherical waves, while $h_l^{(2)}$ represent *converging* spherical waves. Thus, the plasmons are *radiative* for a spherical cavity in a bulk metal and *nonradiative* for a metal sphere. In a more general case, where both the sphere and the surrounding material are absorbing, the plasmons can be either radiative or nonradiative, depending on the structural area where the energy dissipation prevails.

2.1.4 Scattering by a sphere

Now, let us consider the scattering of plane waves by a spherical particle in a non-absorbing environment, where $\mathrm{Im}(\varepsilon_e) = 0$. This problem has an analytical solution called the *Mie solution*, named after the German physicist Gustav Mie (1868–1957), who solved the problem in 1908 [7]. Here, we derive the Mie solution within the vector spherical-harmonics formalism, which is an equivalent approach to the classical consideration [4, 5] which is based on the vector spherical-wavefunction expansion.

Owing to the spherical symmetry of the problem, the scattering of light is independent of the polarization of the incident wave. For the sake of simplicity, we assume that the incident wave is x-polarized and propagates along the z axis with frequency ω. Then, its electric and magnetic fields can be written as follows:

$$[\boldsymbol{E}_\omega]_{\mathrm{inc}} = (\boldsymbol{e}_r \sin\theta\cos\phi + \boldsymbol{e}_\theta \cos\theta\cos\phi - \boldsymbol{e}_\phi \sin\phi) E_0 \mathrm{e}^{\mathrm{i}k_e r\cos\theta}, \tag{2.24}$$

$$[\boldsymbol{H}_\omega]_{\mathrm{inc}} = \frac{k_e}{k_0} (\boldsymbol{e}_r \sin\theta\sin\phi + \boldsymbol{e}_\theta \cos\theta\sin\phi + \boldsymbol{e}_\phi \cos\phi) E_0 \mathrm{e}^{\mathrm{i}k_e r\cos\theta}, \tag{2.25}$$

where E_0 is the amplitude of the incident time-harmonic wave. According to Eqs. (2.12) and (2.13), the fields of the incident wave can be expanded into the vector spherical harmonics and split into the TM and TE modes. The governing field components of these modes,

$$\left[H_{lm}^{(3)}\right]_{\mathrm{inc}} = \int_\Omega [\boldsymbol{H}_\omega]_{\mathrm{inc}} \cdot \boldsymbol{Y}_{lm}^{(3)*}\, \mathrm{d}\Omega,$$

$$\left[E_{lm}^{(3)}\right]_{\mathrm{inc}} = \int_\Omega [\boldsymbol{E}_\omega]_{\mathrm{inc}} \cdot \boldsymbol{Y}_{lm}^{(3)*}\, \mathrm{d}\Omega,$$

can be calculated with the expansion of a plane wave into the scalar spherical harmonics,

$$\mathrm{e}^{\mathrm{i}k_e r\cos\theta} = 4\pi \sum_{l=0}^{\infty} \mathrm{i}^l (2l+1)^{1/2} j_l(k_e r) Y_{l0}(\theta,\phi), \tag{2.26}$$

which gives us

$$\left[H_{lm}^{(3)}\right]_{\mathrm{inc}} = -\mathrm{i}\frac{k_e}{k_0} E_{lm} j_l(k_e r), \tag{2.27}$$

$$\left[E_{lm}^{(3)}\right]_{\mathrm{inc}} = E_{lm} j_l(k_e r), \tag{2.28}$$

where

$$E_{lm} = \mathrm{i}^{l+1}\sqrt{\pi(2l+1)}\delta_{m,\pm 1} E_0.$$

Thus, any plane wave in the vector spherical harmonics has only two nonzero components with $m = \pm 1$ for every $l > 0$.

The overall electromagnetic field outside of the particle, $r \geq R$, is given as a superposition of the incident and scattered fields,

$$\boldsymbol{E}_{\text{ext}} = \boldsymbol{E}_{\text{inc}} + \boldsymbol{E}_{\text{sca}}, \qquad \boldsymbol{H}_{\text{ext}} = \boldsymbol{H}_{\text{inc}} + \boldsymbol{H}_{\text{sca}}. \tag{2.29}$$

The governing field components for the scattered light can be written in the same manner as Eqs. (2.20) and (2.21) in the eigenvalue problem,

$$\left[H_{lm}^{(3)}\right]_{\text{sca}} = \mathrm{i}\frac{k_{\text{e}}}{k_0} a_{lm} E_{lm} h_l(k_{\text{e}}r), \tag{2.30}$$

$$\left[E_{lm}^{(3)}\right]_{\text{sca}} = -b_{lm} E_{lm} h_l(k_{\text{e}}r), \tag{2.31}$$

where the coefficients a_{lm} and b_{lm} (called the *Mie scattering coefficients*) characterize the amplitude of the scattered fields, as compared with the incident wave fields. Thus, for the external fields we get

$$\left[H_{lm}^{(3)}\right]_{\text{ext}} = -\mathrm{i}\frac{k_{\text{e}}}{k_0} E_{lm}[j_l(k_{\text{e}}r) - a_{lm} h_l(k_{\text{e}}r)], \tag{2.32}$$

$$\left[E_{lm}^{(3)}\right]_{\text{ext}} = E_{lm}[j_l(k_{\text{e}}r) - b_{lm} h_l(k_{\text{e}}r)]. \tag{2.33}$$

At the same time, the internal fields induced inside the particle can be written as follows:

$$\left[H_{lm}^{(3)}\right]_{\text{int}} = -\mathrm{i}\frac{k_{\text{i}}}{k_0} d_{lm} E_{lm} j_l(k_{\text{i}}r), \tag{2.34}$$

$$\left[E_{lm}^{(3)}\right]_{\text{int}} = c_{lm} E_{lm} j_l(k_{\text{i}}r). \tag{2.35}$$

Now, we can apply the boundary conditions for the internal and external fields that require the continuity of the tangential components $H_{lm}^{(3)}$, $E_{lm}^{(2)}$ (for the TM modes) and $E_{lm}^{(3)}$, $H_{lm}^{(2)}$ (for the TE modes) on the surface of the sphere in order to determine the unknown coefficients. For the TM modes, this gives us

$$a_{lm} = \frac{q_{\text{i}}\psi_l(q_{\text{i}})\psi_l'(q_{\text{e}}) - q_{\text{e}}\psi_l(q_{\text{e}})\psi_l'(q_{\text{i}})}{q_{\text{i}}\psi_l(q_{\text{i}})\xi_l'(q_{\text{e}}) - q_{\text{e}}\xi_l(q_{\text{e}})\psi_l'(q_{\text{i}})}, \tag{2.36}$$

$$d_{lm} = \frac{q_{\text{i}}\psi_l(q_{\text{e}})\xi_l'(q_{\text{e}}) - q_{\text{i}}\xi_l(q_{\text{e}})\psi_l'(q_{\text{e}})}{q_{\text{i}}\psi_l(q_{\text{i}})\xi_l'(q_{\text{e}}) - q_{\text{e}}\xi_l(q_{\text{e}})\psi_l'(q_{\text{i}})}, \tag{2.37}$$

while for the TE modes it results in

$$b_{lm} = \frac{q_{\text{e}}\psi_l(q_{\text{i}})\psi_l'(q_{\text{e}}) - q_{\text{i}}\psi_l(q_{\text{e}})\psi_l'(q_{\text{i}})}{q_{\text{e}}\psi_l(q_{\text{i}})\xi_l'(q_{\text{e}}) - q_{\text{i}}\xi_l(q_{\text{e}})\psi_l'(q_{\text{i}})}, \tag{2.38}$$

$$c_{lm} = \frac{q_{\text{i}}\psi_l(q_{\text{e}})\xi_l'(q_{\text{e}}) - q_{\text{i}}\xi_l(q_{\text{e}})\psi_l'(q_{\text{e}})}{q_{\text{e}}\psi_l(q_{\text{i}})\xi_l'(q_{\text{e}}) - q_{\text{i}}\xi_l(q_{\text{e}})\psi_l'(q_{\text{i}})}. \tag{2.39}$$

The coefficients obtained do not depend on the azimuthal index m, similarly to the eigenfrequencies of the plasmons discussed in Section 2.1.3. To indicate this independence, we will omit the index m in the notation of the derived coefficients: $a_l \equiv a_{lm}$, $b_l \equiv b_{lm}$, $c_l \equiv c_{lm}$, $d_l \equiv d_{lm}$.

Note that the denominators of the scattering coefficients contain either plasmon dispersion relation (2.22) or (2.23). This may give a wrong impression that the coefficients a_l and d_l can grow infinitely at the TM plasmon frequencies $\omega_{ln}^{\rm TM}$, while b_l and c_l can become unlimited at the TE plasmon frequencies $\omega_{ln}^{\rm TE}$. In reality, the scattering coefficients and the plasmon dispersion equations contain different types of spherical Hankel function – for the scattering problem $h_l \equiv h_l^{(1)}$, since $k_{\rm e}$ is purely real, while for the eigenvalue problem of a metal sphere $h_l \equiv h_l^{(2)}$, since $\mathrm{Im}(k_{\rm e}) < 0$. In addition, the plasmon eigenfrequencies ω_{ln} are complex, while the frequency of the incident light ω is real. Eventually, all this makes exact matching between the plasmons and photons *impossible.*

Nonetheless, at photon energies $\hbar\omega$ close to the plasmon ones E_{ln}, one can observe growth of the scattering coefficients caused by the resonant coupling with the plasmons. Especially strong coupling takes place for subwavelength particles with $q_{\rm e} \ll 1$, for which the mismatch induced by the Hankel functions is small enough,

$$\frac{h_l^{(1)\prime}(q_{\rm e})}{h_l^{(1)}(q_{\rm e})} = \frac{h_l^{(2)\prime}(q_{\rm e})}{h_l^{(2)}(q_{\rm e})} + O(q_{\rm e}^{2l}).$$

In this case, the photons exhibit quite efficient coupling especially to long-lived plasmons that feature $|\omega_{ln}''| \ll \omega_{ln}'$.

2.1.5 Cross-sections

After the spatial distribution of the electromagnetic fields inside and outside a spherical particle has been derived, we can proceed to a discussion of the scattering and absorption of the incident light. The amount of the scattered light is given by the scattered power, which can be calculated as the closed surface integral over the particle surface,

$$P_{\rm sca} = \int_0^{4\pi} (\boldsymbol{S}_{\rm sca} \cdot \boldsymbol{e}_r) R^2 \, \mathrm{d}\Omega, \tag{2.40}$$

where the scattered Poynting vector is considered as averaged over time,

$$\boldsymbol{S}_{\rm sca} = \frac{c}{8\pi} \mathrm{Re}[\boldsymbol{E}_{\rm sca} \times \boldsymbol{H}_{\rm sca}^*]\Big|_{r=R}.$$

For the absorption, we should evaluate the power

$$P_{\rm abs} = -\int_0^{4\pi} (\boldsymbol{S}_{\rm ext} \cdot \boldsymbol{e}_r) R^2 \, \mathrm{d}\Omega, \tag{2.41}$$

which is based on the time-averaged total Poynting vector

$$\boldsymbol{S}_{\rm ext} = \frac{c}{8\pi} \mathrm{Re}[\boldsymbol{E}_{\rm ext} \times \boldsymbol{H}_{\rm ext}^*]\Big|_{r=R}.$$

The different signs in Eqs. (2.40) and (2.41) indicate that the absorbed and scattered energies go in different directions with respect to the particle surface-normal vector $\boldsymbol{e}_r$ – the power P_{abs} is absorbed inside the particle (in the inward direction), while P_{sca} is scattered outside of it (in the outward direction).

Owing to the orthogonality of the vector spherical harmonics, the closed surface integral of the normal component for the Poynting vector can be calculated as

$$\int_0^{4\pi} (\boldsymbol{S} \cdot \boldsymbol{e}_r) R^2 \, \mathrm{d}\Omega = \frac{cR^2}{8\pi} \sum_{l,m} \mathrm{Re} \left[E_{lm}^{(2)} H_{lm}^{(3)*} - E_{lm}^{(3)*} H_{lm}^{(2)} \right], \tag{2.42}$$

where the first term in the sum gives the contribution of the TM modes, while the second term is the contribution of the TE modes. Using this expression, we can derive

$$P_{\text{sca}} = \frac{c|E_0|^2}{4k_0 k_{\text{e}}} \sum_{l=1}^{\infty} (2l+1)(|a_l|^2 + |b_l|^2), \tag{2.43}$$

$$P_{\text{abs}} = \frac{c|E_0|^2}{4k_0 k_{\text{e}}} \sum_{l=1}^{\infty} (2l+1) \left[\mathrm{Re}(a_l + b_l) - (|a_l|^2 + |b_l|^2) \right]. \tag{2.44}$$

Eventually, this gives us the scattering and absorption cross-sections

$$\sigma_{\text{sca}} = \frac{P_{\text{sca}}}{S_{\text{inc}}} = \frac{2\pi}{k_{\text{e}}^2} \sum_{l=1}^{\infty} (2l+1)(|a_l|^2 + |b_l|^2), \tag{2.45}$$

$$\sigma_{\text{abs}} = \frac{P_{\text{abs}}}{S_{\text{inc}}} = \frac{2\pi}{k_{\text{e}}^2} \sum_{l=1}^{\infty} (2l+1) \left[\mathrm{Re}(a_l + b_l) - (|a_l|^2 + |b_l|^2) \right], \tag{2.46}$$

where $S_{\text{inc}} = \sqrt{\varepsilon_{\text{e}}} c |E_0|^2 / (8\pi)$ is the intensity of the incident wave.

For some cases, it is more convenient to operate with the extinction cross-section that characterizes the total losses caused by both scattering and absorption,

$$\sigma_{\text{ext}} = \sigma_{\text{sca}} + \sigma_{\text{abs}}.$$

In terms of the coefficients a_l and b_l, the cross-section σ_{ext} can be written as

$$\sigma_{\text{ext}} = \frac{2\pi}{k_{\text{e}}^2} \sum_{l=1}^{\infty} (2l+1) \mathrm{Re}(a_l + b_l). \tag{2.47}$$

Following the expressions derived for the scattering and extinction cross-sections, we can draw a conclusion about the Mie scattering coefficients – the real parts of a_l and b_l describe the total losses caused by scattering and absorption by the TM and TE modes, while their square moduli give the respective contribution of the scattering.

As mentioned before, the scattering coefficients can increase close to the plasmon energies. This effect becomes apparent if we plot the spectral cross-sections of both scattering and absorption. Figures 2.3 and 2.4 clearly show the distinct peaks that appear in the spectra of silver nanoparticles as a result of the excitation of the first three orbital modes with $l = 1, 2, 3$. It is noteworthy that the scattering and absorption

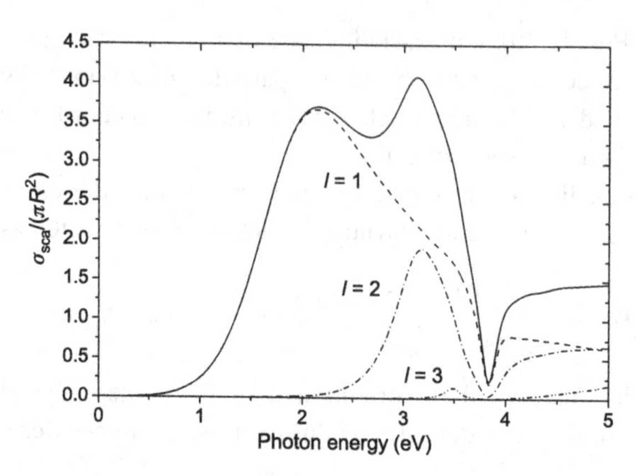

Figure 2.3 The spectral scattering cross-section of a silver nanoparticle of radius $R = 100$ nm. The dashed lines show the contributions of the orbital modes with $l = 1, 2, 3$. The optical data are from Ref. [3].

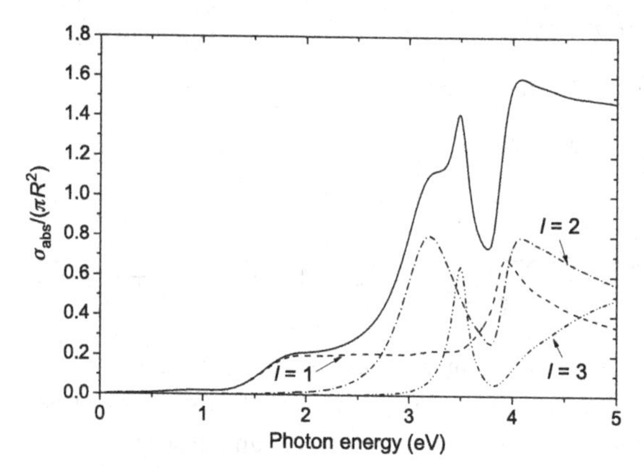

Figure 2.4 The spectral absorption cross-section of a silver nanoparticle of radius $R = 100$ nm. The dashed lines show the contributions of the orbital modes with $l = 1, 2, 3$. The optical data are from Ref. [3].

cross-sections at the plasmon energies may be even higher than the geometric cross-section of the particle, $\sigma_{geo} = \pi R^2$. Hence, one can scatter and absorb more light than can impinge upon the particle surface according to geometrical optics [5]. In other words, through the coupling with plasmons, one can attract more photons and enhance the light–matter interaction. In addition, the scattering and absorption spectra can be controlled by changing the energies of plasmons. This can be done by tuning the size (Fig. 2.5) or material (Fig. 2.6) of the particle. Another factor that can be used to control the plasmon energy is the permittivity of the surrounding medium, ε_e. Figure 2.7 shows the change of the scattering and absorption cross-sections of a silver nanoparticle with $R = 50$ nm for different values of ε_e.

Figures 2.5–2.7 demonstrate that surface plasmons are particularly sensitive to the structural parameters. In addition, they provide very efficient interaction with photons.

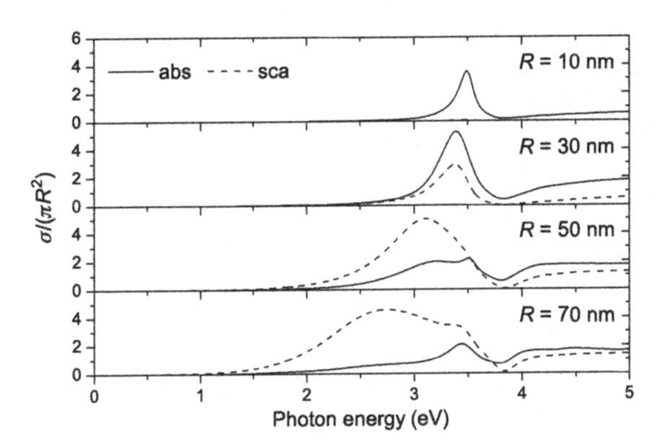

Figure 2.5 Scattering and absorption cross-sections of silver nanoparticles with different radii R. The optical data are from Ref. [3].

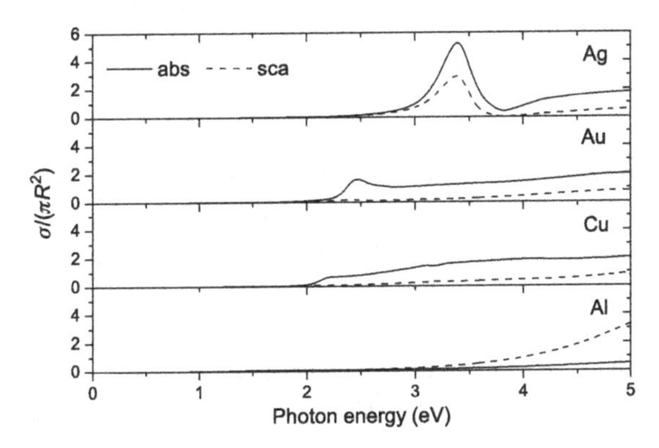

Figure 2.6 Scattering and absorption cross-sections of nanoparticles of different metals. The radius of the nanoparticles is fixed to $R = 30$ nm. The optical data are from Ref. [8].

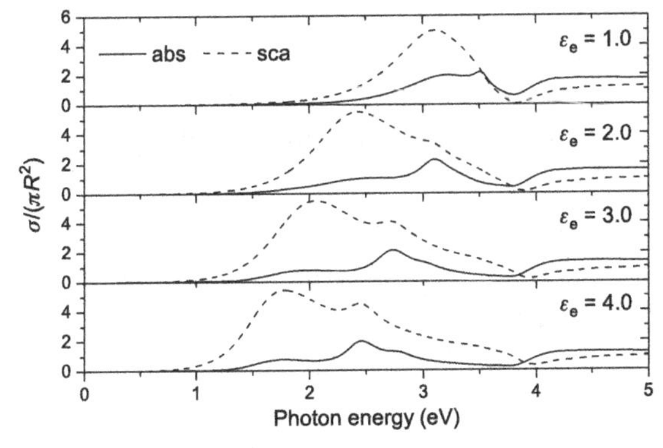

Figure 2.7 Scattering and absorption cross-sections of a silver nanoparticle with $R = 50$ nm embedded in different materials. The optical data are from Ref. [3].

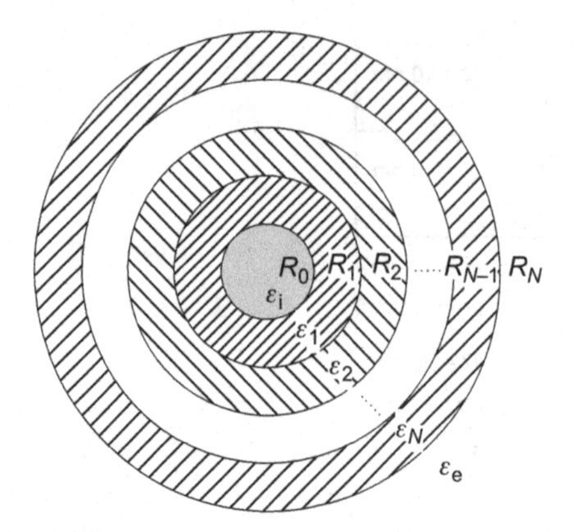

Figure 2.8 A sketch of a multilayer sphere with an N-layer coating.

Thus surface plasmons are a very powerful tool for optical applications that require strongly coupled light–matter interaction [9].

2.1.6 A multilayer sphere

It is obvious that the presence of a coating on the sphere surface affects the eigenfrequencies of the plasmons. Moreover, it can significantly change the spectral characteristics of scattering and absorption. To study the coating effect, we consider the general case of a multilayer sphere with an N-layer coating as shown in Fig. 2.8. The scattering coefficients of such a structure can be obtained with the *transfer-matrix method.* To demonstrate the method, let us consider a general form of the governing fields excited in every layer of the coating. For the TM and TE polarizations, they are given by

$$H_{lm}^{(3)} = -\mathrm{i}\frac{E_{lm}}{k_0 r}[e_{lm}\psi_l(\sqrt{\varepsilon}k_0 r) + f_{lm}\xi_l(\sqrt{\varepsilon}k_0 r)],$$

$$E_{lm}^{(3)} = \frac{E_{lm}}{\sqrt{\varepsilon}k_0 r}[g_{lm}\psi_l(\sqrt{\varepsilon}k_0 r) + h_{lm}\xi_l(\sqrt{\varepsilon}k_0 r)],$$

where e_{lm}, f_{lm}, g_{lm}, and h_{lm} are unknown constants. Using the governing fields, we can compose a vector $\boldsymbol{F}_j$ of the tangential fields in the jth spherical layer, $R_{j-1} \le r \le R_j$, of the coating,

$$\boldsymbol{F}_j(r) = \begin{bmatrix} H_{lm}^{(3)}(r) \\ E_{lm}^{(2)}(r) \\ E_{lm}^{(3)}(r) \\ H_{lm}^{(2)}(r) \end{bmatrix}_j .$$

In matrix form, it can be written as follows:

$$\boldsymbol{F}_j(r) = \boldsymbol{N}_j(r) \cdot \boldsymbol{C}_j, \tag{2.48}$$

where $\boldsymbol{C}_j$ is the vector of unknown constants,

$$\boldsymbol{C}_j = \begin{bmatrix} e_{lm} \\ f_{lm} \\ g_{lm} \\ h_{lm} \end{bmatrix}_j,$$

and the matrix $\boldsymbol{N}_j(r)$ is composed of the Riccati–Bessel functions,

$$\boldsymbol{N}_j(r) = \frac{E_{lm}}{k_0 k_j r} \begin{bmatrix} -\mathrm{i}k_j \psi_l(k_j r) & -\mathrm{i}k_j \xi_l(k_j r) & 0 & 0 \\ -k_0 \psi_l'(k_j r) & -k_0 \xi_l'(k_j r) & 0 & 0 \\ 0 & 0 & k_0 \psi_l(k_j r) & k_0 \xi_l(k_j r) \\ 0 & 0 & \mathrm{i}k_j \psi_l'(k_j r) & \mathrm{i}k_j \xi_l'(k_j r) \end{bmatrix},$$

with $k_j = \sqrt{\varepsilon_j} k_0$ being the wavenumber in the jth layer of the coating.

The vector of unknown constants $\boldsymbol{C}_j$ in Eq. (2.48) can be eliminated with the vector of tangential fields at the inner boundary R_{j-1},

$$\boldsymbol{C}_j = \boldsymbol{N}_j^{-1}(R_{j-1}) \cdot \boldsymbol{F}_j(R_{j-1}).$$

Thus, one can get the relation for the vectors of tangential fields at R_{j-1} and R_j,

$$\boldsymbol{F}_j(R_j) = \boldsymbol{M}_j \cdot \boldsymbol{F}_j(R_{j-1}), \tag{2.49}$$

where $\boldsymbol{M}_j$ is the transfer matrix of the jth spherical layer,

$$\boldsymbol{M}_j = \boldsymbol{N}_j(R_j) \cdot \boldsymbol{N}_j^{-1}(R_{j-1}),$$

that translates the tangential fields from one boundary of the layer to another.

As the boundary conditions between the adjacent layers require the tangential fields to be continuous, we can write

$$\boldsymbol{F}_N(R_N) = \boldsymbol{M} \cdot \boldsymbol{F}_1(R_0), \tag{2.50}$$

where $\boldsymbol{M}$ is the *transfer matrix* of the N-layer system given by

$$\boldsymbol{M} = \boldsymbol{M}_N \cdot \boldsymbol{M}_{N-1} \cdot \dots \cdot \boldsymbol{M}_2 \cdot \boldsymbol{M}_1.$$

Thus, the optical properties of any spherical coating can be described with a 4×4 transfer matrix $\boldsymbol{M}$.

Now, we can require the continuity of the tangential fields at $r = R_0$ and $r = R_N$ with the internal and external fields. This gives us

$$\boldsymbol{F}_{\mathrm{ext}}(R_N) = \boldsymbol{M} \cdot \boldsymbol{F}_{\mathrm{int}}(R_0). \tag{2.51}$$

In general, the vectors of internal and external fields can be written using the vector of scattering coefficients $\boldsymbol{C}_{\mathrm{sca}}$,

$$\boldsymbol{F}_{\mathrm{int}}(r) = \boldsymbol{N}_{\mathrm{int}}(r) \cdot \boldsymbol{C}_{\mathrm{sca}}, \tag{2.52}$$

$$\boldsymbol{F}_{\mathrm{ext}}(r) = \boldsymbol{N}_{\mathrm{sca}}(r) \cdot \boldsymbol{C}_{\mathrm{sca}} + \boldsymbol{F}_{\mathrm{inc}}(r), \tag{2.53}$$

where

$$N_{\text{int}}(r) = \frac{E_{lm}}{k_0 k_{\text{i}} r}\begin{bmatrix} 0 & 0 & 0 & -\text{i}k_{\text{i}}\psi_l(k_{\text{i}}r) \\ 0 & 0 & 0 & -k_0\psi_l'(k_{\text{i}}r) \\ 0 & 0 & k_0\psi_l(k_{\text{i}}r) & 0 \\ 0 & 0 & \text{i}k_{\text{i}}\psi_l'(k_{\text{i}}r) & 0 \end{bmatrix},$$

$$N_{\text{sca}}(r) = \frac{E_{lm}}{k_0 k_{\text{e}} r}\begin{bmatrix} \text{i}k_{\text{e}}\xi_l(k_{\text{e}}r) & 0 & 0 & 0 \\ k_0\xi_l'(k_{\text{e}}r) & 0 & 0 & 0 \\ 0 & -k_0\xi_l(k_{\text{e}}r) & 0 & 0 \\ 0 & -\text{i}k_{\text{e}}\xi_l'(k_{\text{e}}r) & 0 & 0 \end{bmatrix},$$

$$F_{\text{inc}}(r) = \frac{E_{lm}}{k_0 k_{\text{e}} r}\begin{bmatrix} -\text{i}k_{\text{e}}\psi_l(k_{\text{e}}r) \\ -k_0\psi_l'(k_{\text{e}}r) \\ k_0\psi_l(k_{\text{e}}r) \\ \text{i}k_{\text{e}}\psi_l'(k_{\text{e}}r) \end{bmatrix}, \quad C_{\text{sca}} = \begin{bmatrix} a_{lm} \\ b_{lm} \\ c_{lm} \\ d_{lm} \end{bmatrix}.$$

Thus, Eq. (2.51) gives us a linear equation for C_{sca},

$$N_{\text{sca}}(R_N)\cdot C_{\text{sca}} + F_{\text{inc}}(R_N) = M\cdot N_{\text{int}}(R_0)\cdot C_{\text{sca}}, \tag{2.54}$$

the solution of which is given by

$$C_{\text{sca}} = [M\cdot N_{\text{int}}(R_0) - N_{\text{sca}}(R_N)]^{-1}\cdot F_{\text{inc}}(R_N). \tag{2.55}$$

In terms of the transfer-matrix elements M_{ij}, the scattering coefficients a_{lm} and b_{lm} can be written as follows:

$$a_{lm} = \frac{k_{\text{e}}\psi_l(k_{\text{e}}R_N)Z_2 + \text{i}k_0\psi_l'(k_{\text{e}}R_N)Z_1}{k_{\text{e}}\xi_l(k_{\text{e}}R_N)Z_2 + \text{i}k_0\xi_l'(k_{\text{e}}R_N)Z_1}, \tag{2.56}$$

$$b_{lm} = \frac{k_{\text{e}}\psi_l'(k_{\text{e}}R_N)Z_3 + \text{i}k_0\psi_l(k_{\text{e}}R_N)Z_4}{k_{\text{e}}\xi_l'(k_{\text{e}}R_N)Z_3 + \text{i}k_0\xi_l(k_{\text{e}}R_N)Z_4}, \tag{2.57}$$

where

$$Z_1 = k_{\text{i}}M_{11}\psi_l(k_{\text{i}}R_0) - \text{i}k_0M_{12}\psi_l'(k_{\text{i}}R_0),$$

$$Z_2 = k_{\text{i}}M_{21}\psi_l(k_{\text{i}}R_0) - \text{i}k_0M_{22}\psi_l'(k_{\text{i}}R_0),$$

$$Z_3 = k_{\text{i}}M_{34}\psi_l'(k_{\text{i}}R_0) - \text{i}k_0M_{33}\psi_l(k_{\text{i}}R_0),$$

$$Z_4 = k_{\text{i}}M_{44}\psi_l'(k_{\text{i}}R_0) - \text{i}k_0M_{43}\psi_l(k_{\text{i}}R_0).$$

As one can see, the scattering coefficients for a coated sphere do not depend on the azimuthal index m, similarly to the case of an uncoated sphere. Thus, we can omit the index m in the notation of the scattering coefficients: $a_l \equiv a_{lm}$ and $b_l \equiv b_{lm}$.

The expressions obtained for a_l and b_l allow us to get the dispersion relations of the TM and TE plasmons that exist in a multilayer sphere. These relations can be derived by requiring

$$M\cdot N_{\text{int}}(R_0) - N_{\text{sca}}(R_N) = 0, \tag{2.58}$$

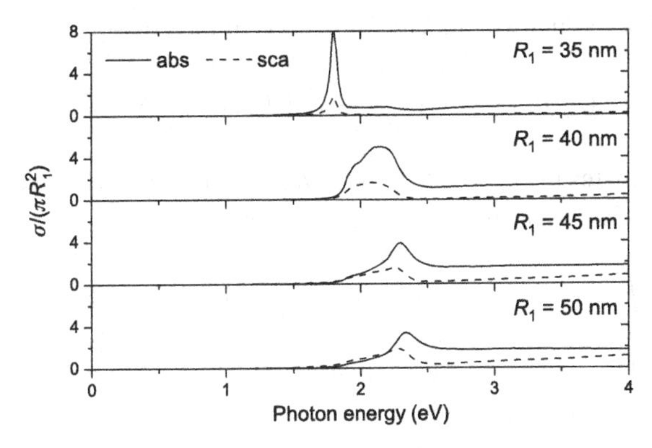

Figure 2.9 Scattering and absorption cross-sections of spherical SiO_2 nanoparticles coated with gold. The radius of the SiO_2 core is fixed to $R_0 = 30$ nm; the thickness of the Au coating varies from 5 to 20 nm. The optical data are from Ref. [8].

when the denominators of a_l or b_l formally become zero, with the subsequent change of h_l in the external fields to the appropriate spherical Hankel function $h_l^{(1)}$ or $h_l^{(2)}$ depending on the imaginary part of k_e. Following this simple procedure, we can write the dispersion relations as

$$k_e \xi_l(k_e R_N) Z_2 + \mathrm{i} k_0 \xi_l'(k_e R_N) Z_1 = 0, \tag{2.59}$$

$$k_e \xi_l'(k_e R_N) Z_3 + \mathrm{i} k_0 \xi_l(k_e R_N) Z_4 = 0 \tag{2.60}$$

for the TM and TE plasmons, respectively.

Figure 2.9 shows the scattering and absorption cross-sections of SiO_2–Au core–shell particles obtained with the transfer-matrix method. It reveals the spectral change of the cross-sections with varying thickness of the gold coating. The peaks seen in Fig. 2.9 are attributed to the surface plasmons of the metal nanoshell, which provide very efficient heating of the structure. The most prominent effect of the nanoshell plasmons is observed for thin Au coatings, for which the resonant coupling with the surface plasmons results in a huge absorption of the incident light with a high frequency selectivity [10].

2.2 Plasmonic modes in cylindrical geometry

In this section, we consider the transverse normal modes of an infinitely long cylinder. On the basis of this study, we analyze the plasmon polaritons of a metal cylinder and a cylindrical cavity in a bulk metal. In addition, we consider the scattering of plane waves by a metal nanowire and discuss the case of multilayer cylindrical structures.

2.2.1 Cylindrical harmonics

To describe the normal modes of cylindrical structures, we introduce two-dimensional *scalar cylindrical harmonics*

$$V_m(k_z; \theta, z) = \frac{\mathrm{e}^{\mathrm{i}(m\theta + k_z z)}}{2\pi} \tag{2.61}$$

with integer m and real k_z, which are the eigenfunctions of the Poisson equation

$$\nabla^2 V_m(k_z;\theta,z) = -\left(\frac{m^2}{r^2}+k_z^2\right) V_m(k_z;\theta,z) \tag{2.62}$$

in the cylindrical coordinates (r,θ,z). The scalar cylindrical harmonics give us a complete set of orthonormal functions defined in the (θ,z) space,

$$\int_0^{2\pi} \mathrm{d}\theta \int_{-\infty}^{\infty} \mathrm{d}z\, V_m(k_z;\theta,z)V_{m'}^*(k_z';\theta,z) = \delta_{mm'}\delta(k_z-k_z'), \tag{2.63}$$

where $\delta(k_z-k_z')$ is the Dirac delta function. By using $V_m(k_z;\theta,z)$, any arbitrary scalar function $g(\theta,z)$ can be expanded as follows:

$$g(\theta,z) = \sum_{m=-\infty}^{\infty} \int_{-\infty}^{\infty} \mathrm{d}k_z\, A_m(k_z)V_m(k_z;\theta,z), \tag{2.64}$$

with the expansion coefficients given by

$$A_m(k_z) = \int_0^{2\pi} \mathrm{d}\theta \int_{-\infty}^{\infty} \mathrm{d}z\, g(\theta,z)V_m^*(k_z;\theta,z). \tag{2.65}$$

By performing a vector extension of the scalar cylindrical harmonics, we can obtain the orthogonal *vector cylindrical harmonics*

$$\begin{aligned} \boldsymbol{V}_m^r(k_z;\theta,z) &= \boldsymbol{e}_r V_m(k_z;\theta,z), \\ \boldsymbol{V}_m^\theta(k_z;\theta,z) &= \boldsymbol{e}_\theta V_m(k_z;\theta,z), \\ \boldsymbol{V}_m^z(k_z;\theta,z) &= \boldsymbol{e}_z V_m(k_z;\theta,z). \end{aligned} \tag{2.66}$$

With the vector harmonics, any arbitrary vector field $\boldsymbol{G}(\theta,z)$ can be expanded as follows:

$$\boldsymbol{G} = \sum_{m=-\infty}^{\infty} \int_{-\infty}^{\infty} \mathrm{d}k_z \left[G_m^r(k_z)\boldsymbol{V}_m^r + G_m^\theta(k_z)\boldsymbol{V}_m^\theta + G_m^z(k_z)\boldsymbol{V}_m^z\right], \tag{2.67}$$

with the expansion coefficients given by

$$\begin{aligned} G_m^r(k_z) &= \int_0^{2\pi} \mathrm{d}\theta \int_{-\infty}^{\infty} \mathrm{d}z\, \boldsymbol{G}\cdot\boldsymbol{V}_m^{r*}(k_z;\theta,z), \\ G_m^\theta(k_z) &= \int_0^{2\pi} \mathrm{d}\theta \int_{-\infty}^{\infty} \mathrm{d}z\, \boldsymbol{G}\cdot\boldsymbol{V}_m^{\theta*}(k_z;\theta,z), \\ G_m^z(k_z) &= \int_0^{2\pi} \mathrm{d}\theta \int_{-\infty}^{\infty} \mathrm{d}z\, \boldsymbol{G}\cdot\boldsymbol{V}_m^{z*}(k_z;\theta,z). \end{aligned} \tag{2.68}$$

2.2.2 Electromagnetic fields in vector cylindrical harmonics

The vector cylindrical harmonics provide the complete tangential dependence of any vector field on the θ and z coordinates. Thus, by decomposing the electric $\boldsymbol{E}_\omega$ and magnetic $\boldsymbol{H}_\omega$ fields into the vector cylindrical harmonics, we can separate their radial dependences from the tangential ones,

$$\boldsymbol{E}_\omega = \sum_{m=-\infty}^{\infty} \int_{-\infty}^{\infty} \mathrm{d}k_z \left[E_m^r(r)\boldsymbol{V}_m^r + E_m^\theta(r)\boldsymbol{V}_m^\theta + E_m^z(r)\boldsymbol{V}_m^z \right], \tag{2.69}$$

$$\boldsymbol{H}_\omega = \sum_{m=-\infty}^{\infty} \int_{-\infty}^{\infty} \mathrm{d}k_z \left[H_m^r(r)\boldsymbol{V}_m^r + H_m^\theta(r)\boldsymbol{V}_m^\theta + H_m^z(r)\boldsymbol{V}_m^z \right], \tag{2.70}$$

and derive the radial distributions from the Maxwell equations for transverse electromagnetic fields.

In general, the normal modes in cylindrical geometry are *hybrid* – their TM and TE polarizations are no longer independent, except in a very few particular cases that we will discuss below. Formally, we still can consider the two polarizations separately, but finally they become coupled through the boundary conditions. For the TM polarization with $H_m^z = 0$, the electromagnetic field components

$$E_m^r = \frac{\mathrm{i}k_z}{k_0^2\varepsilon - k_z^2}\frac{\mathrm{d}E_m^z}{\mathrm{d}r}, \qquad E_m^\theta = -\frac{m}{r}\frac{k_z}{k_0^2\varepsilon - k_z^2}E_m^z, \tag{2.71}$$

$$H_m^r = \frac{m}{r}\frac{k_0\varepsilon}{k_0^2\varepsilon - k_z^2}E_m^z, \qquad H_m^\theta = \frac{\mathrm{i}k_0\varepsilon}{k_0^2\varepsilon - k_z^2}\frac{\mathrm{d}E_m^z}{\mathrm{d}r} \tag{2.72}$$

are defined through the E_m^z field, which obeys the differential equation

$$\frac{\mathrm{d}^2E_m^z}{\mathrm{d}r^2} + \frac{1}{r}\frac{\mathrm{d}E_m^z}{\mathrm{d}r} - \left(\frac{m^2}{r^2} - k_0^2\varepsilon + k_z^2\right)E_m^z = 0. \tag{2.73}$$

At the same time, in the TE polarization with $E_m^z = 0$, the field components can be written as follows:

$$E_m^r = -\frac{m}{r}\frac{k_0}{k_0^2\varepsilon - k_z^2}H_m^z, \qquad E_m^\theta = -\frac{\mathrm{i}k_0}{k_0^2\varepsilon - k_z^2}\frac{\mathrm{d}H_m^z}{\mathrm{d}r}, \tag{2.74}$$

$$H_m^r = \frac{\mathrm{i}k_z}{k_0^2\varepsilon - k_z^2}\frac{\mathrm{d}H_m^z}{\mathrm{d}r}, \qquad H_m^\theta = -\frac{m}{r}\frac{k_z}{k_0^2\varepsilon - k_z^2}H_m^z, \tag{2.75}$$

where the governing magnetic field H_m^z satisfies the equation

$$\frac{\mathrm{d}^2H_m^z}{\mathrm{d}r^2} + \frac{1}{r}\frac{\mathrm{d}H_m^z}{\mathrm{d}r} - \left(\frac{m^2}{r^2} - k_0^2\varepsilon + k_z^2\right)H_m^z = 0. \tag{2.76}$$

Depending on the region under consideration, the solutions of Eqs. (2.73) and (2.76) are given by two different cylindrical functions. For an internal region containing the origin $r = 0$, the solution is given by the cylindrical Bessel functions,

$$\left[E_m^z\right]_{\mathrm{int}},\ \left[H_m^z\right]_{\mathrm{int}} \propto J_m(k_\perp r),$$

where $k_\perp = \sqrt{k_0^2\varepsilon - k_z^2}$. For an external part including infinity, it is given by the cylindrical Hankel functions,

$$\left[E_m^z\right]_{\mathrm{ext}},\ \left[H_m^z\right]_{\mathrm{ext}} \propto H_m(k_\perp r),$$

the type of which is defined by the complex value $k_\perp$ in accordance with the requirement for finiteness of the solution at $r \to \infty$,

$$H_m(k_\perp r) = \begin{cases} H_m^{(1)}(k_\perp r), & \text{if } \mathrm{Im}(k_\perp) \geq 0, \\ H_m^{(2)}(k_\perp r), & \text{if } \mathrm{Im}(k_\perp) < 0. \end{cases}$$

For intermediate regions, one should consider a linear combination of the cylindrical functions $J_m(k_\perp r)$ and $H_m(k_\perp r)$.

2.2.3 Cylindrical plasmon polaritons

Now, let us consider a uniform infinite cylinder of the radius R and dielectric permittivity $\varepsilon = \varepsilon_\mathrm{i}$ embedded in the material with $\varepsilon = \varepsilon_\mathrm{e}$. The governing fields of the TM and TE polarizations inside the cylinder, $r \leq R$, can be written as

$$\left[E_m^z\right]_{\mathrm{int}} = d_m J_m(k_{\mathrm{i}\perp} r), \tag{2.77}$$

$$\left[H_m^z\right]_{\mathrm{int}} = -\mathrm{i}\frac{k_\mathrm{i}}{k_0} c_m J_m(k_{\mathrm{i}\perp} r), \tag{2.78}$$

while for the surrounding medium, $r \geq R$, they are given by

$$\left[E_m^z\right]_{\mathrm{ext}} = -a_m H_m(k_{\mathrm{e}\perp} r), \tag{2.79}$$

$$\left[H_m^z\right]_{\mathrm{ext}} = \mathrm{i}\frac{k_\mathrm{e}}{k_0} b_m H_m(k_{\mathrm{e}\perp} r), \tag{2.80}$$

where a_m, b_m, c_m, and d_m are unknown constants; and $k_{\mathrm{i}\perp} = (k_0^2\varepsilon_\mathrm{i} - k_z^2)^{1/2}$ and $k_{\mathrm{e}\perp} = (k_0^2\varepsilon_\mathrm{e} - k_z^2)^{1/2}$ are the wavenumbers $k_\perp$ in the internal and external media, respectively.

For the eigenmodes, the unknown constants a_m, b_m, c_m, and d_m in Eqs. (2.77)–(2.80) must satisfy the boundary conditions that imply the continuity of the tangential components E_m^θ, E_m^z, H_m^θ, and H_m^z on the cylinder boundary, $r = R$. These conditions give us the dispersion relation for the eigenfrequencies ω of the plasmon polaritons supported by the structure as a function of k_z

$$D_m^2(\omega, k_z) = W_m^{\mathrm{TM}}(\omega, k_z) W_m^{\mathrm{TE}}(\omega, k_z), \tag{2.81}$$

where

$$D_m(\omega, k_z) = \frac{m k_z}{k_0}\left(\frac{1}{\eta_\mathrm{e}^2} - \frac{1}{\eta_\mathrm{i}^2}\right) J_m(\eta_\mathrm{i}) H_m(\eta_\mathrm{e}),$$

$$W_m^{\mathrm{TM}}(\omega, k_z) = \frac{\varepsilon_\mathrm{e}}{\eta_\mathrm{e}} J_m(\eta_\mathrm{i}) H_m'(\eta_\mathrm{e}) - \frac{\varepsilon_\mathrm{i}}{\eta_\mathrm{i}} J_m'(\eta_\mathrm{i}) H_m(\eta_\mathrm{e}),$$

$$W_m^{\mathrm{TE}}(\omega, k_z) = \frac{1}{\eta_\mathrm{e}} J_m(\eta_\mathrm{i}) H_m'(\eta_\mathrm{e}) - \frac{1}{\eta_\mathrm{i}} J_m'(\eta_\mathrm{i}) H_m(\eta_\mathrm{e}),$$

with $\eta_\mathrm{i} = k_{\mathrm{i}\perp} R$ and $\eta_\mathrm{e} = k_{\mathrm{e}\perp} R$.

As we mentioned above, the character of the cylindrical plasmon polaritons is hybrid. Their TM and TE components are coupled through the boundary conditions. However, there are two cases in which the polaritons split into two independent polarizations.

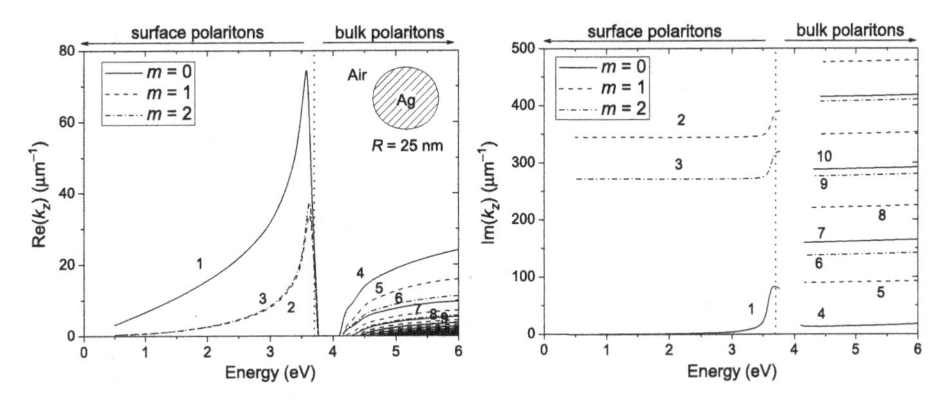

Figure 2.10 The dependence of the complex wavenumbers k_z on the energy $\hbar\omega_{mn}$ for plasmon polaritons with azimuthal numbers $m = 0, 1, 2$ propagating in a silver nanowire of radius $R = 25$ nm. The vertical dotted lines show the level of the plasma energy $\hbar\omega_{\mathrm{p}}$ in silver. The optical data are from Ref. [3].

One can see from the dispersion equation that $D_m(\omega, k_z)$ vanishes, if $m = 0$ or $k_z = 0$. In these particular cases, the dispersion equation gives us $W_m^{\mathrm{TM}}(\omega, k_z) = 0$ for the TM modes and $W_m^{\mathrm{TE}}(\omega, k_z) = 0$ for the TE modes, respectively. For any other m and k_z, the plasmon polaritons are hybrid, featuring quite complex field patterns with all six nonzero field components.

Note that there is an infinite number of plasmon polaritons supported by cylindrical structures for every azimuthal index m. Dispersion equation (2.81) provides the relation for their eigenfrequencies $\omega = \omega_{mn}(k_z)$, where n denotes the mode sequence number. Owing to the energy dissipation occurring in a metal, all plasmon polaritons attenuate. Their attenuation is described by a complex $\omega_{mn} = \omega'_{mn} + \mathrm{i}\omega''_{mn}$ for real k_z. This is the so-called *temporal solution* derived under the assumption that the polaritons attenuate in time in accordance with the plasmon lifetime $\tau_{mn} = -1/(2\omega''_{mn})$. However, for waveguiding applications it is more natural to consider the *spatial solution* of the problem, for which ω_{mn} is real, but $k_z = k'_z + \mathrm{i}k''_z$ is complex, i.e. the polaritons attenuate in space during their propagation. These two solutions are the particular cases of the general *spatio-temporal solution*, when the propagating polaritons attenuate both in space and in time according to the spatio-temporal relation for the attenuation rates [11, 12],

$$\omega''_{mn} + V_{\mathrm{gr}} k''_z = 0, \tag{2.82}$$

where $V_{\mathrm{gr}} = \partial\omega'_{mn}/\partial k'_z$ is the group velocity that characterizes the energy transfer by the polaritons.

To demonstrate the typical dispersion curves for the complex wavenumber k_z of the plasmon polaritons supported by cylindrical structures, we consider two cases, namely those of a metal nanowire (Fig. 2.10) and a cavity in a bulk metal (Fig. 2.11). In general, the frequency bands of those plasmonic modes are similar to the bands in spherical geometry. The low-frequency polaritons with energies below the plasma level, $E_{mn} < \hbar\omega_{\mathrm{p}}$, are *surface* polaritons. They are formed by the propagating surface plasmons that feature large imaginary parts for $k_{\mathrm{i}\perp}$. As a result, the electromagnetic field

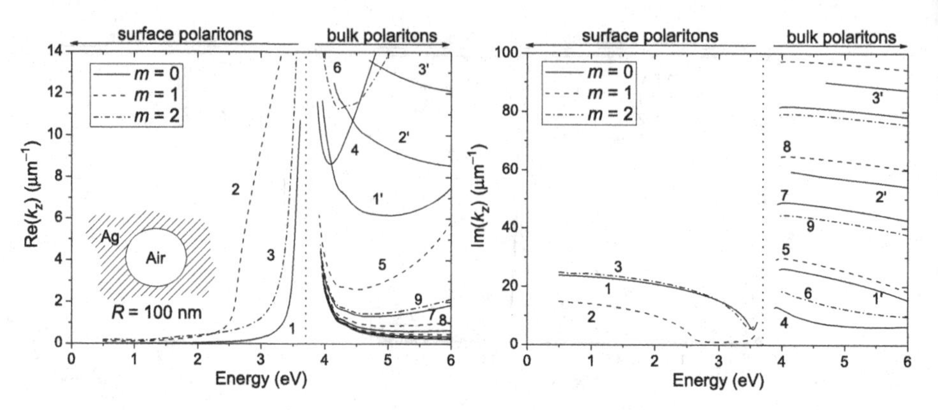

Figure 2.11 The dependence of the complex wavenumbers k_z on the energy $\hbar\omega_{mn}$ for plasmon polaritons with azimuthal numbers $m = 0, 1, 2$ propagating in a cylindrical cavity of radius $R = 100$ nm made in bulk silver. The vertical dotted lines show the level of the plasma energy $\hbar\omega_p$ in silver. The optical data are from Ref. [3].

of such polaritons decays away from the boundary $r = R$. For energies above the plasma level, $E_{mn} > \hbar\omega_p$, the plasmon polaritons are *bulk*. They exhibit high real parts of $k_{i\perp}$ and form oscillating field patterns similar to standing waves at $r < R$. Thus, for every azimuthal index m, there is an infinite number of allowed bulk polaritons and only one mode is a surface polariton.

Despite the similarities in frequency bands, the plasmon polaritons of complementary structures behave differently [13]. It is especially clear for the bulk plasmons, $\mathrm{Re}(k_z)$ of which grows with ω_{mn} in the vicinity of the plasma frequency ω_p for the nanowire (Fig. 2.10) and decreases in the cavity case (Fig. 2.11). For the surface polaritons, the difference manifests itself through different trends of $\mathrm{Im}(k_z)$. Such a difference in modal behavior is determined by the region where the plasmon energy dissipation occurs – for the nanowire, it is the inner part, while for the cavity, it is the outer part. As a result, the radial power flux at $r > R$ is *inward* for the former case and *outward* for the latter. Mathematically, it is given by different cylindrical Hankel functions used in the solution for the external medium (see Eqs. (2.79) and (2.80)). For the nanowire, $H_m \equiv H_m^{(2)}$ (this solution corresponds to an *incoming evanescent* cylindrical wave), while for the cavity case, $H_m \equiv H_m^{(1)}$ (which gives an *outgoing attenuating* cylindrical wave).

2.2.4 Scattering by a cylinder

Now, we consider the scattering of plane waves by an infinitely long cylinder in a non-absorbing environment, for which $\mathrm{Im}(\varepsilon_e) = 0$. We assume that the incident wave propagates in the x–z plane, and its wavevector $\boldsymbol{k}_{\text{inc}}$ forms the angle α with the z direction, i.e.

$$\boldsymbol{k}_{\text{inc}} = k_e(\boldsymbol{e}_x \sin\alpha + \boldsymbol{e}_z \cos\alpha). \tag{2.83}$$

Depending on the orientation of the electric field, the incident wave is either TM- or TE-polarized. For the TM polarization, the electric field is parallel to the x–z plane, and for the TE polarization, it is perpendicular. Below we consider separately all these cases.

TM incident wave

First, we consider the case of TM polarization with the electric field parallel to the x–z plane, for which

$$[\boldsymbol{E}_\omega]_{\rm inc} = (\boldsymbol{e}_z \sin\alpha - \boldsymbol{e}_x \cos\alpha) E_0 {\rm e}^{{\rm i}k_{\rm e}(r \sin\alpha \cos\theta + z\cos\alpha)}, \tag{2.84}$$

$$[\boldsymbol{H}_\omega]_{\rm inc} = -\frac{k_{\rm e}}{k_0}\boldsymbol{e}_y E_0 {\rm e}^{{\rm i}k_{\rm e}(r \sin\alpha \cos\theta + z\cos\alpha)}, \tag{2.85}$$

where E_0 is the amplitude of the incident time-harmonic wave. In this case, the governing fields are given by

$$\left[E_m^z\right]_{\rm inc} = E_m J_m(k_{{\rm e}\perp} r), \quad \left[H_m^z\right]_{\rm inc} = 0, \tag{2.86}$$

where $E_m = 2\pi {\rm i}^m E_0 \delta(k_z - k_{\rm e}\cos\alpha)\sin\alpha$. Thus, the governing electric field is nonzero only for $k_z = k_{\rm e}\cos\alpha$.

Despite the fact that the incident wave is purely TM-polarized, the scattered light is composed of both polarizations due to their coupling by the boundary conditions. We write them by using the scattering coefficients a_m and b_m for the TM and TE polarizations, respectively,

$$\left[E_m^z\right]_{\rm sca} = -a_m E_m H_m(k_{{\rm e}\perp} r), \tag{2.87}$$

$$\left[H_m^z\right]_{\rm sca} = {\rm i}\frac{k_{\rm e}}{k_0} b_m E_m H_m(k_{{\rm e}\perp} r). \tag{2.88}$$

Similarly, we can write the internal TM and TE governing fields as follows:

$$\left[E_m^z\right]_{\rm int} = d_m E_m J_m(k_{{\rm i}\perp} r), \tag{2.89}$$

$$\left[H_m^z\right]_{\rm int} = -{\rm i}\frac{k_{\rm i}}{k_0} c_m E_m J_m(k_{{\rm i}\perp} r). \tag{2.90}$$

Now, we can apply the boundary conditions for the incident, scattered, and internal fields by requiring continuity of the tangential components E_m^θ, E_m^z, H_m^θ, and H_m^z on the cylinder interface. This results in the following scattering coefficients:

$$a_m = a_m^{\rm TM} = \frac{D_m B_m - V_m^{\rm TM} W_m^{\rm TE}}{D_m^2 - W_m^{\rm TM} W_m^{\rm TE}}, \tag{2.91}$$

$$b_m = b_m^{\rm TM} = \frac{k_0 (D_m V_m^{\rm TM} - B_m W_m^{\rm TM})}{k_{\rm e}(D_m^2 - W_m^{\rm TM} W_m^{\rm TE})}, \tag{2.92}$$

where

$$B_m = \frac{m k_z}{k_0}\left(\frac{1}{\eta_{\rm e}^2} - \frac{1}{\eta_{\rm i}^2}\right) J_m(\eta_{\rm i}) J_m(\eta_{\rm e}),$$

$$V_m^{\rm TM} = \frac{\varepsilon_{\rm e}}{\eta_{\rm e}} J_m(\eta_{\rm i}) J_m'(\eta_{\rm e}) - \frac{\varepsilon_{\rm i}}{\eta_{\rm i}} J_m'(\eta_{\rm i}) J_m(\eta_{\rm e}).$$

Analysis of the coefficients obtained shows that with respect to the azimuthal index m they obey the following symmetry:

$$a_{-m}^{\mathrm{TM}} = a_m^{\mathrm{TM}}, \quad b_{-m}^{\mathrm{TM}} = -b_m^{\mathrm{TM}}, \quad b_0^{\mathrm{TM}} = 0. \tag{2.93}$$

Thus, one can use it to simplify the summation of harmonics over the index m.

TE incident wave

For the second case of incident-wave polarization, the electric field $\boldsymbol{E}_{\mathrm{inc}}$ is perpendicular to the x–z plane, representing a purely TE-polarized wave with

$$[\boldsymbol{E}_\omega]_{\mathrm{inc}} = -\mathrm{i}\boldsymbol{e}_y E_0 \mathrm{e}^{\mathrm{i}k_{\mathrm{e}}(r\sin\alpha\cos\theta + z\cos\alpha)}, \tag{2.94}$$

$$[\boldsymbol{H}_\omega]_{\mathrm{inc}} = -\mathrm{i}\frac{k_{\mathrm{e}}}{k_0}(\boldsymbol{e}_z\sin\alpha - \boldsymbol{e}_x\cos\alpha)E_0\mathrm{e}^{\mathrm{i}k_{\mathrm{e}}(r\sin\alpha\cos\theta + z\cos\alpha)}. \tag{2.95}$$

In this case, the governing fields of the incident light are

$$\left[E_m^z\right]_{\mathrm{inc}} = 0, \quad \left[H_m^z\right]_{\mathrm{inc}} = -\mathrm{i}\frac{k_{\mathrm{e}}}{k_0}E_m J_m(k_{\mathrm{e}\perp}r), \tag{2.96}$$

where E_m is the same quantity as for TM-polarized light.

Applying the boundary conditions for the incident, scattered, and internal fields on the cylinder surface gives the scattering coefficients

$$a_m = a_m^{\mathrm{TE}} = \frac{k_{\mathrm{e}}(D_m V_m^{\mathrm{TE}} - B_m W_m^{\mathrm{TE}})}{k_0(D_m^2 - W_m^{\mathrm{TM}} W_m^{\mathrm{TE}})}, \tag{2.97}$$

$$b_m = b_m^{\mathrm{TE}} = \frac{D_m B_m - V_m^{\mathrm{TE}} W_m^{\mathrm{TM}}}{D_m^2 - W_m^{\mathrm{TM}} W_m^{\mathrm{TE}}}, \tag{2.98}$$

where

$$V_m^{\mathrm{TE}} = \frac{1}{\eta_{\mathrm{e}}}J_m(\eta_{\mathrm{i}})J_m'(\eta_{\mathrm{e}}) - \frac{1}{\eta_{\mathrm{i}}}J_m'(\eta_{\mathrm{i}})J_m(\eta_{\mathrm{e}}).$$

With respect to the azimuthal index m, these coefficients show the following symmetry:

$$a_{-m}^{\mathrm{TE}} = -a_m^{\mathrm{TE}}, \quad a_0^{\mathrm{TE}} = 0, \quad b_{-m}^{\mathrm{TE}} = b_m^{\mathrm{TE}}, \tag{2.99}$$

which will help us later in the summation of harmonics over the index m.

Note that, although the denominators of the scattering coefficients in (2.91) and (2.92), as well as those of (2.97) and (2.98), contain the dispersion equation (2.81) of the plasmon polaritons, exact matching between the polaritons and the incident photons is *never* fulfilled. This is so because different types of cylindrical Hankel functions are used in the solution for the electromagnetic field distribution at $r > R$. For the scattering problem with $k_z'' = 0$ and $\omega'' = 0$, one should use Hankel functions of the first type, $H_m \equiv H_m^{(1)}$ (this type corresponds to the cylindrically diverging waves of the scattered light), while for the eigenvalue problem of a metallic cylinder with $k_z'' = 0$ and $\omega_{mn}'' < 0$, one should take Hankel functions of the second type, $H_m \equiv H_m^{(2)}$ (that give the cylindrically converging waves with zero radiation losses for the plasmon polaritons). In other words, the actual dispersion equation of the plasmon polaritons and the denominators of the scattering coefficients differ from each other, insofar as they contain different sets of functions H_m. In addition, the finite lifetime of the plasmons results in complex eigenfrequencies $\omega_{mn}(k_z)$ of the polaritons for real wavenumbers k_z, making exact frequency matching with the incident photons, for which $\omega = \omega_{mn}(k_e \cos\alpha)$, impossible.

Nonetheless, close to the plasmon polariton energies, $E_{mn} = \hbar\omega_{mn}'(k_e \cos\alpha)$, one can observe growth of the scattering coefficients caused by the resonant photon–plasmon coupling. Especially strong interaction is detected in smaller cylinders, for which the difference between the two sets of cylindrical Hankel functions is small enough. In addition, the photons exhibit better coupling to long-lived plasmon polaritons with $\omega_{mn}' \gg |\omega_{mn}''|$, for which the mismatch induced by $\omega_{mn}'' \neq 0$ is smaller.

2.2.5 Cross-sections per unit length

Note that the scattering and absorption cross-section of an infinite cylinder is infinite. However, the amount of light scattered and absorbed per unit length is finite. The power scattered per unit length can be calculated as

$$P_{\text{sca}} = \lim_{L\to\infty} \frac{1}{L} \int_{-L/2}^{L/2} \mathrm{d}z \int_0^{2\pi} \mathrm{d}\theta (\boldsymbol{S}_{\text{sca}} \cdot \boldsymbol{e}_r) R, \tag{2.100}$$

where the time-averaged scattered Poynting vector is taken on the cylinder surface,

$$\boldsymbol{S}_{\text{sca}} = \frac{c}{8\pi} \operatorname{Re}[\boldsymbol{E}_{\text{sca}} \times \boldsymbol{H}_{\text{sca}}^*]\Big|_{r=R}.$$

For the power absorbed per unit length, we should evaluate

$$P_{\text{abs}} = -\lim_{L\to\infty} \frac{1}{L} \int_{-L/2}^{L/2} \mathrm{d}z \int_0^{2\pi} \mathrm{d}\theta (\boldsymbol{S}_{\text{ext}} \cdot \boldsymbol{e}_r) R \tag{2.101}$$

with the total Poynting vector on the cylinder surface

$$\boldsymbol{S}_{\text{ext}} = \frac{c}{8\pi} \operatorname{Re}[\boldsymbol{E}_{\text{ext}} \times \boldsymbol{H}_{\text{ext}}^*]\Big|_{r=R}.$$

Performing the integration in P_{sca} gives

$$P_{\text{sca}} = \frac{c|E_0|^2}{2\pi k_0} \sum_{m=-\infty}^{\infty} (|a_m|^2 + |b_m|^2). \tag{2.102}$$

For the cross-section per unit length, this gives us

$$\sigma_{\text{sca}} = \frac{P_{\text{sca}}}{S_{\text{inc}}} = \frac{4}{k_e} \sum_{m=-\infty}^{\infty} (|a_m|^2 + |b_m|^2). \tag{2.103}$$

Taking into account symmetry conditions (2.93) and (2.99) of the scattering coefficients, we can write

$$\sigma_{\text{sca}}^{\text{TM}} = \frac{4}{k_{\text{e}}}\left[\left|a_0^{\text{TM}}\right|^2 + 2\sum_{m=1}^{\infty}\left(\left|a_m^{\text{TM}}\right|^2 + \left|b_m^{\text{TM}}\right|^2\right)\right], \tag{2.104}$$

$$\sigma_{\text{sca}}^{\text{TE}} = \frac{4}{k_{\text{e}}}\left[\left|b_0^{\text{TE}}\right|^2 + 2\sum_{m=1}^{\infty}\left(\left|a_m^{\text{TE}}\right|^2 + \left|b_m^{\text{TE}}\right|^2\right)\right] \tag{2.105}$$

for the TM and TE incident waves.

For the absorption, P_{abs} is given by

$$P_{\text{abs}}^{\text{TM}} = \frac{c|E_0|^2}{2\pi k_0}\sum_{m=-\infty}^{\infty}\left[\text{Re}(a_m^{\text{TM}}) - \left|a_m^{\text{TM}}\right|^2 - \left|b_m^{\text{TM}}\right|^2\right], \tag{2.106}$$

$$P_{\text{abs}}^{\text{TE}} = \frac{c|E_0|^2}{2\pi k_0}\sum_{m=-\infty}^{\infty}\left[\text{Re}(b_m^{\text{TE}}) - \left|a_m^{\text{TE}}\right|^2 - \left|b_m^{\text{TE}}\right|^2\right], \tag{2.107}$$

resulting in the following absorption cross-sections per unit length:

$$\sigma_{\text{abs}}^{\text{TM}} = \frac{4}{k_{\text{e}}}\left[\text{Re}(a_0^{\text{TM}}) + 2\sum_{m=1}^{\infty}\text{Re}(a_m^{\text{TM}})\right] - \sigma_{\text{sca}}^{\text{TM}}, \tag{2.108}$$

$$\sigma_{\text{abs}}^{\text{TE}} = \frac{4}{k_{\text{e}}}\left[\text{Re}(b_0^{\text{TE}}) + 2\sum_{m=1}^{\infty}\text{Re}(b_m^{\text{TE}})\right] - \sigma_{\text{sca}}^{\text{TE}}. \tag{2.109}$$

Thus, the extinction cross-sections per unit length for the TM and TE incident waves can be written as

$$\sigma_{\text{ext}}^{\text{TM}} = \frac{4}{k_{\text{e}}}\left[\text{Re}(a_0^{\text{TM}}) + 2\sum_{m=1}^{\infty}\text{Re}(a_m^{\text{TM}})\right], \tag{2.110}$$

$$\sigma_{\text{ext}}^{\text{TE}} = \frac{4}{k_{\text{e}}}\left[\text{Re}(b_0^{\text{TE}}) + 2\sum_{m=1}^{\infty}\text{Re}(b_m^{\text{TE}})\right]. \tag{2.111}$$

In general, scattering and absorption of light by cylindrical structures are polarization-sensitive. The cross-sections for the TM and TE incident light may differ greatly. Therefore, for arbitrarily polarized incident light, one should consider the TM- and TE-polarized components separately. In the case of unpolarized incident light, the overall cross-sections can be calculated as the average value of the TM and TE cross-sections,

$$\sigma = (\sigma^{\text{TM}} + \sigma^{\text{TE}})/2.$$

Figure 2.12 shows the spectral cross-sections per unit length normalized with $2R$ (this being the geometrical cross-section of the cylinder per unit length) for scattering and absorption by a silver nanowire of $R = 100$ nm for two polarizations of light incident at $\alpha = 15°$. It clearly identifies the peaks that appear in the spectra as a result of the excitation of modes with different azimuthal wavenumbers m. Similarly to the coupling with spherical plasmons (that was considered in Section 2.1.5), resonant excitation of the

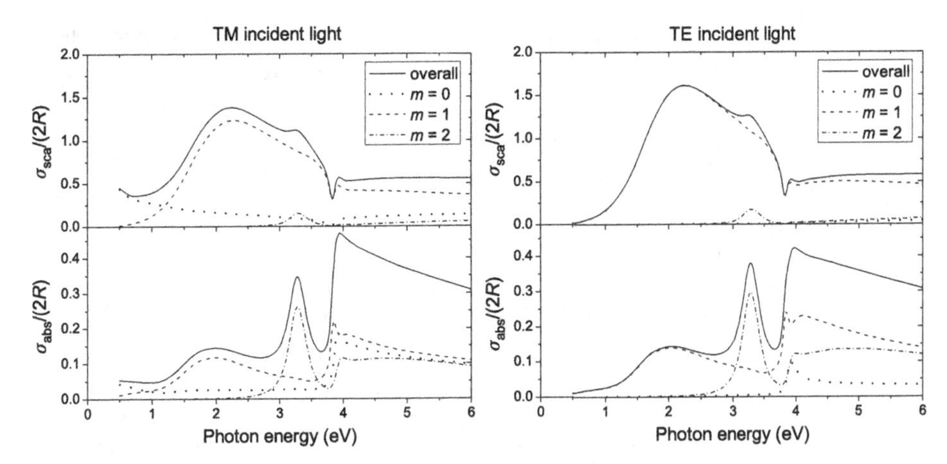

Figure 2.12 Spectral scattering and absorption cross-sections per unit length of a silver nanowire of radius $R = 100$ nm for different polarizations of light incident at $\alpha = 15°$. The dashed lines show the contributions of the azimuthal modes with $m = 0, 1, 2$. The optical data are from Ref. [3].

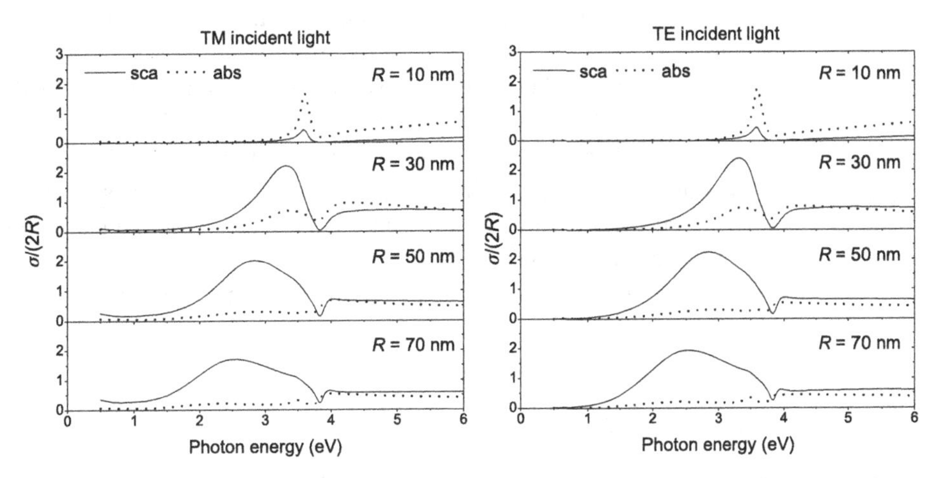

Figure 2.13 Spectral scattering and absorption cross-sections per unit length of silver nanowires with different radii R for TM- and TE-polarized light incident at $\alpha = 15°$. The optical data are from Ref. [3].

cylindrical plasmon polaritons enhances scattering and absorption of the incident light. The most prominent effect is observed for surface polaritons of smaller structures that provide stronger interactions with higher frequency selectivity, as shown in Fig. 2.13.

Interestingly, incident photons of any polarization never couple to surface plasmon polaritons with $m = 0$. This is because the high wavenumbers of such polaritons, $k_z > k_e$, render them inaccessible to photons [13]. For surface polaritons with $m \neq 0$, this limitation is not so strict. It allows coupling in structures with small R, where the wavenumber matching is met. However, as the wavenumber k_z of surface polaritons grows with the cylinder radius R for every m, the prohibition may be extended to higher

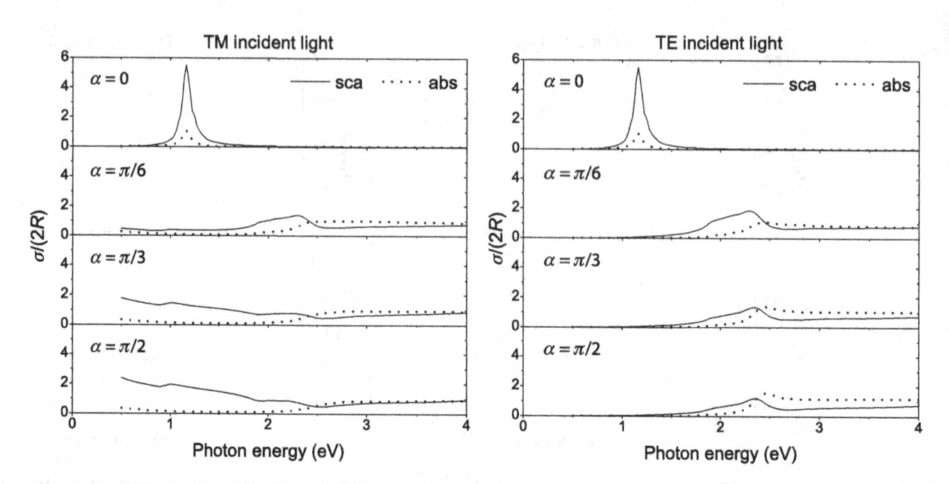

Figure 2.14 The influence of the incidence angle α on the scattering and absorption spectra of a gold nanowire with $R = 30$ nm embedded in silicon dioxide for two polarizations of the incident light. The optical data are from Ref. [8].

azimuthal indexes and imposed on the coupling with the corresponding surface plasmon polaritons.

The efficiency of coupling with the plasmon polaritons is also defined by the angle of incidence and the polarization state of photons, as shown in Fig. 2.14. For $\alpha = 0°$, the polarization state does not affect scattering and absorption, whereas for larger angles, the difference in interaction of the TM- and TE-polarized light becomes more pronounced. Finally, at normal incidence, the difference is maximal, since the plasmon polaritons with $k_z = 0$ split into independent TM and TE modes, and, hence, the scattering of the TM-polarized light becomes completely independent from that of the TE-polarized light.

2.2.6 Multilayer cylinder

To study the effect of a coating on the cylinder surface, let us consider a multilayer cylindrical structure with an N-layer coating as shown in Fig. 2.15. As in spherical geometry, the scattering coefficients can be obtained with the *transfer-matrix method*. However, since the TE and TM modes in cylindrical geometry are generally coupled, the transfer matrix and its analysis for a multilayer cylinder are more complicated.

First, we consider a general form of the polarization governing fields E^z_m and H^z_m excited in every layer of the coating,

$$E^z_m = E_m[e_m J_m(k_\perp r) + f_m H_m(k_\perp r)],$$
$$H^z_m = -\mathrm{i}\sqrt{\varepsilon} E_m[g_m J_m(k_\perp r) + h_m H_m(k_\perp r)],$$

where e_m, f_m, g_m, and h_m are unknown constants. Using these fields, we can compose a vector $\boldsymbol{F}_j$ of the tangential fields in the jth cylindrical layer, $R_{j-1} \leq r \leq R_j$, of the

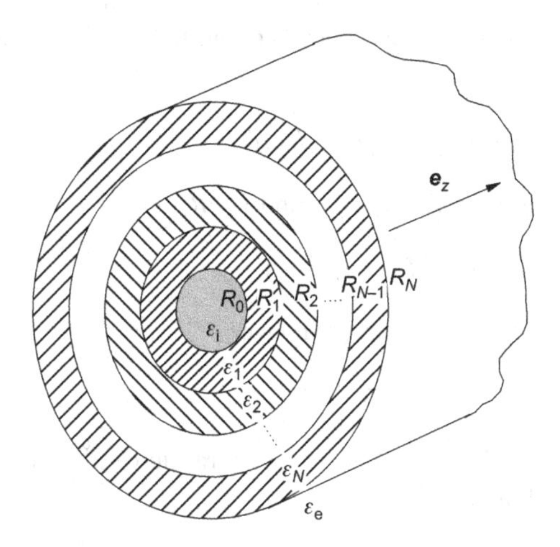

Figure 2.15 A sketch of a multilayer cylinder with an N-layer coating.

coating. In matrix form, it can be written as follows:

$$\boldsymbol{F}_j(r) = \boldsymbol{N}_j(r) \cdot \boldsymbol{C}_j, \tag{2.112}$$

where

$$\boldsymbol{F}_j(r) = \begin{bmatrix} E_m^\theta(r) \\ E_m^z(r) \\ H_m^\theta(r) \\ H_m^z(r) \end{bmatrix}_j, \quad \boldsymbol{C}_j = \begin{bmatrix} e_m \\ f_m \\ g_m \\ h_m \end{bmatrix}_j$$

are the vectors of tangential field components and unknown constants. The matrix $\boldsymbol{N}_j$ can be written as a superposition of two polarizations,

$$\boldsymbol{N}_j(r) = \boldsymbol{N}_j^{\mathrm{TM}}(r) + \boldsymbol{N}_j^{\mathrm{TE}}(r), \tag{2.113}$$

where

$$\boldsymbol{N}_j^{\mathrm{TM}}(r) = \frac{E_m}{k_0 k_{j\perp}^2 r} \begin{bmatrix} -mk_0 k_z J_m(k_{j\perp} r) & -mk_0 k_z H_m(k_{j\perp} r) & 0 & 0 \\ k_0 k_{j\perp}^2 r J_m(k_{j\perp} r) & k_0 k_{j\perp}^2 r H_m(k_{j\perp} r) & 0 & 0 \\ \mathrm{i} k_j^2 k_{j\perp} r J_m'(k_{j\perp} r) & \mathrm{i} k_j^2 k_{j\perp} r H_m'(k_{j\perp} r) & 0 & 0 \\ 0 & 0 & 0 & 0 \end{bmatrix},$$

$$\boldsymbol{N}_j^{\mathrm{TE}}(r) = \frac{E_m}{k_0 k_{j\perp}^2 r} \begin{bmatrix} 0 & 0 & -k_0 k_j k_{j\perp} r J_m'(k_{j\perp} r) & -k_0 k_j k_{j\perp} r H_m'(k_{j\perp} r) \\ 0 & 0 & 0 & 0 \\ 0 & 0 & \mathrm{i} m k_j k_z J_m(k_{j\perp} r) & \mathrm{i} m k_j k_z H_m(k_{j\perp} r) \\ 0 & 0 & -\mathrm{i} k_j k_{j\perp}^2 r J_m(k_{j\perp} r) & -\mathrm{i} k_j k_{j\perp}^2 r H_m(k_{j\perp} r) \end{bmatrix},$$

with $k_j = \sqrt{\varepsilon_j} k_0$ and $k_{j\perp} = \sqrt{k_0^2 \varepsilon_j - k_z^2}$ being the wavenumbers in the jth cylindrical layer.

After elimination of the vector of unknown constants $\boldsymbol{C}_j$ with the vector of tangential fields at the inner boundary R_{j-1},

$$\boldsymbol{C}_j = \boldsymbol{N}_j(R_{j-1})^{-1} \cdot \boldsymbol{F}_j(R_{j-1}),$$

we can get the relation connecting the fields $\boldsymbol{F}_j(R_j)$ and $\boldsymbol{F}_j(R_{j-1})$,

$$\boldsymbol{F}_j(R_j) = \boldsymbol{M}_j \cdot \boldsymbol{F}_j(R_{j-1}), \tag{2.114}$$

where $\boldsymbol{M}_j$ is the transfer matrix of the jth cylindrical layer,

$$\boldsymbol{M}_j = \boldsymbol{N}_j(R_j) \cdot \boldsymbol{N}_j^{-1}(R_{j-1}).$$

Applying the boundary conditions for the adjacent layers in the coating allows us to write a similar relation for the whole coating,

$$\boldsymbol{F}_N(R_N) = \boldsymbol{M} \cdot \boldsymbol{F}_1(R_0), \tag{2.115}$$

where $\boldsymbol{M}$ is the *transfer matrix* of the N-layer system given by

$$\boldsymbol{M} = \boldsymbol{M}_N \cdot \boldsymbol{M}_{N-1} \cdot \cdots \cdot \boldsymbol{M}_2 \cdot \boldsymbol{M}_1.$$

After the transfer matrix of the coating has been derived, we can apply the boundary conditions to the tangential fields at $r = R_0$ and $r = R_N$ by requiring their continuity with the internal and external fields. In this way, we obtain

$$\boldsymbol{F}_{\text{ext}}(R_N) = \boldsymbol{M} \cdot \boldsymbol{F}_{\text{int}}(R_0). \tag{2.116}$$

Upon introducing the vector of scattering coefficients

$$\boldsymbol{C}_{\text{sca}} = \begin{bmatrix} a_m \\ b_m \\ c_m \\ d_m \end{bmatrix}, \tag{2.117}$$

the vectors of the internal and external fields can be written as

$$\boldsymbol{F}_{\text{int}}(r) = \boldsymbol{N}_{\text{int}}(r) \cdot \boldsymbol{C}_{\text{sca}}, \tag{2.118}$$

$$\boldsymbol{F}_{\text{ext}}(r) = \boldsymbol{N}_{\text{sca}}(r) \cdot \boldsymbol{C}_{\text{sca}} + \boldsymbol{F}_{\text{inc}}(r), \tag{2.119}$$

where the matrices $\boldsymbol{N}_{\text{int}}(r)$ and $\boldsymbol{N}_{\text{sca}}(r)$ are given by

$$\boldsymbol{N}_{\text{int}}(r) = \frac{E_m}{k_0 k_{\text{i}\perp}^2 r} \begin{bmatrix} 0 & 0 & -k_0 k_{\text{i}} k_{\text{i}\perp} r J_m'(k_{\text{i}\perp} r) & -m k_0 k_z J_m(k_{\text{i}\perp} r) \\ 0 & 0 & 0 & k_0 k_{\text{i}\perp}^2 r J_m(k_{\text{i}\perp} r) \\ 0 & 0 & \text{i} m k_{\text{i}} k_z J_m(k_{\text{i}\perp} r) & \text{i} k_{\text{i}}^2 k_{\text{i}\perp} r J_m'(k_{\text{i}\perp} r) \\ 0 & 0 & -\text{i} k_{\text{i}} k_{\text{i}\perp}^2 r J_m(k_{\text{i}\perp} r) & 0 \end{bmatrix},$$

$$\boldsymbol{N}_{\text{sca}}(r) = \frac{E_m}{k_0 k_{\text{e}\perp}^2 r} \begin{bmatrix} m k_0 k_z H_m(k_{\text{e}\perp} r) & k_0 k_{\text{e}} k_{\text{e}\perp} r H_m'(k_{\text{e}\perp} r) & 0 & 0 \\ -k_0 k_{\text{e}\perp}^2 r H_m(k_{\text{e}\perp} r) & 0 & 0 & 0 \\ -\text{i} k_{\text{e}}^2 k_{\text{e}\perp} r H_m'(k_{\text{e}\perp} r) & -\text{i} m k_{\text{e}} k_z H_m(k_{\text{e}\perp} r) & 0 & 0 \\ 0 & \text{i} k_{\text{e}} k_{\text{e}\perp}^2 r H_m(k_{\text{e}\perp} r) & 0 & 0 \end{bmatrix}.$$

The vector $\boldsymbol{F}_{\text{inc}}$ is defined by the polarization of the incident wave, and can be written in the following forms for the TM and TE incident light:

$$\boldsymbol{F}_{\text{inc}}^{\text{TM}}(r) = \frac{E_m}{k_0 k_{\text{e}\perp}^2 r} \begin{bmatrix} -m k_0 k_z J_m(k_{\text{e}\perp} r) \\ k_0 k_{\text{e}\perp}^2 r J_m(k_{\text{e}\perp} r) \\ \text{i} k_{\text{e}}^2 k_{\text{e}\perp} r J_m'(k_{\text{e}\perp} r) \\ 0 \end{bmatrix},$$

$$\boldsymbol{F}_{\text{inc}}^{\text{TE}}(r) = \frac{E_m}{k_0 k_{\text{e}\perp}^2 r} \begin{bmatrix} -k_0 k_{\text{e}} k_{\text{e}\perp} r J_m'(k_{\text{e}\perp} r) \\ 0 \\ \text{i} m k_{\text{e}} k_z J_m(k_{\text{e}\perp} r) \\ -\text{i} k_{\text{e}} k_{\text{e}\perp}^2 r J_m(k_{\text{e}\perp} r) \end{bmatrix}.$$

Thus, Eq. (2.116) gives us a linear equation for the vector of scattering coefficients,

$$\boldsymbol{N}_{\text{sca}}(R_N) \cdot \boldsymbol{C}_{\text{sca}} + \boldsymbol{F}_{\text{inc}}(R_N) = \boldsymbol{M} \cdot \boldsymbol{N}_{\text{int}}(R_0) \cdot \boldsymbol{C}_{\text{sca}}, \tag{2.120}$$

the solution of which can be written as

$$\boldsymbol{C}_{\text{sca}} = [\boldsymbol{M} \cdot \boldsymbol{N}_{\text{int}}(R_0) - \boldsymbol{N}_{\text{sca}}(R_N)]^{-1} \cdot \boldsymbol{F}_{\text{inc}}(R_N). \tag{2.121}$$

The analysis of the solution obtained is quite complicated due to the coupling of the TE and TM polarizations. Nonetheless, one can easily use the vector $\boldsymbol{C}_{\text{sca}}$ for numerical calculation of the scattering coefficients for arbitrary coatings. The same scheme can also be used for analysis of the plasmon polaritons supported in a multilayer cylindrical system. The corresponding dispersion equation can be derived by requiring

$$\boldsymbol{M} \cdot \boldsymbol{N}_{\text{int}}(R_0) - \boldsymbol{N}_{\text{sca}}(R_N) = 0, \tag{2.122}$$

when the denominator of $\boldsymbol{C}_{\text{sca}}$ formally becomes zero, with the subsequent change of the type of cylindrical Hankel functions used for the external fields to the appropriate one in accordance with the sign of $\text{Im}(k_{\text{e}\perp})$.

The transfer-matrix method allows us to investigate more complex cylindrical structures. For example, Fig. 2.16 shows the results obtained with this method for a silicon dioxide nanowire of radius $R_0 = 30$ nm coated with gold. It shows how the scattering and absorption of TM- and TE-polarized light incident at $\alpha = 30^\circ$ vary with the thickness of the gold coating. The peaks observed in Fig. 2.16 are attributed to the plasmon polaritons supported by the metal nanotube. It is noteworthy that the strongest coupling is provided by thinner coatings, where the light effectively interacts with the surface plasmon polaritons.

2.3 Plasmonic modes in planar geometry

In this section, we consider the transverse normal modes of an infinitely long planar slab. We study the peculiarities of plasmon polaritons in metal–insulator–metal and insulator–metal–insulator waveguides by solving the eigenvalue problem. Also, we analyze reflection and transmission by a metal slab, and discuss the coupling of photons with transverse plasmon polaritons of the slab. Finally, using the transfer-matrix method, we study the general case of a planar multilayer structure.

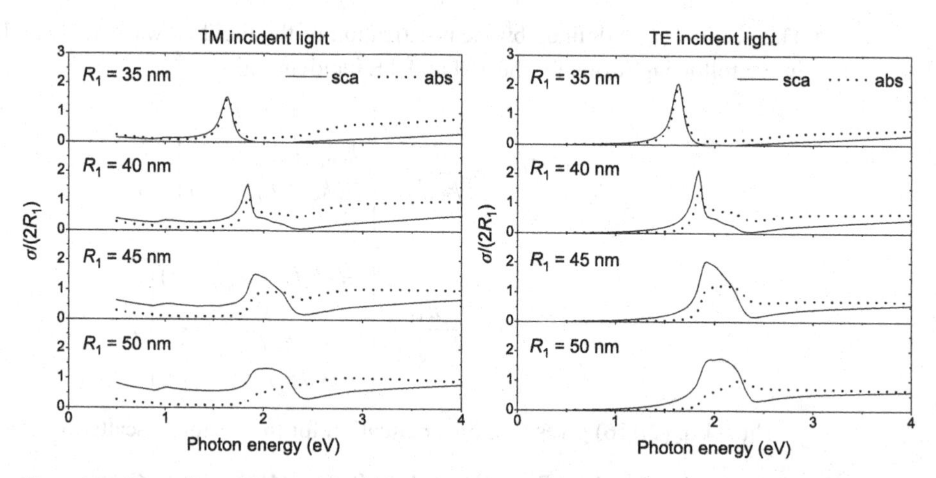

Figure 2.16 Scattering and absorption cross-sections per unit length of an SiO_2 nanowire coated with gold at $\alpha = \pi/6$. The radius of the SiO_2 nanowire is fixed to $R_0 = 30$ nm, while the thickness of the Au coating varies from 5 to 20 nm. The optical data are from Ref. [8].

2.3.1 Planar harmonics

To describe the normal modes in planar geometry, we introduce *scalar planar harmonics*

$$X(\boldsymbol{k}_\parallel; y, z) = \frac{e^{i\boldsymbol{k}_\parallel \cdot \boldsymbol{r}}}{2\pi}, \tag{2.123}$$

which are the two-dimensional eigenfunctions of the Poisson equation

$$\nabla^2 X(\boldsymbol{k}_\parallel; y, z) = -k_\parallel^2 X(\boldsymbol{k}_\parallel; y, z) \tag{2.124}$$

in the Cartesian coordinates (x, y, z), with $\boldsymbol{k}_\parallel = \boldsymbol{e}_y k_y + \boldsymbol{e}_z k_z$ being a two-dimensional real wavevector on the y–z plane. The wavevector $\boldsymbol{k}_\parallel$ is the only parameter of the harmonics that completely defines their spatial distribution. The crucial properties of the scalar planar harmonics are their completeness and orthogonality,

$$\iint_{-\infty}^{\infty} X(\boldsymbol{k}_\parallel; y, z) X^*(\boldsymbol{k}'_\parallel; y, z) \mathrm{d}y\, \mathrm{d}z = \delta(\boldsymbol{k}_\parallel - \boldsymbol{k}'_\parallel). \tag{2.125}$$

They enable us to expand any arbitrary function $g(y, z)$ as

$$g(y, z) = \iint_{-\infty}^{\infty} A(\boldsymbol{k}_\parallel) X(\boldsymbol{k}_\parallel; y, z) \mathrm{d}^2 k_\parallel, \tag{2.126}$$

where $\mathrm{d}^2 k_\parallel = \mathrm{d}k_y\, \mathrm{d}k_z$, with the expansion image $A(\boldsymbol{k}_\parallel)$ given by

$$A(\boldsymbol{k}_\parallel) = \iint_{-\infty}^{\infty} g(y, z) X^*(\boldsymbol{k}_\parallel; y, z) \mathrm{d}y\, \mathrm{d}z. \tag{2.127}$$

Thus, the expansion into the planar harmonics $X(\boldsymbol{k}_\parallel; y, z)$ represents the two-dimensional Fourier transform from the real space (y, z) to the space of wavenumbers (k_y', k_z).

By using the scalar planar harmonics $X(\boldsymbol{k}_\parallel; y, z)$, we can define *vector planar harmonics*. They can be written with the use of the $\boldsymbol{e}_x$ unit vector in the following way:

$$\begin{aligned}
\boldsymbol{X}_1(\boldsymbol{k}_\parallel; y, z) &= \boldsymbol{e}_x X(\boldsymbol{k}_\parallel; y, z), \\
\boldsymbol{X}_2(\boldsymbol{k}_\parallel; y, z) &= \frac{\nabla X(\boldsymbol{k}_\parallel; y, z)}{k_\parallel}, \\
\boldsymbol{X}_3(\boldsymbol{k}_\parallel; y, z) &= \frac{\boldsymbol{e}_x \times \nabla X(\boldsymbol{k}_\parallel; y, z)}{k_\parallel}.
\end{aligned} \tag{2.128}$$

The vectors generated are orthogonal in the real space,

$$\begin{aligned}
\boldsymbol{X}_1(\boldsymbol{k}_\parallel; y, z) \cdot \boldsymbol{X}_2(\boldsymbol{k}_\parallel; y, z) &= 0, \\
\boldsymbol{X}_1(\boldsymbol{k}_\parallel; y, z) \cdot \boldsymbol{X}_3(\boldsymbol{k}_\parallel; y, z) &= 0, \\
\boldsymbol{X}_2(\boldsymbol{k}_\parallel; y, z) \cdot \boldsymbol{X}_3(\boldsymbol{k}_\parallel; y, z) &= 0,
\end{aligned} \tag{2.129}$$

as well as exhibiting orthogonality in the Hilbert space,

$$\iint\limits_{-\infty}^{\infty} \boldsymbol{X}_j(\boldsymbol{k}_\parallel; y, z) \cdot \boldsymbol{X}_j^*(\boldsymbol{k}'_\parallel; y, z) \mathrm{d}y\, \mathrm{d}z = \delta(\boldsymbol{k}_\parallel - \boldsymbol{k}'_\parallel), \quad j = 1, 2, 3. \tag{2.130}$$

Using the vector planar harmonics, any vector field $\boldsymbol{G}(y, z)$ can be expanded as

$$\boldsymbol{G}(y, z) = \sum_{j=1}^{3} \iint\limits_{-\infty}^{\infty} G_j(\boldsymbol{k}_\parallel) \boldsymbol{X}_j(\boldsymbol{k}_\parallel; y, z) \mathrm{d}^2 k_\parallel, \tag{2.131}$$

with the expansion images given by

$$G_j(\boldsymbol{k}_\parallel) = \iint\limits_{-\infty}^{\infty} \boldsymbol{G}(y, z) \cdot \boldsymbol{X}_j^*(\boldsymbol{k}_\parallel; y, z) \mathrm{d}y\, \mathrm{d}z, \quad j = 1, 2, 3. \tag{2.132}$$

Thus, the image $G_1(\boldsymbol{k}_\parallel)$ gives the normal x component of the field $\boldsymbol{G}$, while the images $G_2(\boldsymbol{k}_\parallel)$ and $G_3(\boldsymbol{k}_\parallel)$ are the tangential field components on the y–z plane.

2.3.2 Electromagnetic fields in vector planar harmonics

The vector planar harmonics provide the complete tangential dependence of any vector field on the y–z plane. Thus, we can decompose electric $\boldsymbol{E}_\omega$ and magnetic $\boldsymbol{H}_\omega$ fields into $\boldsymbol{X}_j(\boldsymbol{k}_\parallel; y, z)$ to separate their tangential (on y and z) and normal (on x) dependences,

$$\boldsymbol{E}_\omega = \sum_{j=1}^{3} \iint\limits_{-\infty}^{\infty} E_j(x) \boldsymbol{X}_j(\boldsymbol{k}_\parallel; y, z) \mathrm{d}^2 k_\parallel, \tag{2.133}$$

$$\boldsymbol{H}_\omega = \sum_{j=1}^{3} \iint\limits_{-\infty}^{\infty} H_j(x) \boldsymbol{X}_j(\boldsymbol{k}_\parallel; y, z) \mathrm{d}^2 k_\parallel. \tag{2.134}$$

In a similar way, we can expand the Maxwell equations for transverse fields by taking into account that, for any arbitrary field $\boldsymbol{G}$,

$$\nabla \times \boldsymbol{G} = \iint\limits_{-\infty}^{\infty} \left[-k_\parallel G_3 \boldsymbol{X}_1(\boldsymbol{k}_\parallel; y, z) - \frac{\mathrm{d}G_3}{\mathrm{d}x} \boldsymbol{X}_2(\boldsymbol{k}_\parallel; y, z) + \left(-k_\parallel G_1 + \frac{\mathrm{d}G_2}{\mathrm{d}x} \right) \boldsymbol{X}_3(\boldsymbol{k}_\parallel; y, z) \right] \mathrm{d}^2 k_\parallel .$$

After that, the Maxwell equations split into two independent sets of equations that describe the TM and TE polarizations. The TM modes have two nonzero electric components,

$$E_1 = -\frac{\mathrm{i}k_\parallel}{k_0 \varepsilon} H_3, \quad E_2 = -\frac{\mathrm{i}}{k_0 \varepsilon} \frac{\mathrm{d}H_3}{\mathrm{d}x}, \tag{2.135}$$

and feature the transverse magnetic field H_3 that obeys the equation

$$\frac{\mathrm{d}^2 H_3}{\mathrm{d}x^2} + \left(k_0^2 \varepsilon - k_\parallel^2\right) H_3 = 0. \tag{2.136}$$

The TE polarization exhibits the other two magnetic field components,

$$H_1 = \frac{\mathrm{i}k_\parallel}{k_0} E_3, \quad H_2 = \frac{\mathrm{i}}{k_0} \frac{\mathrm{d}E_3}{\mathrm{d}x}, \tag{2.137}$$

and is described with the transverse electric field E_3 that satisfies the equation

$$\frac{\mathrm{d}^2 E_3}{\mathrm{d}x^2} + \left(k_0^2 \varepsilon - k_\parallel^2\right) E_3 = 0. \tag{2.138}$$

The solutions of Eqs. (2.136) and (2.138) are given by the two different exponential functions $\exp(\pm \mathrm{i}k_x x)$, where $k_x = (k_0^2 \varepsilon - k_\parallel^2)^{1/2}$. For an external region containing $x = -\infty$, the solutions are written as

$$[H_3]^-_{\mathrm{ext}}, \; [E_3]^-_{\mathrm{ext}} \propto \exp(-\mathrm{i}\zeta k_x x),$$

while for an external part including $x = +\infty$, they are given by

$$[H_3]^+_{\mathrm{ext}}, \; [E_3]^+_{\mathrm{ext}} \propto \exp(+\mathrm{i}\zeta k_x x),$$

where the parameter ζ defines the type of the exponential functions in accordance with the requirement of finiteness of the field at $x \to \pm\infty$,

$$\zeta = \begin{cases} 1, & \text{if } \mathrm{Im}(k_x) \geq 0, \\ -1, & \text{if } \mathrm{Im}(k_x) < 0. \end{cases}$$

For intermediate regions, one should consider a linear combination of the exponential functions $\exp(\pm \mathrm{i}k_x x)$.

2.3.3 Planar plasmon polaritons

Let us consider a slab of thickness L and dielectric permittivity ε_i embedded in a material with permittivity ε_e. The governing fields of the TM and TE modes in the slab,

$-L/2 \leq x \leq L/2$, can be written as

$$[H_3]_{\text{int}} = -\mathrm{i}\frac{k_\mathrm{i}}{k_0}(d_1 \mathrm{e}^{\mathrm{i}k_{\mathrm{i}x}x} + d_2 \mathrm{e}^{-\mathrm{i}k_{\mathrm{i}x}x}), \tag{2.139}$$

$$[E_3]_{\text{int}} = c_1 \mathrm{e}^{\mathrm{i}k_{\mathrm{i}x}x} + c_2 \mathrm{e}^{-\mathrm{i}k_{\mathrm{i}x}x}, \tag{2.140}$$

where $k_{\mathrm{i}x} = (k_0^2 \varepsilon_\mathrm{i} - k_\parallel^2)^{1/2}$. For the surrounding medium, the fields are given by

$$[H_3]_{\text{ext}}^- = -\mathrm{i}\frac{k_\mathrm{e}}{k_0} a_1 \mathrm{e}^{-\mathrm{i}\zeta k_{\mathrm{e}x}x}, \tag{2.141}$$

$$[E_3]_{\text{ext}}^- = b_1 \mathrm{e}^{-\mathrm{i}\zeta k_{\mathrm{e}x}x} \tag{2.142}$$

for the range $x \leq -L/2$, and

$$[H_3]_{\text{ext}}^+ = -\mathrm{i}\frac{k_\mathrm{e}}{k_0} a_2 \mathrm{e}^{\mathrm{i}\zeta k_{\mathrm{e}x}x}, \tag{2.143}$$

$$[E_3]_{\text{ext}}^+ = b_2 \mathrm{e}^{\mathrm{i}\zeta k_{\mathrm{e}x}x} \tag{2.144}$$

for $x \geq L/2$, where $k_{\mathrm{e}x} = (k_0^2 \varepsilon_\mathrm{e} - k_\parallel^2)^{1/2}$ and $a_{1,2}$, $b_{1,2}$, $c_{1,2}$, and $d_{1,2}$ are unknown constants.

For the eigenmodes, the electromagnetic fields must satisfy the boundary conditions that imply the continuity of the tangential components H_3, E_2 (for the TM modes) and E_3, H_2 (for the TE modes) on the slab boundaries, $x = \pm L/2$. These conditions give us the dispersion relations for the eigenfrequencies ω and wavenumbers $k_\parallel$ of the polaritons supported by a slab. For the TM modes, the boundary conditions result in the dispersion relation

$$\left(\frac{k_{\mathrm{e}x}\varepsilon_\mathrm{i} + \zeta k_{\mathrm{i}x}\varepsilon_\mathrm{e}}{k_{\mathrm{e}x}\varepsilon_\mathrm{i} - \zeta k_{\mathrm{i}x}\varepsilon_\mathrm{e}}\right)^2 = \exp(2\mathrm{i}k_{\mathrm{i}x}L), \tag{2.145}$$

which can be split into two independent equations for *even* and *odd* TM modes,

$$\frac{k_{\mathrm{e}x}\varepsilon_\mathrm{i}}{k_{\mathrm{i}x}\varepsilon_\mathrm{e}} = -\mathrm{i}\zeta \cot(k_{\mathrm{i}x}L/2), \quad \frac{k_{\mathrm{e}x}\varepsilon_\mathrm{i}}{k_{\mathrm{i}x}\varepsilon_\mathrm{e}} = \mathrm{i}\zeta \tan(k_{\mathrm{i}x}L/2). \tag{2.146}$$

Similarly, the dispersion equation of the TE modes,

$$\left(\frac{k_{\mathrm{e}x} + \zeta k_{\mathrm{i}x}}{k_{\mathrm{e}x} - \zeta k_{\mathrm{i}x}}\right)^2 = \exp(2\mathrm{i}k_{\mathrm{i}x}L), \tag{2.147}$$

can be split into two independent equations,

$$\frac{k_{\mathrm{e}x}}{k_{\mathrm{i}x}} = -\mathrm{i}\zeta \cot(k_{\mathrm{i}x}L/2), \quad \frac{k_{\mathrm{e}x}}{k_{\mathrm{i}x}} = \mathrm{i}\zeta \tan(k_{\mathrm{i}x}L/2), \tag{2.148}$$

for the modes with even and odd field distributions.

In general, an infinite number of branches for plasmon polaritons is supported in planar metal structures for both polarizations. Dispersion equations (2.145) and (2.147) give us the relation between their eigenfrequencies $\omega = \omega_n$ and the parallel wavenumber $k_\parallel$. At energies above the plasma energy, $E_n > \hbar\omega_\mathrm{p}$, the polaritons are bulk, featuring high real parts of $k_{\mathrm{i}x}$ and spreading the energy of the plasmons out over the slab volume.

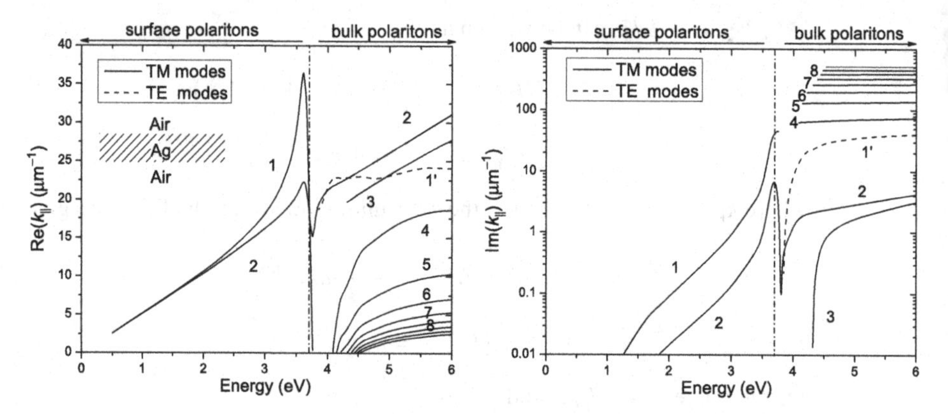

Figure 2.17 The dependence of the complex wavenumbers $k_{||}$ on the energy $\hbar\omega_n$ for the plasmon polaritons propagating in an air/silver/air waveguide with $L = 50$ nm. The vertical dash–dotted lines show the level of the plasma energy $\hbar\omega_p$ in silver. The optical data are from Ref. [3].

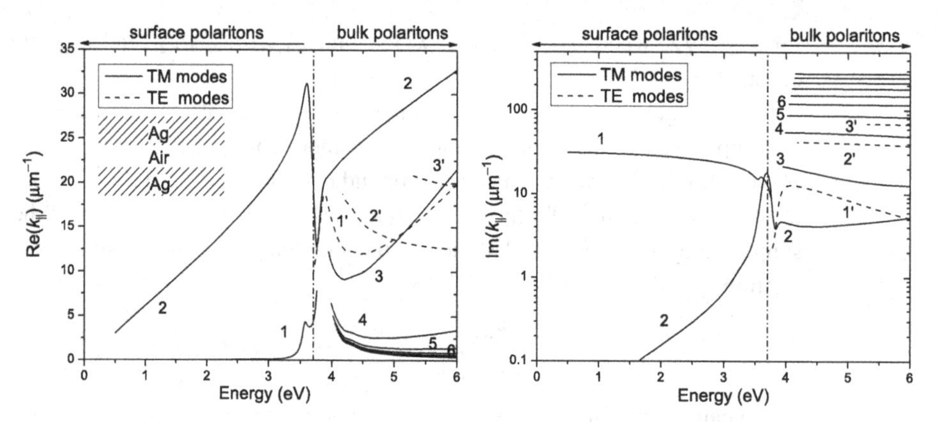

Figure 2.18 Dependence of the complex wavenumbers $k_{||}$ on the energy $\hbar\omega_n$ for the plasmon polaritons propagating in a silver/air/silver waveguide with $L = 100$ nm. The vertical dash–dotted lines show the level of the plasma energy $\hbar\omega_p$ in silver. The optical data are taken from Ref. [3].

They form the *bulk polariton band* that consists of an infinite number of TM and TE branches. Below the plasma level, $E_n < \hbar\omega_p$, the polaritons exhibit high imaginary parts of k_{ix}, and, thus, concentrate the plasmon energy close to the slab boundaries. In contrast to the cases of spherical and cylindrical geometries, only *two* branches of surface modes are allowed in planar structures. These branches correspond to TM-polarized plasmon polaritons with even and odd field distributions across the slab [13, 14].

Figures 2.17 and 2.18 show the dependences of the complex wavenumber $k_{||}$ on the energy E_n of the plasmon polaritons supported in insulator–metal–insulator (IMI) and metal–insulator–metal (MIM) waveguides. The main differences in behavior of the plasmon polaritons maintained by those waveguides are defined by different power flux distributions that occur as a result of the energy dissipation. In both cases, the plasmon energy is dissipated in the metal. Therefore, the power flux at $|x| > L/2$ is

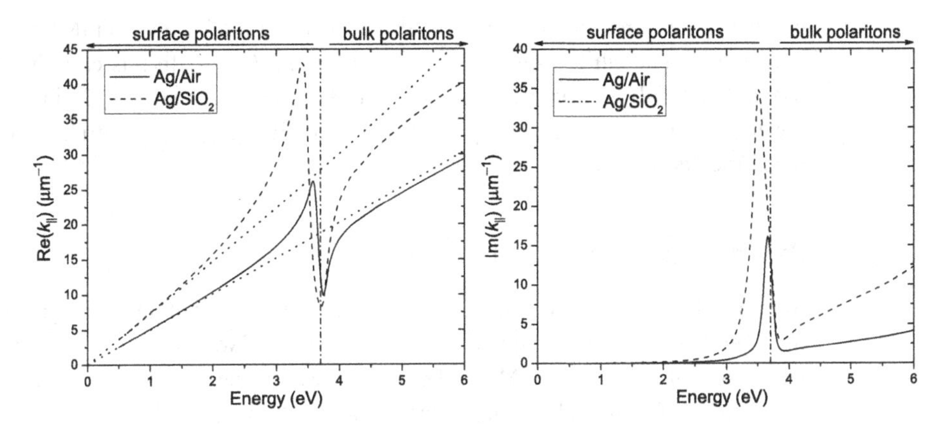

Figure 2.19 The complex wavenumbers $k_\|$ of the plasmon polaritons maintained by single silver/air and silver/silica interfaces. The dotted lines depict the corresponding light lines $\sqrt{\varepsilon}k_0$ in the dielectrics. The vertical dash–dotted lines show the level of the plasma energy $\hbar\omega_\mathrm{p}$ in silver. The optical data are from Ref. [8].

incoming with respect to the slab for an IMI waveguide and *outgoing* for MIM structure. Mathematically, it corresponds to the different signs of ζ used in the field solution at $|x| > L/2$. For the IMI waveguide, $\mathrm{Im}(k_{\mathrm{e}x})$ is always negative, which gives an *incoming evanescent* wave solution with $\zeta = -1$,

$$[H_3]^{\pm}_{\mathrm{ext}},\ [E_3]^{\pm}_{\mathrm{ext}} \propto \exp(\mp \mathrm{i}k_x x),$$

whereas for the MIM waveguide, $\mathrm{Im}(k_{\mathrm{e}x})$ is always positive, which results in an *outgoing attenuating* wave with $\zeta = 1$,

$$[H_3]^{\pm}_{\mathrm{ext}},\ [E_3]^{\pm}_{\mathrm{ext}} \propto \exp(\pm \mathrm{i}k_{\mathrm{e}x} x).$$

Surface plasmon polaritons of a plane boundary

Notably, to guide plasmon polaritons, a single metal/dielectric interface suffices. It is the simplest waveguide structure that supports the propagation of plasmon polaritons. The dispersion equation of those polaritons can be derived directly from Eqs. (2.145) and (2.147), if we put $L \to \infty$. In this way, we get

$$k_{\mathrm{e}x}\varepsilon_\mathrm{i} + \zeta k_{\mathrm{i}x}\varepsilon_\mathrm{e} = 0, \tag{2.149}$$

$$k_{\mathrm{e}x} + \zeta k_{\mathrm{i}x} = 0 \tag{2.150}$$

for the TM and TE polarizations, respectively.

The dispersion equation for the TE polarization has no solution. This means that a single metal/dielectric interface does not support any wave of TE type. At the same time, the equation for the TM polaritons has one solution, which is given by

$$k_\| = k_0 \left(\frac{\varepsilon_\mathrm{i}\varepsilon_\mathrm{e}}{\varepsilon_\mathrm{i} + \varepsilon_\mathrm{e}} \right)^{1/2}. \tag{2.151}$$

In other words, a single metal interface supports only *one* branch of plasmon polaritons, as shown in Fig. 2.19. Depending on the energy, it gives us either surface or bulk plasmon polaritons. The same branch also appears for thick MIM and IMI waveguides, if L is large enough to satisfy the condition $\mathrm{Im}(k_{ix})L \gg 1$. In this case, it represents the *degenerated solution* for even and odd plasmon polaritons [9, 13] in addition to the infinite number of closely packed TM and TE bulk modes extrinsic to a single metal/dielectric interface.

2.3.4 Reflection and transmission by a slab

Since the TM and TE polarizations are completely independent in planar geometry, any incident wave can be considered as a superposition of waves of the two polarizations. Below we consider purely TM and TE incident waves that impinge on an infinitely long metal slab from a non-absorbing material, for which $\mathrm{Im}(\varepsilon_e) = 0$. The wavevector of the incident waves is considered to be on the x–z plane,

$$\boldsymbol{k}_{\mathrm{inc}} = k_e(\boldsymbol{e}_x \sin\alpha + \boldsymbol{e}_z \cos\alpha), \tag{2.152}$$

where α is the incidence angle formed by $\boldsymbol{k}_{\mathrm{inc}}$ and $\boldsymbol{e}_z$.

The TM incident wave

For TM polarization, the incident electric and magnetic fields can be written as follows:

$$[\boldsymbol{E}_\omega]_{\mathrm{inc}} = (\boldsymbol{e}_z \sin\alpha - \boldsymbol{e}_x \cos\alpha)E_0 \mathrm{e}^{\mathrm{i}k_e(x\sin\alpha + z\cos\alpha)}, \tag{2.153}$$

$$[\boldsymbol{H}_\omega]_{\mathrm{inc}} = -\frac{k_e}{k_0}\boldsymbol{e}_y E_0 \mathrm{e}^{\mathrm{i}k_e(x\sin\alpha + z\cos\alpha)}, \tag{2.154}$$

where E_0 is the amplitude of the incident time-harmonic wave. In this case, the incident light is described with the governing field

$$[H_3]_{\mathrm{inc}} = -\mathrm{i}\frac{k_e}{k_0}\tilde{E}\mathrm{e}^{\mathrm{i}k_e x\sin\alpha}, \tag{2.155}$$

where $\tilde{E} = 4\pi^2 E_0 \delta(k_y)\delta(k_z - k_e\cos\alpha)$. Thus, $[H_3]_{\mathrm{inc}}$ is nonzero only for $k_y = 0$ and $k_z = k_e \cos\alpha$.

The light reflected from the slab can be written as

$$[H_3]_{\mathrm{ref}} = -\mathrm{i}\frac{k_e}{k_0}a_1\tilde{E}\mathrm{e}^{-\mathrm{i}k_{ex}x}, \tag{2.156}$$

where a_1 is an unknown TM reflection coefficient. Thus, the overall field H_3 in the semi-infinite region $x \le -L/2$ is given by a superposition of the incident and reflected waves,

$$[H_3]^-_{\mathrm{ext}} = -\mathrm{i}\frac{k_e}{k_0}\tilde{E}\left(\mathrm{e}^{\mathrm{i}k_{ex}x} + a_1\mathrm{e}^{-\mathrm{i}k_{ex}x}\right). \tag{2.157}$$

Similarly, we can write the transmitted light in the region $x \ge L/2$,

$$[H_3]^+_{\mathrm{ext}} = -\mathrm{i}\frac{k_e}{k_0}a_2\tilde{E}\mathrm{e}^{\mathrm{i}k_{ex}x}, \tag{2.158}$$

where a_2 is the TM transmission coefficient. For the internal fields induced inside the slab, $-L/2 \leq x \leq L/2$, we have

$$[H_3]_{\rm int} = -{\rm i}\frac{k_{\rm i}}{k_0}\tilde{E}\left(d_1 {\rm e}^{{\rm i}k_{\rm ix}x} + d_2 {\rm e}^{-{\rm i}k_{\rm ex}x}\right), \tag{2.159}$$

where d_1 and d_2 are unknown constants.

To find the reflection and transmission coefficients, we should apply the boundary conditions for the external and internal fields which require the continuity of the tangential components E_2 and H_3 on the slab surfaces $x = \pm L/2$. This gives us the following coefficients:

$$a_1 = \frac{{\rm e}^{{\rm i}(k_{\rm ix}-k_{\rm ex})L}({\rm e}^{{\rm i}k_{\rm ix}L} - {\rm e}^{-{\rm i}k_{\rm ix}L})(k_{\rm ex}^2\varepsilon_{\rm i}^2 - k_{\rm ix}^2\varepsilon_{\rm e}^2)}{{\rm e}^{2{\rm i}k_{\rm ix}L}(k_{\rm ex}\varepsilon_{\rm i} - k_{\rm ix}\varepsilon_{\rm e})^2 - (k_{\rm ex}\varepsilon_{\rm i} + k_{\rm ix}\varepsilon_{\rm e})^2}, \tag{2.160}$$

$$a_2 = -\frac{4{\rm e}^{{\rm i}(k_{\rm ix}-k_{\rm ex})L}k_{\rm ex}k_{\rm ix}\varepsilon_{\rm e}\varepsilon_{\rm i}}{{\rm e}^{2{\rm i}k_{\rm ix}L}(k_{\rm ex}\varepsilon_{\rm i} - k_{\rm ix}\varepsilon_{\rm e})^2 - (k_{\rm ex}\varepsilon_{\rm i} + k_{\rm ix}\varepsilon_{\rm e})^2}. \tag{2.161}$$

For normal incidence with $\alpha = \pi/2$, the reflection and transmission coefficients are simpler, and can be written as follows:

$$a_1(\pi/2) = \frac{{\rm e}^{{\rm i}(k_{\rm i}-k_{\rm e})L}({\rm e}^{{\rm i}k_{\rm i}L} - {\rm e}^{-{\rm i}k_{\rm i}L})(k_{\rm i}^2 - k_{\rm e}^2)}{{\rm e}^{2{\rm i}k_{\rm i}L}(k_{\rm e} - k_{\rm i})^2 - (k_{\rm e} + k_{\rm i})^2}, \tag{2.162}$$

$$a_2(\pi/2) = -\frac{4{\rm e}^{{\rm i}(k_{\rm i}-k_{\rm e})L}k_{\rm e}k_{\rm i}}{{\rm e}^{2{\rm i}k_{\rm i}L}(k_{\rm e} - k_{\rm i})^2 - (k_{\rm e} + k_{\rm i})^2}. \tag{2.163}$$

The TE incident wave

For TE polarization, the incident fields are given by

$$[\boldsymbol{E}_\omega]_{\rm inc} = -{\rm i}\boldsymbol{e}_y E_0 {\rm e}^{{\rm i}k_{\rm e}(x\sin\alpha + z\cos\alpha)}, \tag{2.164}$$

$$[\boldsymbol{H}_\omega]_{\rm inc} = -{\rm i}\frac{k_{\rm e}}{k_0}(\boldsymbol{e}_z\sin\alpha - \boldsymbol{e}_x\cos\alpha)E_0 {\rm e}^{{\rm i}k_{\rm e}(x\sin\alpha + z\cos\alpha)}, \tag{2.165}$$

where E_0 is the amplitude of the time-harmonic wave. After expansion of the electric field into the vector planar harmonics, we derive

$$[E_3]_{\rm inc} = \tilde{E}{\rm e}^{{\rm i}k_{\rm e}x\sin\alpha}, \tag{2.166}$$

where $\tilde{E}$ is the same quantity as for TM polarization.

Similarly to the case of TM polarization, the reflected light can be written as

$$[E_3]_{\rm ref} = b_1\tilde{E}{\rm e}^{-{\rm i}k_{\rm ex}x}, \tag{2.167}$$

where b_1 is an unknown reflection coefficient for TE polarization. Thus, the overall field in the region $x \leq -L/2$ is given by

$$[E_3]_{\rm ext}^- = \tilde{E}\left({\rm e}^{{\rm i}k_{\rm ex}x} + b_1{\rm e}^{-{\rm i}k_{\rm ex}x}\right). \tag{2.168}$$

For the transmitted light in the region $x \geq L/2$, we can write

$$[E_3]_{\rm ext}^+ = b_2\tilde{E}{\rm e}^{{\rm i}k_{\rm ex}x}, \tag{2.169}$$

where b_2 is a transmission coefficient, while the electric field E_3 induced inside the slab, $-L/2 \leq x \leq L/2$, is given by

$$[E_3]_{\text{int}} = \tilde{E}\left(c_1 e^{ik_{ix}x} + c_2 e^{-ik_{ex}x}\right), \tag{2.170}$$

with c_1 and c_2 being unknown constants.

After applying the boundary conditions for the tangential field components H_2 and E_3 on the slab surfaces $x = \pm L/2$, we get the reflection and transmission coefficients

$$b_1 = \frac{e^{i(k_{ix}-k_{ex})L}(e^{ik_{ix}L} - e^{-ik_{ix}L})(k_{ex}^2 - k_{ix}^2)}{e^{2ik_{ix}L}(k_{ex} - k_{ix})^2 - (k_{ex} + k_{ix})^2}, \tag{2.171}$$

$$b_2 = -\frac{4e^{i(k_{ix}-k_{ex})L}k_{ex}k_{ix}}{e^{2ik_{ix}L}(k_{ex} - k_{ix})^2 - (k_{ex} + k_{ix})^2}. \tag{2.172}$$

For the particular case of normally incident light, they result in

$$b_1(\pi/2) = \frac{e^{i(k_i-k_e)L}(e^{ik_iL} - e^{-ik_iL})(k_e^2 - k_i^2)}{e^{2ik_iL}(k_e - k_i)^2 - (k_e + k_i)^2}, \tag{2.173}$$

$$b_2(\pi/2) = -\frac{4e^{i(k_i-k_e)L}k_ek_i}{e^{2ik_iL}(k_e - k_i)^2 - (k_e + k_i)^2}. \tag{2.174}$$

Thus, the reflection and transmission coefficients for the TM and TE polarizations at normal incidence are no longer independent and feature

$$b_1(\pi/2) = -a_1(\pi/2), \quad b_2(\pi/2) = a_2(\pi/2).$$

It is worth noting that the denominators of the reflection and transmission coefficients in both polarizations look quite similar to the dispersion equations of the plasmon polaritons given by Eqs. (2.145) and (2.147). The only difference is the parameter ζ that appears in the dispersion equations. Recall that for a metal slab $\zeta = -1$. This makes a big difference between the plasmon polariton dispersion equations and the reflection and transmission coefficient denominators. As a result, the reflection and transmission coefficients in planar geometry *do not increase* at the plasmon resonance conditions, in contrast to what happens in spherical and cylindrical geometries, where the scattering coefficients may grow with the photon–plasmon coupling.

2.3.5 Reflectance, transmittance, and absorptance

To quantify the reflection and transmission by a slab, let us consider the normal components of the Poynting vector for the reflected and transmitted light,

$$[S_n]_{\text{ref}} = -[S_x]_{\text{ref}} = \frac{c}{8\pi}\,\text{Re}\left\{[E_2]_{\text{ref}}[H_3]_{\text{ref}}^* - [E_3]_{\text{ref}}[H_2]_{\text{ref}}^*\right\},$$

$$[S_n]_{\text{ext}}^+ = [S_x]_{\text{ext}}^+ = \frac{c}{8\pi}\,\text{Re}\left\{[E_3]_{\text{ext}}^+[H_2]_{\text{ext}}^{+*} - [E_2]_{\text{ext}}^+[H_3]_{\text{ext}}^{+*}\right\}.$$

By taking into account the expressions for the scattered and transmitted fields, we can derive

$$[S_n]_{\text{ref}}^{\text{TM}} = |a_1|^2[S_n]_{\text{inc}}^{\text{TM}}, \quad [S_n]_{\text{ext}}^{+\text{TM}} = |a_2|^2[S_n]_{\text{inc}}^{\text{TM}} \tag{2.175}$$

for the TM incident waves and

$$[S_n]^{\text{TE}}_{\text{ref}} = |b_1|^2[S_n]^{\text{TE}}_{\text{inc}}, \quad [S_n]^{+\text{TE}}_{\text{ext}} = |b_2|^2[S_n]^{\text{TE}}_{\text{inc}} \tag{2.176}$$

for the TE incident waves. Thus, we can introduce the reflectance R and transmittance T as the fractions of the incident power $[S_n]_{\text{inc}}$ which are reflected and transmitted by the slab. In terms of the reflection and transmission coefficients, they can be written as

$$R^{\text{TM}} = |a_1|^2, \quad T^{\text{TM}} = |a_2|^2 \tag{2.177}$$

for the TM polarization and as

$$R^{\text{TE}} = |b_1|^2, \quad T^{\text{TE}} = |b_2|^2 \tag{2.178}$$

for the TE incident waves.

Owing to the energy conservation, we can also introduce the absorptance

$$A = 1 - R - T,$$

namely the fraction of the incident light which is absorbed in the slab. For the TM and TE polarizations, it is given by

$$A^{\text{TM}} = 1 - |a_1|^2 - |a_2|^2, \quad A^{\text{TE}} = 1 - |b_1|^2 - |b_2|^2. \tag{2.179}$$

Note that, for normally incident light, the reflectance, transmittance, and absorptance are independent of the polarization state, since we have

$$|a_1(\pi/2)|^2 = |b_1(\pi/2)|^2, \quad |a_2(\pi/2)|^2 = |b_2(\pi/2)|^2.$$

However, they become polarization-dependent for oblique incidence, so one should consider separately the TM- and TE-polarized components in order to derive the overall characteristics. For the particular case of unpolarized incident light, one can use

$$R = (R^{\text{TM}} + R^{\text{TE}})/2, \quad T = (T^{\text{TM}} + T^{\text{TE}})/2, \quad A = (A^{\text{TM}} + A^{\text{TE}})/2.$$

The introduced reflectance, transmittance, and absorptance of the slab provide a complete description of the energy transfer that occurs in the light–matter interaction. Hence, they should contain traces of the coupling between photons and plasmon polaritons. The most informative parameter is the absorptance, insofar as the coupling with eigenmodes improves the energy confinement in the metal and, thus, increases the dissipation in the slab. Figure 2.20 shows the spectral absorptance of a silver slab of thickness $L = 50$ nm calculated for different incidence angles α. It reveals two distinct spikes of the optical absorption that arise around $\hbar\omega = 3.8$ eV at small incidence angles. One spike is attributed to a bulk TM polariton, while the other is caused by a bulk TE plasmon polariton. The respective branches of the plasmon polaritons are labeled in Fig. 2.17 as 2 and 1$'$. The behavior of those spikes with α exactly follows the dispersion curves, $\omega_n(k_e \cos\alpha)$. They all slightly shift to lower energies and then disappear – the TE spike first and the TM one second.

It is notable that the incident photons *do not couple* with the surface plasmon polaritons of a metal slab [15]. This is so because of the high parallel wavenumbers of those polaritons, $k_\parallel > k_e$ (Fig. 2.17), which are inaccessible to photons. As a result, spatial

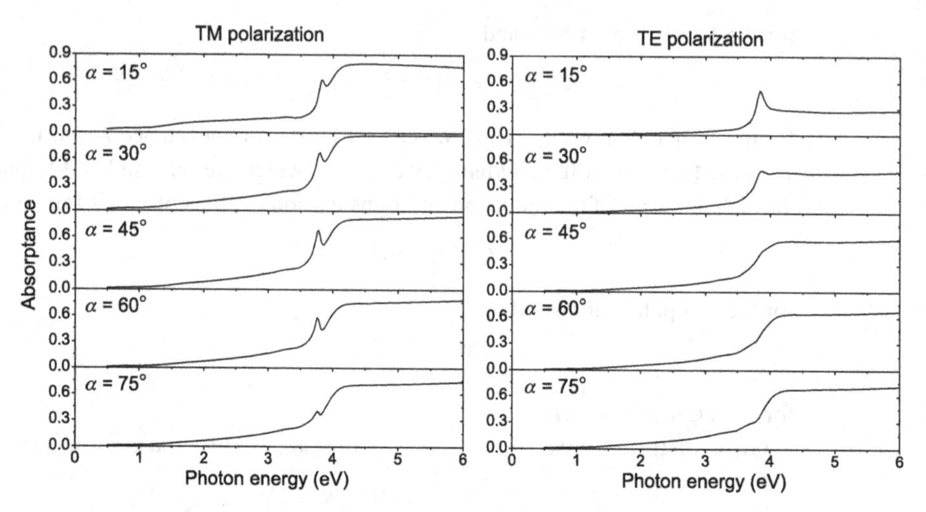

Figure 2.20 The absorptance of a silver slab of thickness $L = 50$ nm for different angles of incidence α and light polarizations. The optical data are from Ref. [3].

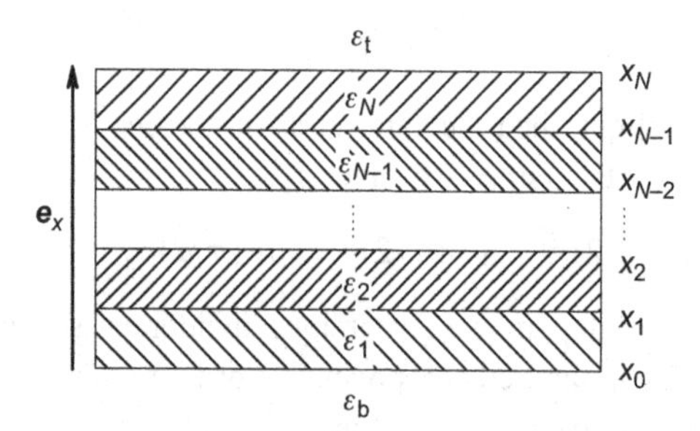

Figure 2.21 A sketch of an N-layer slab.

synchronism of the photons and surface plasmon polaritons becomes impossible. However, this does not mean that their coupling is impossible in general. There are several schemes that enable spatial synchronism between photons and surface polaritons in planar geometry. To show them, we generalize the scattering problem and solve it for the case of a multilayer slab.

2.3.6 A multilayer slab

The transmission and reflection coefficients of a multilayer slab can be obtained with the *transfer-matrix method*. Below, we consider the case of an N-layer slab as shown in Fig. 2.21. To implement this method in planar geometry, let us write a general form of the governing fields induced in every layer of the structure. For the TM and TE polarizations, they are given by

$$H_3 = -\mathrm{i}\sqrt{\varepsilon}\tilde{E}[d_1 \mathrm{e}^{\mathrm{i}k_x x} + d_2 \mathrm{e}^{-\mathrm{i}k_x x}],$$

$$E_3 = \tilde{E}[c_1 \mathrm{e}^{\mathrm{i}k_x x} + c_2 \mathrm{e}^{-\mathrm{i}k_x x}],$$

where c_1, c_2, d_1, and d_2 are unknown constants. Using the governing fields, we can compose a vector $\boldsymbol{F}_j$ of the tangential field components in the jth layer, $x_{j-1} \leq x \leq x_j$, of the slab. In matrix form, it can be written as

$$\boldsymbol{F}_j(x) = \boldsymbol{N}_j(x) \cdot \boldsymbol{C}_j, \tag{2.180}$$

where, depending on the polarization, the vectors of tangential components and unknown constants are given by

$$\boldsymbol{F}_j^{\mathrm{TM}}(x) = \begin{bmatrix} H_3(x) \\ E_2(x) \end{bmatrix}_j, \quad \boldsymbol{F}_j^{\mathrm{TE}}(x) = \begin{bmatrix} E_3(x) \\ H_2(x) \end{bmatrix}_j$$

and

$$\boldsymbol{C}_j^{\mathrm{TM}} = \begin{bmatrix} d_1 \\ d_2 \end{bmatrix}_j, \quad \boldsymbol{C}_j^{\mathrm{TE}} = \begin{bmatrix} c_1 \\ c_2 \end{bmatrix}_j,$$

and the matrix $\boldsymbol{N}_j(x)$ takes the forms

$$\boldsymbol{N}_j^{\mathrm{TM}}(x) = \frac{\tilde{E}}{k_0 k_j} \begin{bmatrix} -\mathrm{i}k_j^2 \mathrm{e}^{\mathrm{i}k_{jx}x} & -\mathrm{i}k_j^2 \mathrm{e}^{-\mathrm{i}k_{jx}x} \\ -\mathrm{i}k_0 k_{jx} \mathrm{e}^{\mathrm{i}k_{jx}x} & \mathrm{i}k_0 k_{jx} \mathrm{e}^{-\mathrm{i}k_{jx}x} \end{bmatrix},$$

$$\boldsymbol{N}_j^{\mathrm{TE}}(x) = \frac{\tilde{E}}{k_0 k_j} \begin{bmatrix} k_0 k_j \mathrm{e}^{\mathrm{i}k_{jx}x} & k_0 k_j \mathrm{e}^{-\mathrm{i}k_{jx}x} \\ -k_j k_{jx} \mathrm{e}^{\mathrm{i}k_{jx}x} & k_j k_{jx} \mathrm{e}^{-\mathrm{i}k_{jx}x} \end{bmatrix}$$

with $k_j = k_0(\varepsilon_j)^{1/2}$ and $k_{jx} = (k_0^2 \varepsilon_j - k_\parallel^2)^{1/2}$ being the wavenumbers in the jth layer.

The vector of unknown constants $\boldsymbol{C}_j$ can be eliminated in Eq. (2.180) with the vector of tangential fields at the boundary x_{j-1},

$$\boldsymbol{C}_j = \boldsymbol{N}_j^{-1}(x_{j-1}) \cdot \boldsymbol{F}_j(x_{j-1}).$$

Thus, one can get the relation for the vectors of tangential fields at $x = x_{j-1}$ and $x = x_j$,

$$\boldsymbol{F}_j(x_j) = \boldsymbol{M}_j \cdot \boldsymbol{F}_j(x_{j-1}), \tag{2.181}$$

where $\boldsymbol{M}_j$ is the transfer matrix of the jth layer,

$$\boldsymbol{M}_j = \boldsymbol{N}_j(x_j) \cdot \boldsymbol{N}_j^{-1}(x_{j-1}),$$

that translates the tangential fields from one boundary of the layer to another.

Since the boundary conditions require the tangential fields to be continuous between the adjacent layers, we can write that

$$\boldsymbol{F}_N(x_N) = \boldsymbol{M} \cdot \boldsymbol{F}_1(x_0), \tag{2.182}$$

where $\boldsymbol{M}$ is the *transfer matrix* of the N-layer slab given by

$$\boldsymbol{M} = \boldsymbol{M}_N \cdot \boldsymbol{M}_{N-1} \cdot \cdots \cdot \boldsymbol{M}_2 \cdot \boldsymbol{M}_1.$$

Thus, the optical properties of the N-layer slab can be described with two 2×2 transfer matrices $\boldsymbol{M}^{\mathrm{TM}}$ and $\boldsymbol{M}^{\mathrm{TE}}$.

Now, we can require the continuity of the tangential fields at $x = x_0$ and $x = x_N$ with the external fields. Thus, we get

$$\boldsymbol{F}^{+}_{\text{ext}}(x_N) = \boldsymbol{M} \cdot \boldsymbol{F}^{-}_{\text{ext}}(x_0). \tag{2.183}$$

On introducing the vector of reflection and transmission coefficients $\boldsymbol{C}$ for the TM and TE polarizations,

$$\boldsymbol{C}^{\text{TM}} = \begin{bmatrix} a_1 \\ a_2 \end{bmatrix}, \quad \boldsymbol{C}^{\text{TE}} = \begin{bmatrix} b_1 \\ b_2 \end{bmatrix},$$

the vectors of external fields can be written as follows:

$$\boldsymbol{F}^{-}_{\text{ext}}(x) = \boldsymbol{N}_{\text{ref}}(x) \cdot \boldsymbol{C} + \boldsymbol{F}_{\text{inc}}(x), \tag{2.184}$$

$$\boldsymbol{F}^{+}_{\text{ext}}(x) = \boldsymbol{N}_{\text{tra}}(x) \cdot \boldsymbol{C}. \tag{2.185}$$

Here, the matrices for the reflected and transmitted light are

$$\boldsymbol{N}^{\text{TM}}_{\text{ref}}(x) = \frac{\tilde{E}}{k_0 k_{\text{b}}} \begin{bmatrix} -\mathrm{i}k_{\text{b}}^2 \mathrm{e}^{-\mathrm{i}k_{\text{b}x}x} & 0 \\ \mathrm{i}k_0 k_{\text{b}x} \mathrm{e}^{-\mathrm{i}k_{\text{b}x}x} & 0 \end{bmatrix}, \quad \boldsymbol{N}^{\text{TE}}_{\text{ref}}(x) = \frac{\tilde{E}}{k_0 k_{\text{b}}} \begin{bmatrix} k_0 k_{\text{b}} \mathrm{e}^{-\mathrm{i}k_{\text{b}x}x} & 0 \\ k_{\text{b}} k_{\text{b}x} \mathrm{e}^{-\mathrm{i}k_{\text{b}x}x} & 0 \end{bmatrix},$$

$$\boldsymbol{N}^{\text{TM}}_{\text{tra}}(x) = \frac{\tilde{E}}{k_0 k_{\text{t}}} \begin{bmatrix} 0 & -\mathrm{i}k_{\text{t}}^2 \mathrm{e}^{\mathrm{i}k_{\text{t}x}x} \\ 0 & -\mathrm{i}k_0 k_{\text{t}x} \mathrm{e}^{\mathrm{i}k_{\text{t}x}x} \end{bmatrix}, \quad \boldsymbol{N}^{\text{TE}}_{\text{tra}}(x) = \frac{\tilde{E}}{k_0 k_{\text{t}}} \begin{bmatrix} 0 & k_0 k_{\text{t}} \mathrm{e}^{\mathrm{i}k_{\text{t}x}x} \\ 0 & -k_{\text{t}} k_{\text{t}x} \mathrm{e}^{\mathrm{i}k_{\text{t}x}x} \end{bmatrix},$$

where the wavenumber indexes b and t were introduced to differentiate the bottom and top external media that occupy the semi-infinite regions $x \le x_0$ and $x \ge x_N$, as shown in Fig. 2.21. Depending on the polarization of the incident light, the vector of incident fields $\boldsymbol{F}_{\text{inc}}$ can be written in the following forms:

$$\boldsymbol{F}^{\text{TM}}_{\text{inc}}(x) = \frac{\tilde{E}}{k_0 k_{\text{b}}} \begin{bmatrix} -\mathrm{i}k_{\text{b}}^2 \mathrm{e}^{\mathrm{i}k_{\text{b}x}x} \\ -\mathrm{i}k_0 k_{\text{b}x} \mathrm{e}^{\mathrm{i}k_{\text{b}x}x} \end{bmatrix}, \quad \boldsymbol{F}^{\text{TE}}_{\text{inc}}(x) = \frac{\tilde{E}}{k_0 k_{\text{b}}} \begin{bmatrix} k_0 k_{\text{b}} \mathrm{e}^{\mathrm{i}k_{\text{b}x}x} \\ -k_{\text{b}} k_{\text{b}x} \mathrm{e}^{\mathrm{i}k_{\text{b}x}x} \end{bmatrix}.$$

Thus, Eq. (2.183) gives us a linear equation for the vector of reflection and transmission coefficients $\boldsymbol{C}$,

$$\boldsymbol{N}_{\text{tra}}(x_N) \cdot \boldsymbol{C} = \boldsymbol{M} \cdot [\boldsymbol{N}_{\text{ref}}(x_0) \cdot \boldsymbol{C} + \boldsymbol{F}_{\text{inc}}(x_0)], \tag{2.186}$$

the solution of which is

$$\boldsymbol{C} = [\boldsymbol{N}_{\text{tra}}(x_N) - \boldsymbol{M} \cdot \boldsymbol{N}_{\text{ref}}(x_0)]^{-1} \cdot \boldsymbol{M} \cdot \boldsymbol{F}_{\text{inc}}(x_0). \tag{2.187}$$

In terms of the the transfer matrix elements M_{ij}, the TM reflection and transmission coefficients can be written as

$$a_1 = \frac{k_0 k_{\text{b}x}(k_{\text{t}}^2 M^{\text{TM}}_{22} - k_0 k_{\text{t}x} M^{\text{TM}}_{12}) + k_{\text{b}}^2(k_{\text{t}}^2 M^{\text{TM}}_{21} - k_0 k_{\text{t}x} M^{\text{TM}}_{11})}{k_0 k_{\text{b}x}(k_{\text{t}}^2 M^{\text{TM}}_{22} - k_0 k_{\text{t}x} M^{\text{TM}}_{12}) - k_{\text{b}}^2(k_{\text{t}}^2 M^{\text{TM}}_{21} - k_0 k_{\text{t}x} M^{\text{TM}}_{11})} \mathrm{e}^{2\mathrm{i}k_{\text{b}x}x_0},$$

$$a_2 = \frac{2k_0 k_{\text{b}} k_{\text{t}} k_{\text{b}x}(M^{\text{TM}}_{11} M^{\text{TM}}_{22} - M^{\text{TM}}_{12} M^{\text{TM}}_{21}) \mathrm{e}^{\mathrm{i}(k_{\text{b}x}x_0 - k_{\text{t}x}x_N)}}{k_0 k_{\text{b}x}(k_{\text{t}}^2 M^{\text{TM}}_{22} - k_0 k_{\text{t}x} M^{\text{TM}}_{12}) - k_{\text{b}}^2(k_{\text{t}}^2 M^{\text{TM}}_{21} - k_0 k_{\text{t}x} M^{\text{TM}}_{11})},$$

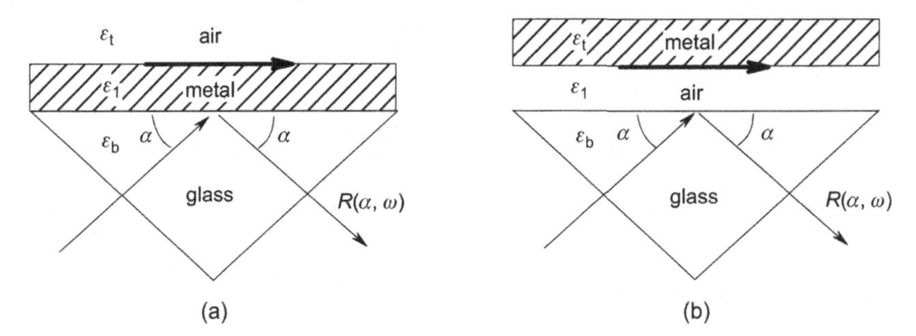

Figure 2.22 The Kretschmann (a) and Otto (b) configurations for coupling with the surface plasmon polaritons propagating at a metal/dielectric interface.

while the TE coefficients are given by

$$b_1 = \frac{k_{bx}(k_{tx}M_{12}^{TE} + k_0 M_{22}^{TE}) - k_0(k_{tx}M_{11}^{TE} + k_0 M_{21}^{TE})}{k_{bx}(k_{tx}M_{12}^{TE} + k_0 M_{22}^{TE}) + k_0(k_{tx}M_{11}^{TE} + k_0 M_{21}^{TE})} e^{2ik_{bx}x_0},$$

$$b_2 = \frac{2k_0 k_{bx}(M_{11}^{TE}M_{22}^{TE} - M_{12}^{TE}M_{21}^{TE})e^{i(k_{bx}x_0 - k_{tx}x_N)}}{k_{bx}(k_{tx}M_{12}^{TE} + k_0 M_{22}^{TE}) + k_0(k_{tx}M_{11}^{TE} + k_0 M_{21}^{TE})}.$$

The reflection and transmission coefficients obtained also allow us to get the dispersion relations of the TM and TE plasmon polaritons supported in a multilayer slab. These relations can be derived by requiring

$$\boldsymbol{N}_{tra}(x_N) - \boldsymbol{M} \cdot \boldsymbol{N}_{ref}(x_0) = 0, \tag{2.188}$$

with the subsequent changes $k_{bx} \to \zeta_b k_{bx}$ and $k_{tx} \to \zeta_t k_{tx}$, where ζ_b and ζ_t follow the definition of ζ and take the values ± 1 depending on the sign of $\mathrm{Im}(k_b)$ and $\mathrm{Im}(k_t)$, respectively. In this way, the dispersion equations can be written as

$$\zeta_b k_0 k_{bx}(k_t^2 M_{22}^{TM} - \zeta_t k_0 k_{tx} M_{12}^{TM}) - k_b^2(k_t^2 M_{21}^{TM} - \zeta_t k_0 k_{tx} M_{11}^{TM}) = 0,$$
$$\zeta_b k_{bx}(\zeta_b k_{tx} M_{12}^{TE} + k_0 M_{22}^{TE}) + k_0(\zeta_t k_{tx} M_{11}^{TE} + k_0 M_{21}^{TE}) = 0$$

for the TM and TE plasmon polaritons, respectively.

Coupling with surface plasmon polaritons

As we mentioned above, the coupling with the surface plasmon polaritons is impossible in the simple case of a metal slab due to the high polariton wavenumbers $k_\parallel > k_e$, which are inaccessible to photons propagating in the external medium with $\varepsilon = \varepsilon_e$. However, if we consider a structure in which the external media below and above the metal slab are different, with $\varepsilon = \varepsilon_b$ and $\varepsilon = \varepsilon_t$, as shown in Fig. 2.22(a), then the photons propagating in the bottom external medium can access higher wavenumbers in the top external medium, if $\varepsilon_b > \varepsilon_t$. In this way, the parallel wavenumbers of the photons can reach $k_b \cos\alpha > k_t$ and couple to the surface plasmon polaritons that propagate at the top metal interface with $k_\parallel > k_t$.

Note that the condition $k_b \cos\alpha > k_t$ is the well-known condition of the *attenuated total reflection* at the interface between the media with ε_b and ε_t [12]. It occurs for

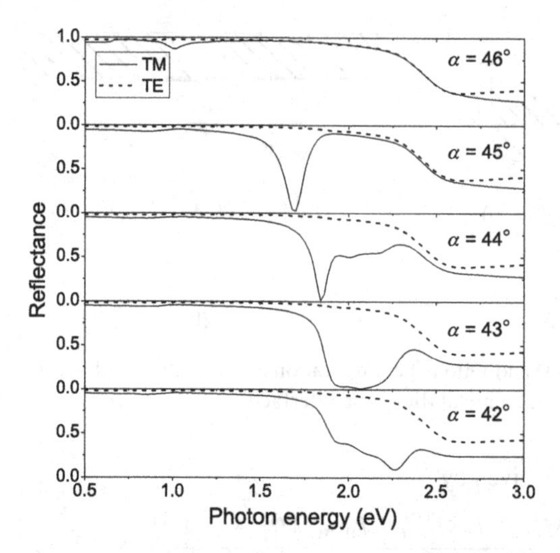

Figure 2.23 The spectral reflectance of a gold film of thickness 50 nm in the Kretschmann configuration SiO_2/Au/air. For the energies shown, the critical angle α_{cr} changes within the range $46.4° \pm 0.7°$. The optical data are from Ref. [8].

$\varepsilon_b > \varepsilon_t$ only, if the incidence angle is smaller than the critical angle, $\alpha < \alpha_{cr}$, where

$$\alpha_{cr} = \arccos(\sqrt{\varepsilon_t/\varepsilon_b}).$$

As a result of the higher parallel wavenumbers, the photons are not allowed to penetrate into the medium with ε_t under the attenuated total reflection conditions. They form evanescent fields in the top external medium that propagate with an overcritical parallel wavenumber similar to those of the surface plasmon polaritons supported by the interface between the metal slab and the top external medium. This allows us to couple and transform the photons incident in one medium (with ε_b) to the surface polaritons propagating in another (with ε_t).

The basis of the attenuated-total-reflection method lies in the two configurations used for the coupling with surface plasmon polaritons shown in Fig. 2.22. In the *Kretschmann configuration* [16], a metal film is deposited directly onto a prism with relatively high ε, whereas in the *Otto configuration* [17], they are separated by a dielectric of low ε. Both configurations provide coupling with the surface polaritons that propagate on the interface between the metal and the low-index dielectric. The coupling is realized at incidence angles below α_{cr} and for TM-polarized light only. In general, it increases the slab absorption at the resonance frequencies. However, since direct experimental measurement of the absorptance is impossible, the coupling is generally detected through the minimal reflectance R which appears as a result of the enhanced absorption [15].

Figure 2.23 shows typical reflectance spectra obtained in the Kretschmann configuration for different incidence angles α. The spectra exhibit the dips that appear for TM-polarized light only and strictly follow the energy of the surface polaritons at the metal/air interface. Similar minima appear for the reflectance in the Otto configuration, as shown in Fig. 2.24. They follow the same dispersion curve as that of the surface polaritons on a single interface, but for a broader range of the incidence angle α.

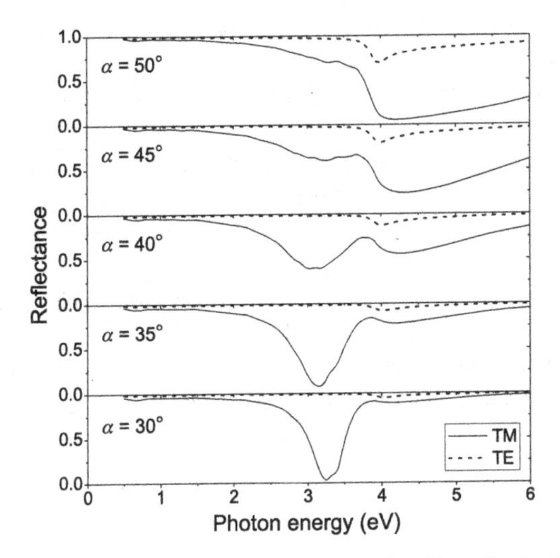

Figure 2.24 The spectral reflectance of a silver film in the Otto configuration SiO_2/air (100 nm)/Ag. For the energies shown, the critical angle α_{cr} changes within the range 46.6° ± 3.0°. The optical data are from Ref. [8].

In general, these two configurations provide similar efficiency for the coupling with the surface plasmon polaritons. However, the Kretschmann scheme usually shows better coupling at low photon energies, while the Otto configuration is commonly more appropriate for the coupling at higher photon energies.

References

[1] J. D. Jackson, *Classical Electrodynamics*, 2nd ed. New York: Wiley, 1975.

[2] R. G. Barrera, G. A. Estévez, and J. Giraldo, "Vector spherical harmonics and their application to magnetostatics," *Eur. J. Phys.*, vol. 6, pp. 287–294, 1985.

[3] H. J. Hagemann, W. Gudat, and C. Kunz, "Optical constants from the far infrared to the x-ray region: Mg, Al, Cu, Ag, Au, Bi, C, and Al_2O_3," *J. Opt. Soc. Am.*, vol. 65, pp. 742–744, 1975.

[4] J. A. Stratton, *Electromagnetic Theory*. New York: McGraw-Hill Book Company. Inc., 1941.

[5] C. F. Bohren and D. R. Huffman, *Absorption and Scattering of Light by Small Particles*, 2nd ed. New York: Wiley-Interscience, 1998.

[6] R. Fuchs and K. L. Kliewer, "Optical modes of vibration in an ionic crystal sphere," *J. Opt. Soc. Am.*, vol. 58, pp. 319–330, 1968.

[7] G. Mie, "Beiträge zur Optik trüber Medien, speziell kolloidaler Metallösungen," *Ann. Phys.*, vol. 25, pp. 377–445, 1908.

[8] SOPRA database, http://www.sspectra.com/sopra.html.

[9] S. A. Maier, *Plasmonics: Fundamentals and Applications*. New York: Springer, 2007.

[10] C. Loo, A. Lin, L. Hirsch *et al.*, "Nanoshell-enabled photonics-based imaging and therapy of cancer," *Technol. Cancer Res. Treat.*, vol. 3, pp. 33–40, 2004.

[11] V. L. Ginzburg, *The Propagation of Electromagnetic Waves in Plasmas*. New York: Pergamon, 1970.

[12] L. D. Landau, E. M. Lifshitz, and L. P. Pitaevskii, *Electrodynamics of Continuous Media*, 2nd ed. Oxford: Pergamon, 1984.

[13] S. I. Bozhevolnyi, ed., *Plasmonic Nanoguides and Circuits*. Singapore: Pan Stanford Publishing, 2009.

[14] A. V. Zayats, I. I. Smolyaninov, and A. A. Maradudin, "Nano-optics of surface plasmon polaritons," *Phys. Rep.*, vol. 408, pp. 131–314, 2005.

[15] H. Raether, *Surface Plasmons on Smooth and Rough Surfaces and on Gratings*. Berlin: Springer-Verlag, 1988.

[16] E. Kretschmann and H. Raether, "Radiative decay of non-radiative surface plasmons excited by light," *Z. Naturforsch.*, vol. 23a, pp. 2135–2136, 1968.

[17] A. Otto, "Excitation of non-radiative surface plasma waves in silver by the method of frustrated total reflection," *Z. Phys.*, vol. 216, pp. 398–410, 1968.

3 Frequency-domain methods for modeling plasmonics

3.1 Introduction

The rapid development of research on plasmonics in recent years has led to numerous interesting applications, and many plasmonic nanostructures have been designed and fabricated to achieve novel functionalities and/or better performance. For example, optical antennas are used for biochemical sensing [1–4], plasmonic waveguides have been proposed for on-chip optical communications [5], and metamaterials are under consideration for subwavelength imaging [6]. Most of those plasmonic nanostructures are complicated, so we cannot find analytical solutions for them. Therefore numerical modeling methods are the only choice when it comes to device modeling and structural design. Many numerical methods for solving Maxwell's equations have been established. They can be generally categorized into two classes: frequency-domain methods and time-domain methods.

In frequency-domain methods we assume that the electromagnetic wave is a single-frequency harmonic wave with a time dependence term $e^{i\omega t}$ or $e^{-i\omega t}$ (most electrical engineers use $e^{i\omega t}$, while $e^{-i\omega t}$ is more popular among physicists. They are essentially the same, except that, when a medium is lossy, the imaginary parts of its refractive index and permittivity take positive values for $e^{-i\omega t}$ and negative values for $e^{i\omega t}$). Then the time dependence in Maxwell's equations can easily be eliminated and the fields are functions solely of space coordinates. The solutions obtained from frequency-domain methods are generally steady-state solutions. There are many frequency-domain methods available for plasmonic device modeling, among which the finite-element method (FEM) and the method of moments (MoM) are very popular. Both the FEM and the MoM are well-established methods that are capable of handling complicated geometries. However, their implementations are very involved, requiring several major steps such as geometry drawing, meshing, and sparse-matrix solving, each being a separate research area of its own. To model plasmonic devices with the FEM or the MoM, it is better to use commercial software packages, such as COMSOL Multiphysics. Therefore those methods will not be discussed in this chapter. Readers interested in FEM may refer to dedicated literature such as Jianming Jin's *The Finite Element Method in Electromagnetics*, which also covers the basics of the MoM [7].

Time-domain methods are based on the time-dependent Maxwell equations, with the most common ones being the finite-difference time-domain (FDTD) method and the finite-integration technique (FIT). Time-domain methods are good for transient analysis and nonlinear phenomena modeling, but the implementations of material dispersions in the time domain require special treatment and often cause problems. Readers may consult Ref. [8] for more details.

In this chapter we discuss a frequency-domain method called rigorous coupled-wave analysis (RCWA) and show examples of its application to plasmonic device modeling. We also develop a semi-analytical method that has unique capabilities to analyze near-field coupling effects.

3.2 Rigorous coupled-wave analysis

Rigorous coupled-wave analysis (RCWA), also known as the Fourier modal method (FMM) or spatial harmonic analysis (SHA), is a numerical method to rigorously solve Maxwell's equations for periodic structures. The origin of this method dates back to the 1960s, when Burckhardt proposed a two-dimensional (2D) RCWA for modeling sinusoidally modulated diffraction gratings [9]. Later this method was extended to include non-sinusoidal and complex dielectric constants by Kaspar [10], and further generalized by Knop, Moharam, and Gaylord [11–13]. A fast-converging formulation was proposed in 1996 to solve the convergence problem for transverse magnetic (TM) waves by Lalanne and Morris [14], and Granet and Guizal [15], and then the mathematical proof of this formulation was given by Li [16]. Li also provided a stable recursive matrix algorithm for thick multilayer gratings [17, 18]. Three-dimensional (3D) RCWA was first proposed by Noponen and Turunen [19], and was later reformulated using correct Fourier factorization rules by Li [20]. RCWA is a powerful method in modeling nanostructures, and is used as a major tool in many studies. This chapter provides the theoretical formulation of RCWA, and shows how we can use it to model plasmonic nanostructures and metamaterials.

3.2.1 Formulations

In RCWA we solve the differential form of Maxwell's equations for source-free regions:

$$\begin{aligned} \nabla \times \boldsymbol{H} - \frac{\partial}{\partial t}\boldsymbol{D} &= 0, \\ \nabla \times \boldsymbol{E} + \frac{\partial}{\partial t}\boldsymbol{B} &= 0, \\ \nabla \cdot \boldsymbol{D} &= 0, \\ \nabla \cdot \boldsymbol{B} &= 0. \end{aligned} \tag{3.1}$$

Note that in this chapter we use SI units, so Maxwell's equations here look different from those in Chapter 1, Eqs. (1.1) and (1.2). We then assume that the electromagnetic fields are monochromatic and the materials are isotropic, so the following relations are

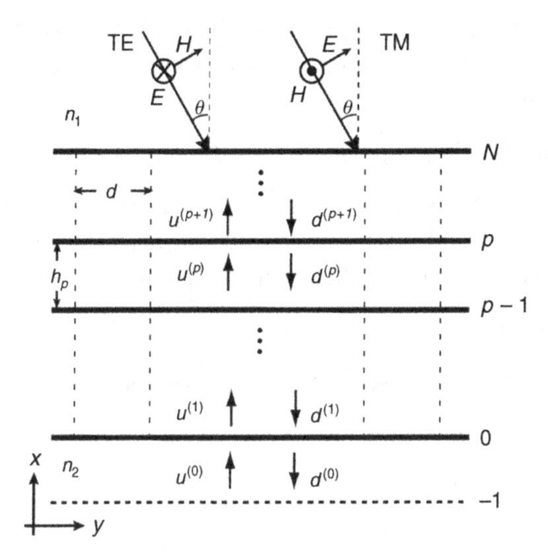

Figure 3.1 A multilayer single-period grating studied by 2D RCWA. The period is d. Each layer has a thickness h_p.

true:

$$\begin{aligned} \boldsymbol{E} &= \boldsymbol{E}(x, y, z)\mathrm{e}^{-\mathrm{i}\omega t}, \\ \boldsymbol{H} &= \boldsymbol{H}(x, y, z)\mathrm{e}^{-\mathrm{i}\omega t}, \\ \boldsymbol{D} &= \varepsilon(x, y, z)\boldsymbol{E}, \\ \boldsymbol{B} &= \mu_0 \boldsymbol{H}. \end{aligned} \tag{3.2}$$

Then Maxwell's equations are

$$\begin{aligned} \partial_y E_z - \partial_z E_y &= \mathrm{i}\omega\mu_0 H_x, \\ \partial_z E_x - \partial_x E_z &= \mathrm{i}\omega\mu_0 H_y, \\ \partial_x E_y - \partial_y E_x &= \mathrm{i}\omega\mu_0 H_z, \\ \partial_y H_z - \partial_z H_y &= -\mathrm{i}\omega\varepsilon E_x, \\ \partial_z H_x - \partial_x H_z &= -\mathrm{i}\omega\varepsilon E_y, \\ \partial_x H_y - \partial_y H_x &= -\mathrm{i}\omega\varepsilon E_z, \end{aligned} \tag{3.3}$$

where $\partial_x = \partial/\partial x$, $\partial_y = \partial/\partial y$, and $\partial_z = \partial/\partial z$.

Two-dimensional RCWA

Two-dimensional (2D) RCWA solves single-period multilayer gratings as illustrated in Fig. 3.1 [21]. In each layer, the permittivity ε is modulated periodically along the y direction with a period of d, and remains invariant along the x and z directions, and the refractive indexes of the incident and output media are n_1 and n_2, respectively. In each layer the Maxwell equations can be converted into an eigenvalue equation. The incident wave is a monochromatic plane wave with a vacuum wavelength λ, and the vacuum wavevector is $k = 2\pi/\lambda$; θ is the angle of incidence. For transverse electric (TE) polarization where the electric field is normal to the incidence plane, the incident electric field can expressed as

$$\boldsymbol{E}_{\text{I}}(\mathbf{r}) = \hat{z}E_{\text{iz}}\exp[\text{i}(\boldsymbol{k}_0 \cdot \boldsymbol{r} - \omega t)], \tag{3.4}$$

where $\boldsymbol{k}_0$ is the incident wavevector, $\boldsymbol{k}_0 = \alpha_0\hat{x} + \beta_0\hat{y}$, where $\alpha_0 = n_1 k\cos\theta$ and $\beta_0 = n_1 k\sin\theta$. For transverse magnetic (TM) polarization, where the magnetic field is normal to the incidence plane, the incident magnetic field can be expressed as

$$\boldsymbol{H}_{\text{I}}(\boldsymbol{r}) = \hat{z}H_{\text{iz}}\exp[\text{i}(\boldsymbol{k}_0 \cdot \boldsymbol{r} - \omega t)]. \tag{3.5}$$

The electromagnetic fields in the modulated region can be expressed as

$$\begin{aligned} \boldsymbol{E} &= \boldsymbol{E}(x, y)\text{e}^{-\text{i}\omega t}, \\ \boldsymbol{H} &= \boldsymbol{H}(x, y)\text{e}^{-\text{i}\omega t}, \\ \boldsymbol{D} &= \varepsilon(y)\boldsymbol{E}, \\ \boldsymbol{B} &= \mu_0\boldsymbol{H}, \end{aligned} \tag{3.6}$$

where $\varepsilon(y)$ is a periodic function of y with a period of d.

Eigenequations for TE polarization

For TE polarization $E_x = E_y = H_z = \partial_z = 0$, so Eq. (3.3) is reduced to

$$\begin{aligned} \partial_y E_z &= \text{i}\omega\mu_0 H_x, \\ -\partial_x E_z &= \text{i}\omega\mu_0 H_y, \\ \partial_x H_y - \partial_y H_x &= -\text{i}\omega\varepsilon E_z, \end{aligned} \tag{3.7}$$

which can be further reduced to

$$\frac{\partial^2 E_z}{\partial x^2} + \frac{\partial^2 E_z}{\partial y^2} + \varepsilon_{\text{r}} k^2 E_z = 0. \tag{3.8}$$

According to the Floquet–Bloch theorem, the solutions can be expressed as

$$E_z(x, y) = \sum_{m,n} E_{zmn}\exp[\text{i}(k_{xm}x + k_{yn}y)], \tag{3.9}$$

where m and n are integers ranging from $-\infty$ to ∞, $k_{yn} = \beta_0 + 2\pi n/d$, and k_{xm} is to be determined.

The permittivity and its inverse can be expressed as a sum of Fourier series,

$$\varepsilon_r(y) = \varepsilon(y)/\varepsilon_0 = \sum_p \varepsilon_p \exp[\text{i}2\pi py/d], \tag{3.10}$$

where ε_p denotes the Fourier coefficients defined as

$$\varepsilon_p = \frac{1}{d}\int_0^d \varepsilon_r(y)\exp[-\text{i}2\pi py/d]\text{d}y. \tag{3.11}$$

On substituting (3.9) and (3.10) into Eq. (3.8) and matching the coefficients of each term of the form $\exp[i(k_{xm}x + k_{yn}y)]$ on both sides of the equation for each particular set of (m, n), we obtain

$$E_{zmn}\left(k_{xm}^2 + k_{yn}^2\right) = k^2 \sum_p \varepsilon_{n-p} E_{zmp}. \tag{3.12}$$

For each m, Eq. (3.12) can be rewritten as

$$\tilde{k}_{xm}^2 \boldsymbol{E}_m = \left(\boldsymbol{\varepsilon} - \tilde{\boldsymbol{k}}_y^2\right) \boldsymbol{E}_m, \tag{3.13}$$

where $\tilde{k}_{xm} = k_{xm}/k$, $\tilde{\boldsymbol{k}}_y = [\tilde{k}_{yn}]$ is a diagonal matrix, $\tilde{k}_{yn} = k_{yn}/k$, $\boldsymbol{E}_m = [E_{zmn}]$, and $\boldsymbol{\varepsilon}$ is composed of the Fourier coefficients ε_p,

$$\boldsymbol{\varepsilon} = \begin{bmatrix} \varepsilon_0 & \varepsilon_{-1} & \varepsilon_{-2} & \cdots & \varepsilon_{-2M} \\ \varepsilon_1 & \varepsilon_0 & \varepsilon_{-1} & \cdots & \varepsilon_{-2M+1} \\ \vdots & \vdots & \vdots & \vdots & \vdots \\ \varepsilon_{2M} & \varepsilon_{2M-1} & \cdots & \varepsilon_1 & \varepsilon_0 \end{bmatrix}. \tag{3.14}$$

This is an eigenvalue equation, with $\tilde{k}_{xm}^2$ being the eigenvalue and $\boldsymbol{E}_m$ being the eigenvector.

Eigenequations for TM polarization

For TM polarization $H_x = H_y = E_z = \partial_z = 0$, so Eq. (3.3) is reduced to

$$\begin{aligned} \partial_y H_z &= -\mathrm{i}\omega\varepsilon E_x, \\ \frac{1}{\varepsilon}\,\partial_x H_z &= \mathrm{i}\omega E_y, \\ \partial_x E_y - \partial_y E_x &= \mathrm{i}\omega\mu_0 H_z. \end{aligned} \tag{3.15}$$

Note that the position of ε in Eq. (3.15) is important: we must place ε on the right-hand side of the first equation and $1/\varepsilon$ on the left-hand side of the second equation in order to achieve fast convergence [16]. Then we write the fields as

$$\begin{aligned} E_x(x, y) &= \sum_{m,n} E_{xmn} \exp[\mathrm{i}(k_{xm}x + k_{yn}y)], \\ E_y(x, y) &= \sum_{m,n} E_{ymn} \exp[\mathrm{i}(k_{xm}x + k_{yn}y)], \\ H_z(x, y) &= \sum_{m,n} H_{zmn} \exp[\mathrm{i}(k_{xm}x + k_{yn}y)]. \end{aligned} \tag{3.16}$$

By substituting Eq. (3.10) and (3.16) into Eq. (3.15), and matching the coefficients for every term of the form $\exp[\mathrm{i}(k_{xm}x + k_{yn}y)]$, we obtain

$$\begin{aligned} -\omega \sum_p \varepsilon_{n-p} E_{xmp} &= k_{yn} H_{zmn}, \\ \omega E_{ymn} &= \sum_l \left(\frac{1}{\varepsilon}\right)_{n-p} k_{xm} H_{zmp}, \\ k_{xm} E_{ymn} - k_{yn} E_{xmn} &= \omega\mu_0 H_{zmn}. \end{aligned} \tag{3.17}$$

For each m, Eq. (3.17) can be rewritten as

$$\begin{aligned} -\omega\varepsilon_0 \boldsymbol{\varepsilon} \boldsymbol{E}_{xm} &= k\tilde{\boldsymbol{k}}_y \boldsymbol{H}_{zm}, \\ \omega\varepsilon_0 \boldsymbol{E}_{ym} &= k\tilde{k}_{xm}\bar{\boldsymbol{\varepsilon}} \boldsymbol{H}_{zm}, \\ k\tilde{k}_{xm} \boldsymbol{E}_{ym} - k\tilde{\boldsymbol{k}}_y \boldsymbol{E}_{xm} &= \omega\mu_0 \boldsymbol{H}_{zm}, \end{aligned} \tag{3.18}$$

where $\boldsymbol{E}_{xm} = [E_{xmn}]$, $\boldsymbol{E}_{ym} = [E_{ymn}]$, and $\boldsymbol{H}_{zm} = [H_{zmn}]$ are vectors, and $\bar{\boldsymbol{\varepsilon}}$ is composed of the Fourier coefficients of $1/\varepsilon$,

$$\bar{\boldsymbol{\varepsilon}} = \begin{bmatrix} (1/\varepsilon)_0 & (1/\varepsilon)_{-1} & (1/\varepsilon)_{-2} & \cdots & (1/\varepsilon)_{-2M} \\ (1/\varepsilon)_1 & (1/\varepsilon)_0 & (1/\varepsilon)_{-1} & \cdots & (1/\varepsilon)_{-2M+1} \\ \vdots & \vdots & \vdots & \vdots & \vdots \\ (1/\varepsilon)_{2M} & (1/\varepsilon)_{2M-1} & \cdots & (1/\varepsilon)_1 & (1/\varepsilon)_0 \end{bmatrix}. \tag{3.19}$$

By eliminating $\boldsymbol{E}_{xm}$ and $\boldsymbol{E}_{ym}$ from Eq. (3.18), we obtain an eigenvalue equation for TM polarization:

$$\tilde{k}_{xm}^2 \bar{\boldsymbol{\varepsilon}} \boldsymbol{H}_{zm} = \left(\boldsymbol{I} - \tilde{\boldsymbol{k}}_y \boldsymbol{\varepsilon}^{-1} \tilde{\boldsymbol{k}}_y\right) \boldsymbol{H}_{zm}. \tag{3.20}$$

This is a generalized eigenvalue equation, with $\tilde{k}_{xm}^2$ being the eigenvalue and $\boldsymbol{H}_{zm}$ being the eigenvector. We can solve it as a standard eigenvalue problem

$$\tilde{k}_{xm}^2 \boldsymbol{H}_{zm} = \bar{\boldsymbol{\varepsilon}}^{-1} \left(\boldsymbol{I} - \tilde{\boldsymbol{k}}_y \boldsymbol{\varepsilon}^{-1} \tilde{\boldsymbol{k}}_y\right) \boldsymbol{H}_{zm}. \tag{3.21}$$

The dimensions of the matrix in Eqs. (3.13) and (3.21) are infinite; however, in practical implementations the matrices have to be truncated.

Three-dimensional RCWA

Three-dimensional (3D) RCWA solves Maxwell's equations for structures like the one shown in Fig. 3.2(a) [19], where the permittivity is modulated periodically along the x and y directions, and remains invariant along the z direction. The incident wave is a monochromatic plane wave with a vacuum wavelength λ, as shown in Fig. 3.2(b). The vacuum wavevector is $k = 2\pi/\lambda$, θ is the angle of incidence, and ϕ is the angle between the x axis and the plane of incidence.

The incident electric field is

$$\boldsymbol{E}_{\mathrm{I}}(\boldsymbol{r}) = \boldsymbol{E}_{\mathrm{i}} \exp[\mathrm{i}(\boldsymbol{k}_0 \cdot \boldsymbol{r} - \omega t)], \tag{3.22}$$

where $\boldsymbol{k}_0$ is the wavevector, $\boldsymbol{k}_0 = \alpha_0\hat{\boldsymbol{x}} + \beta_0\hat{\boldsymbol{y}} + r_{00}\hat{\boldsymbol{z}}$, where $\alpha_0 = n_1 k \sin\theta \cos\phi$, $\beta_0 = n_1 k \sin\theta \sin\phi$, and $r_{00} = n_1 k \cos\theta$. In the case of a linearly polarized incident wave, $\boldsymbol{E}_{\mathrm{i}}$ is

$$\begin{aligned} \boldsymbol{E}_{\mathrm{i}} = {} & (\cos\psi \cos\theta \cos\phi - \sin\psi \sin\phi)\hat{\boldsymbol{x}} \\ & + (\cos\psi \cos\theta \sin\phi + \sin\psi \cos\phi)\hat{\boldsymbol{y}} \\ & - \cos\psi \sin\theta\, \hat{\boldsymbol{z}}, \end{aligned} \tag{3.23}$$

where ψ is the angle between the electric field vector and the plane of incidence.

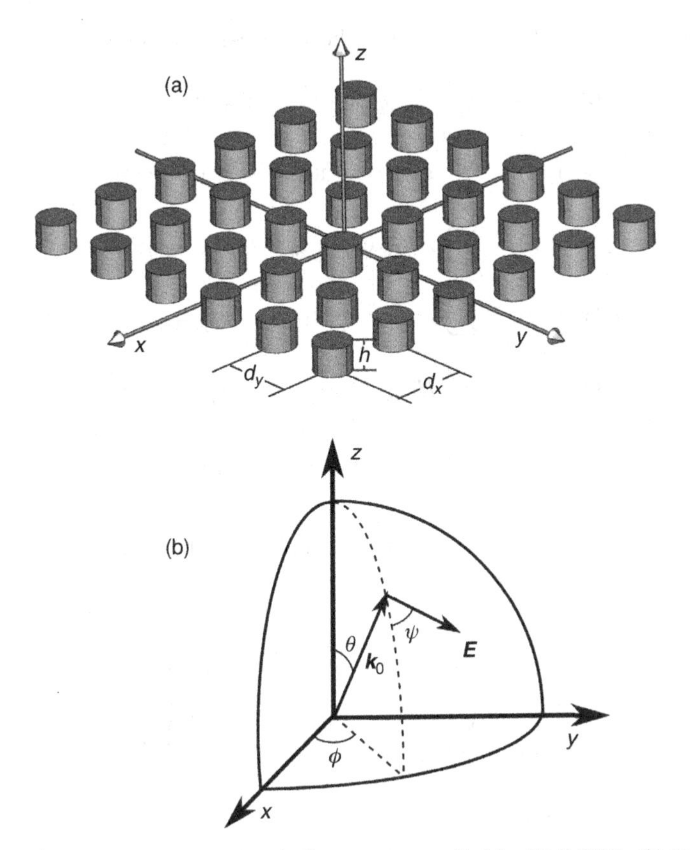

Figure 3.2 (a) A typical periodic structure studied by 3D RCWA. (b) Definitions of the angles.

The electromagnetic fields in the modulated region can be expressed as

$$\begin{aligned} \boldsymbol{E} &= \boldsymbol{E}(x, y, z)\mathrm{e}^{-\mathrm{i}\omega t}, \\ \boldsymbol{H} &= \boldsymbol{H}(x, y, z)\mathrm{e}^{-\mathrm{i}\omega t}, \\ \boldsymbol{D} &= \varepsilon(x, y)\boldsymbol{E}, \\ \boldsymbol{B} &= \mu_0 \boldsymbol{H}, \end{aligned} \tag{3.24}$$

where $\varepsilon(x, y)$ is a periodic function of x and y with the periods being d_x and d_y. By eliminating E_z and H_z using $E_z = [i/(\omega\varepsilon)](\partial_x H_y - \partial_y H_x)$ and $H_z = -[i/(\omega\mu_0)](\partial_x E_y - \partial_y E_x)$ from Eq. (3.3) we get

$$\begin{aligned} \omega\varepsilon_0\, \partial_z E_x &= \mathrm{i}k^2 H_y + \mathrm{i}\, \partial_x \left[\varepsilon_\mathrm{r}^{-1} \left(\partial_x H_y - \partial_y H_x\right)\right], \\ \omega\varepsilon_0\, \partial_z E_y &= -\mathrm{i}k^2 H_x + \mathrm{i}\, \partial_y \left[\varepsilon_\mathrm{r}^{-1} \left(\partial_x H_y - \partial_y H_x\right)\right], \\ \omega\mu_0\, \partial_z H_x &= -\mathrm{i}k^2 \varepsilon_\mathrm{r} E_y - \mathrm{i}\, \partial_x \left[\left(\partial_x E_y - \partial_y E_x\right)\right], \\ \omega\mu_0\, \partial_z H_y &= \mathrm{i}k^2 \varepsilon_\mathrm{r} E_x - \mathrm{i}\, \partial_y \left[\left(\partial_x E_y - \partial_y E_x\right)\right]. \end{aligned} \tag{3.25}$$

According to the Floquet–Bloch theorem, the fields can be expressed as

$$\begin{aligned}
E_x(x,y,z) &= \sum_{l,m,n} E_{xmnl} \exp[\mathrm{i}(\alpha_m x + \beta_n y + \gamma_l z)],\\
E_y(x,y,z) &= \sum_{l,m,n} E_{ymnl} \exp[\mathrm{i}(\alpha_m x + \beta_n y + \gamma_l z)],\\
E_z(x,y,z) &= \sum_{l,m,n} E_{zmnl} \exp[\mathrm{i}(\alpha_m x + \beta_n y + \gamma_l z)],\\
H_x(x,y,z) &= \sum_{l,m,n} H_{xmnl} \exp[\mathrm{i}(\alpha_m x + \beta_n y + \gamma_l z)],\\
H_y(x,y,z) &= \sum_{l,m,n} H_{ymnl} \exp[\mathrm{i}(\alpha_m x + \beta_n y + \gamma_l z)],\\
H_z(x,y,z) &= \sum_{l,m,n} H_{zmnl} \exp[\mathrm{i}(\alpha_m x + \beta_n y + \gamma_l z)],
\end{aligned} \tag{3.26}$$

where $\alpha_m = \alpha_0 + 2\pi m/d_x$, $\beta_n = \beta_0 + 2\pi n/d_y$, and γ_l is to be determined.

The permittivity and its inverse can be expressed as 2D Fourier series,

$$\begin{aligned}
\varepsilon_{\mathrm{r}}(x,y) &= \sum_{p,q} \varepsilon_{pq} \exp\left[\mathrm{i}2\pi\left(\frac{px}{d_x} + \frac{qy}{d_y}\right)\right],\\
\frac{1}{\varepsilon_{\mathrm{r}}(x,y)} &= \sum_{p,q} \varepsilon_{pq}^{-} \exp\left[\mathrm{i}2\pi\left(\frac{px}{d_x} + \frac{qy}{d_y}\right)\right],
\end{aligned} \tag{3.27}$$

where ε_{pq}^{-} and ε_{pq} are Fourier coefficients defined as

$$\begin{aligned}
\varepsilon_{pq} &= \frac{1}{d_x d_y}\int_0^{d_x}\int_0^{d_y} \varepsilon(x,y) \exp\left[-\mathrm{i}2\pi\left(\frac{px}{d_x} + \frac{qy}{d_y}\right)\right] \mathrm{d}x\,\mathrm{d}y,\\
\varepsilon_{pq}^{-} &= \frac{1}{d_x d_y}\int_0^{d_x}\int_0^{d_y} \frac{1}{\varepsilon(x,y)} \exp\left[-\mathrm{i}2\pi\left(\frac{px}{d_x} + \frac{qy}{d_y}\right)\right] \mathrm{d}x\,\mathrm{d}y.
\end{aligned} \tag{3.28}$$

By substituting Eqs. (3.26) and (3.27) into (3.25), and matching the coefficients of the term $\exp[\mathrm{i}(\alpha_m x + \beta_n y + \gamma_l z)]$ on both sides of the equations for each particular set of (l, m, n), we then obtain

$$\begin{aligned}
\omega\varepsilon_0\gamma_l E_{xmnl} &= k^2 H_{ymnl} - \alpha_m \sum_{p,q} \varepsilon_{m-p,n-q}^{-}(\alpha_p H_{ypql} - \beta_q H_{xpql}),\\
\omega\varepsilon_0\gamma_l E_{ymnl} &= -k^2 H_{xmnl} - \beta_n \sum_{p,q} \varepsilon_{m-p,n-q}^{-}(\alpha_p H_{ypql} - \beta_q H_{xpql}),\\
\omega\mu_0\gamma_l H_{xmnl} &= -k^2 \sum_{p,q} \varepsilon_{m-p,n-q} E_{ypql} + \alpha_m(\alpha_m E_{ymnl} - \beta_n E_{xmnl}),\\
\omega\mu_0\gamma_l H_{ymnl} &= k^2 \sum_{p,q} \varepsilon_{m-p,n-q} E_{xpql} + \beta_n(\alpha_m E_{ymnl} - \beta_n E_{xmnl}).
\end{aligned} \tag{3.29}$$

The summations in the above equations are from $-\infty$ to ∞. In practical implementations we need to truncate the m and n, $-M_x \leq m \leq M_x$, $-M_y \leq n \leq M_y$, and let $L_x = 2M_x + 1$, $L_y = 2M_y + 1$, and $L_{\mathrm{t}} = L_x L_y$. Then for each γ_l we construct the vector

$$\boldsymbol{E}_l = \begin{bmatrix} \boldsymbol{E}_{xl} \\ \boldsymbol{E}_{yl} \end{bmatrix},$$

where $\boldsymbol{E}_{xl} = [E_{x1l} \quad E_{x2l} \quad \ldots \quad E_{xul}]^{\mathrm{T}}$ and $\boldsymbol{E}_{yl} = [E_{y1l} \quad E_{y2l} \quad \ldots \quad E_{yul}]^{\mathrm{T}}$. u is obtained from m and n via $u = (m + M_x)(2M_y + 1) + n + M_y + 1$. In the same way

$$\boldsymbol{H}_l = \begin{bmatrix} \boldsymbol{H}_{xl} \\ \boldsymbol{H}_{yl} \end{bmatrix}$$

can be constructed. Then Eq. (3.29) can be rewritten as

$$\begin{aligned} \omega\varepsilon_0\gamma_l \boldsymbol{E}_l &= \boldsymbol{F}\boldsymbol{H}_l, \\ \omega\mu_0\gamma_l \boldsymbol{H}_l &= \boldsymbol{G}\boldsymbol{E}_l. \end{aligned} \tag{3.30}$$

Eliminating $\boldsymbol{H}_l$ from Eq. (3.30) results in an eigenvalue equation,

$$\tilde{\gamma}_l^2 \boldsymbol{E}_l = \tilde{\boldsymbol{F}}\tilde{\boldsymbol{G}}\boldsymbol{E}_l, \tag{3.31}$$

where $\tilde{\gamma}_l = \gamma_l/k$, $\tilde{\boldsymbol{F}} = \boldsymbol{F}/k^2$, and $\tilde{\boldsymbol{G}} = \boldsymbol{G}/k^2$. This is an eigenvalue equation with dimension $2L_{\mathrm{t}}$. The matrices $\tilde{\boldsymbol{F}}$ and $\tilde{\boldsymbol{G}}$ are constructed as

$$\begin{aligned} \tilde{\boldsymbol{F}} &= \begin{bmatrix} \boldsymbol{A}\boldsymbol{E}^{-}\boldsymbol{B} & \boldsymbol{I} - \boldsymbol{A}\boldsymbol{E}^{-}\boldsymbol{A} \\ -\boldsymbol{I} + \boldsymbol{B}\boldsymbol{E}^{-}\boldsymbol{B} & -\boldsymbol{B}\boldsymbol{E}^{-}\boldsymbol{A} \end{bmatrix}, \\ \tilde{\boldsymbol{G}} &= \begin{bmatrix} -\boldsymbol{A}\boldsymbol{B} & \boldsymbol{A}^2 - \boldsymbol{E} \\ -\boldsymbol{B}^2 + \boldsymbol{E} & \boldsymbol{B}\boldsymbol{A} \end{bmatrix}, \end{aligned} \tag{3.32}$$

where $\boldsymbol{A}$ and $\boldsymbol{B}$ are diagonal matrices, $\boldsymbol{A} = [a_{uu}]$, $\boldsymbol{B} = [b_{uu}]$, $a_{uu} = \alpha_m/k$, $b_{uu} = \beta_n/k$, $\boldsymbol{E} = [e_{uv}]$, $\boldsymbol{E}^{-} = [e_{uv}^{-}]$, $e_{uv} = \varepsilon_{m-p,n-q}$, $e_{uv}^{-} = \varepsilon_{m-p,n-q}^{-}$, u is defined as before, and v, p, and q are defined in the same way as u. The details of these matrices are shown below:

$$\boldsymbol{A} = \begin{bmatrix} \boldsymbol{a}_{-M_{\max}} & & & \\ & \boldsymbol{a}_{-M_{\max}+1} & & \\ & & \ddots & \\ & & & \boldsymbol{a}_{M_{\max}} \end{bmatrix}, \text{ where } \boldsymbol{a}_j = \frac{1}{k}\begin{bmatrix} \alpha_j & & & \\ & \alpha_j & & \\ & & \ddots & \\ & & & \alpha_j \end{bmatrix}_{2N_{\max}+1} \tag{3.33}$$

$$\boldsymbol{B} = \begin{bmatrix} \boldsymbol{b} & & & \\ & \boldsymbol{b} & & \\ & & \ddots & \\ & & & \boldsymbol{b} \end{bmatrix}_{2M_{\max}+1}, \text{ where } \boldsymbol{b} = \frac{1}{k}\begin{bmatrix} \beta_{-N_{\max}} & & & \\ & \beta_{-N_{\max}+1} & & \\ & & \ddots & \\ & & & \beta_{N_{\max}} \end{bmatrix} \tag{3.34}$$

$$
\boldsymbol{E} = \begin{bmatrix}
\varepsilon_{0,0} & \varepsilon_{0,-1} & \cdots & \varepsilon_{0,-2N_{\max}} & \varepsilon_{-1,0} & \cdots & \varepsilon_{-2M_{\max},-2N_{\max}} \\
\varepsilon_{0,1} & \varepsilon_{0,0} & \varepsilon_{0,-1} & \cdots & \varepsilon_{0,-2N_{\max}+1} & \cdots & \varepsilon_{-2M_{\max},-2N_{\max}+1} \\
\varepsilon_{0,2} & \varepsilon_{0,1} & \varepsilon_{0,0} & \varepsilon_{0,-1} & \cdots & \cdots & \varepsilon_{-2M_{\max},-2N_{\max}+2} \\
\vdots & \vdots & \vdots & \vdots & \vdots & \vdots & \vdots \\
\varepsilon_{2M_{\max},2N_{\max}} & \varepsilon_{2M_{\max},2N_{\max}-1} & \cdots & \varepsilon_{2M_{\max},0} & \varepsilon_{2M_{\max}-1,2N_{\max}} & \cdots & \varepsilon_{0,0}
\end{bmatrix}, \tag{3.35}
$$

$$
\boldsymbol{E}^{-} = \begin{bmatrix}
\varepsilon^{-}_{0,0} & \varepsilon^{-}_{0,-1} & \cdots & \varepsilon^{-}_{0,-2N_{\max}} & \varepsilon^{-}_{-1,0} & \cdots & \varepsilon^{-}_{-2M_{\max},-2N_{\max}} \\
\varepsilon^{-}_{0,1} & \varepsilon^{-}_{0,0} & \varepsilon^{-}_{0,-1} & \cdots & \varepsilon^{-}_{0,-2N_{\max}+1} & \cdots & \varepsilon^{-}_{-2M_{\max},-2N_{\max}+1} \\
\varepsilon^{-}_{0,2} & \varepsilon^{-}_{0,1} & \varepsilon^{-}_{0,0} & \varepsilon^{-}_{0,-1} & \cdots & \cdots & \varepsilon^{-}_{-2M_{\max},-2N_{\max}+2} \\
\vdots & \vdots & \vdots & \vdots & \vdots & \vdots & \vdots \\
\varepsilon^{-}_{2M_{\max},2N_{\max}} & \varepsilon^{-}_{2M_{\max},2N_{\max}-1} & \cdots & \varepsilon^{-}_{2M_{\max},0} & \varepsilon^{-}_{2M_{\max}-1,2N_{\max}} & \cdots & \varepsilon^{-}_{0,0}
\end{bmatrix}. \tag{3.36}
$$

We can compute $\boldsymbol{J} = \tilde{\boldsymbol{F}}\tilde{\boldsymbol{G}}$ explicitly to further reduce the computation:

$$
\boldsymbol{J} = \tilde{\boldsymbol{F}}\tilde{\boldsymbol{G}} = \begin{bmatrix}
\boldsymbol{E} - \boldsymbol{B}^2 - \boldsymbol{A}\boldsymbol{E}^{-}\boldsymbol{A}\boldsymbol{E} & \boldsymbol{A}\boldsymbol{B} - \boldsymbol{A}\boldsymbol{E}^{-}\boldsymbol{B}\boldsymbol{E} \\
\boldsymbol{A}\boldsymbol{B} - \boldsymbol{B}\boldsymbol{E}^{-}\boldsymbol{A}\boldsymbol{E} & \boldsymbol{E} - \boldsymbol{A}^2 - \boldsymbol{B}\boldsymbol{E}^{-}\boldsymbol{B}\boldsymbol{E}
\end{bmatrix}. \tag{3.37}
$$

By solving the equation

$$
\tilde{\gamma}_l^2 \boldsymbol{E}_l = \boldsymbol{J}\boldsymbol{E}_l, \tag{3.38}
$$

we obtain the electric fields, and then from the electric fields we can calculate the magnetic fields.

We can also calculate $\boldsymbol{H}_l$ using

$$
\tilde{\gamma}_l^2 \boldsymbol{H}_l = \boldsymbol{P}\boldsymbol{H}_l, \tag{3.39}
$$

where

$$
\boldsymbol{P} = \tilde{\boldsymbol{G}}\tilde{\boldsymbol{F}} = \begin{bmatrix}
\boldsymbol{E} - \boldsymbol{A}^2 - \boldsymbol{E}\boldsymbol{B}\boldsymbol{E}^{-}\boldsymbol{B} & \boldsymbol{E}\boldsymbol{B}\boldsymbol{E}^{-}\boldsymbol{A} - \boldsymbol{A}\boldsymbol{B} \\
\boldsymbol{E}\boldsymbol{A}\boldsymbol{E}^{-}\boldsymbol{B} - \boldsymbol{A}\boldsymbol{B} & \boldsymbol{E} - \boldsymbol{B}^2 - \boldsymbol{E}\boldsymbol{A}\boldsymbol{E}^{-}\boldsymbol{A}
\end{bmatrix}. \tag{3.40}
$$

Another formulation is to rewrite Eq. (3.25) as

$$
\begin{aligned}
\omega\varepsilon_0\, \partial_z E_x &= \mathrm{i}k^2 H_y + \mathrm{i}\,\partial_x \left[\varepsilon_{\mathrm{r}}^{-1}(\partial_x H_y - \partial_y H_x)\right], \\
\omega\varepsilon_0\, \partial_z E_y &= -\mathrm{i}k^2 H_x + \mathrm{i}\,\partial_y \left[\varepsilon_{\mathrm{r}}^{-1}(\partial_x H_y - \partial_y H_x)\right], \\
\omega\mu_0\varepsilon_{\mathrm{r}}^{-1}\, \partial_z H_x &= -\mathrm{i}k^2 E_y - \mathrm{i}\varepsilon_{\mathrm{r}}^{-1}\, \partial_x \left[(\partial_x E_y - \partial_y E_x)\right], \\
\omega\mu_0\varepsilon_{\mathrm{r}}^{-1}\, \partial_z H_y &= \mathrm{i}k^2 E_x - \mathrm{i}\varepsilon_{\mathrm{r}}^{-1}\, \partial_y \left[(\partial_x E_y - \partial_y E_x)\right].
\end{aligned} \tag{3.41}
$$

Then Eq. (3.30) is modified accordingly to

$$\begin{aligned} \omega\varepsilon_0\gamma_l \boldsymbol{E}_l &= \boldsymbol{F}\boldsymbol{H}_l, \\ \omega\mu_0\gamma_l \boldsymbol{Q}\boldsymbol{H}_l &= \boldsymbol{G}'\boldsymbol{E}_l, \end{aligned} \tag{3.42}$$

where

$$\boldsymbol{Q} = \begin{bmatrix} \boldsymbol{E}^- & \boldsymbol{0} \\ \boldsymbol{0} & \boldsymbol{E}^- \end{bmatrix}, \qquad \boldsymbol{G}' = \begin{bmatrix} -\boldsymbol{E}^-\boldsymbol{A}\boldsymbol{B} & -\boldsymbol{I} + \boldsymbol{E}^-\boldsymbol{A}^2 \\ \boldsymbol{I} - \boldsymbol{E}^-\boldsymbol{B}^2 & \boldsymbol{E}^-\boldsymbol{B}\boldsymbol{A} \end{bmatrix}.$$

Then we arrive at the generalized eigenvalue problem

$$\tilde{\gamma}_l^2 \boldsymbol{Q}\boldsymbol{H}_l = \boldsymbol{P}_{\mathrm{G}}\boldsymbol{H}_l, \tag{3.43}$$

where

$$\boldsymbol{P}_{\mathrm{G}} = \begin{bmatrix} \boldsymbol{I} - \boldsymbol{B}\boldsymbol{E}^-\boldsymbol{B} - \boldsymbol{E}^-\boldsymbol{A}^2 & \boldsymbol{B}\boldsymbol{E}^-\boldsymbol{A} - \boldsymbol{E}^-\boldsymbol{A}\boldsymbol{B} \\ \boldsymbol{A}\boldsymbol{E}^-\boldsymbol{B} - \boldsymbol{E}^-\boldsymbol{B}\boldsymbol{A} & \boldsymbol{I} - \boldsymbol{A}\boldsymbol{E}^-\boldsymbol{A} - \boldsymbol{E}^-\boldsymbol{B}^2 \end{bmatrix}. \tag{3.44}$$

It was pointed out by Li that, to achieve fast convergence, the above RCWA formulation has to be modified [16, 20]. The matrix $\boldsymbol{E}^-$ in Eq. (3.32) should be replaced by $\boldsymbol{E}^{-1}$, the inverse of $\boldsymbol{E}$. Additionally we introduce the matrices $\boldsymbol{E}_x^-$ and $\boldsymbol{E}_y^-$:

$$\begin{aligned} \left[\boldsymbol{E}_x^-(y)\right]_{mn} &= \frac{1}{d_x}\int_0^{d_x} \frac{1}{\varepsilon}(x, y)\exp\left[-\mathrm{i}2\pi\frac{(m-n)x}{d_x}\right]\mathrm{d}x, \\ \left[\boldsymbol{E}_y^-(x)\right]_{mn} &= \frac{1}{d_y}\int_0^{d_y} \frac{1}{\varepsilon}(x, y)\exp\left[-\mathrm{i}2\pi\frac{(m-n)y}{d_y}\right]\mathrm{d}y. \end{aligned} \tag{3.45}$$

Obviously $\boldsymbol{E}_x^-$ is a function of y and $\boldsymbol{E}_y^-$ is a function of x. Then we can define matrices $\boldsymbol{E}_{xy}^F$ and $\mathbf{E}_{yx}^F$,

$$\begin{aligned} \left[\boldsymbol{E}_{xy}^F\right]_{mn,jl} &= \frac{1}{d_y}\int_0^{d_y} \left[\boldsymbol{E}_x^-(y)\right]_{mj}^{-1} \exp\left[-\mathrm{i}2\pi\frac{(n-l)y}{d_y}\right]\mathrm{d}y, \\ \left[\boldsymbol{E}_{yx}^F\right]_{mn,jl} &= \frac{1}{d_x}\int_0^{d_x} \left[\boldsymbol{E}_y^-(x)\right]_{nl}^{-1} \exp\left[-\mathrm{i}2\pi\frac{(m-j)x}{d_x}\right]\mathrm{d}x, \end{aligned} \tag{3.46}$$

and use them to replace the matrix $\boldsymbol{E}$ in Eq. (3.32). Then Eq. (3.32) is modified to

$$\begin{aligned} \tilde{\boldsymbol{F}} &= \begin{bmatrix} \boldsymbol{A}\boldsymbol{E}^{-1}\boldsymbol{B} & \boldsymbol{I} - \boldsymbol{A}\boldsymbol{E}^{-1}\boldsymbol{A} \\ -\boldsymbol{I} + \boldsymbol{B}\boldsymbol{E}^{-1}\boldsymbol{B} & -\boldsymbol{B}\boldsymbol{E}^{-1}\boldsymbol{A} \end{bmatrix}, \\ \tilde{\boldsymbol{G}} &= \begin{bmatrix} -\boldsymbol{A}\boldsymbol{B} & \boldsymbol{A}^2 - \boldsymbol{E}_{yx}^F \\ -\boldsymbol{B}^2 + \boldsymbol{E}_{xy}^F & \boldsymbol{B}\boldsymbol{A} \end{bmatrix}. \end{aligned} \tag{3.47}$$

Accordingly the matrix for the eigenvalue problem is

$$\boldsymbol{J} = \tilde{\boldsymbol{F}}\tilde{\boldsymbol{G}} = \begin{bmatrix} \boldsymbol{E}_{xy}^F - \boldsymbol{B}^2 - \boldsymbol{A}\boldsymbol{E}^{-1}\boldsymbol{A}\boldsymbol{E}_{xy}^F & \boldsymbol{A}\boldsymbol{B} - \boldsymbol{A}\boldsymbol{E}^{-1}\boldsymbol{B}\boldsymbol{E}_{yx}^F \\ \boldsymbol{A}\boldsymbol{B} - \boldsymbol{B}\boldsymbol{E}^{-1}\boldsymbol{A}\boldsymbol{E}_{xy}^F & \boldsymbol{E}_{yx}^F - \mathbf{A}^2 - \boldsymbol{B}\boldsymbol{E}^{-1}\boldsymbol{B}\boldsymbol{E}_{yx}^F \end{bmatrix}. \tag{3.48}$$

If we solve for magnetic fields first, then Eq. (3.40) becomes

$$P = \tilde{G}\tilde{F} = \begin{bmatrix} E^F_{yx} - A^2 - E^F_{yx} B E^{-1} B & E^F_{yx} B E^{-1} A - AB \\ E^F_{xy} A E^{-1} B - AB & E^F_{xy} - B^2 - E^F_{xy} A E^{-1} A \end{bmatrix}. \tag{3.49}$$

The eigenvalues obtained from those eigenvalue equations are the squares of the wavevectors along the propagation direction, and the eigenvectors determine the wavefronts.

Cascading multilayer periodic structures

The eigenvectors obtained from Eqs. (3.13), (3.21), and (3.38) are not unique since they can be scaled by arbitrary constants, therefore boundary conditions are needed in order to uniquely determine the solutions for the system under study. We use the S-matrix method proposed in [17] to match the tangential electric and magnetic fields for multilayer gratings. In the pth layer each eigenvalue Λ_m corresponds to two propagation constants, namely $\tilde{k}_{xm} = \pm\sqrt{\Lambda_m}$; hence two sets of coefficients, $\boldsymbol{u}^{(p)}$ and $\boldsymbol{d}^{(p)}$ for upward- and downward-propagating waves, respectively, need to be determined (Fig. 3.1). Here $\boldsymbol{u}^{(p)}$ and $\boldsymbol{d}^{(p)}$ are defined at the lower bound of the pth layer. Using the same notations as in [17], the boundary conditions are written as

$$\begin{bmatrix} u^{(p+1)} \\ d^{(0)} \end{bmatrix} = \begin{bmatrix} T^{(p)}_{uu} & R^{(p)}_{ud} \\ R^{(p)}_{du} & T^{(p)}_{dd} \end{bmatrix} \begin{bmatrix} u^{(0)} \\ d^{(p+1)} \end{bmatrix}, \tag{3.50}$$

where $T^{(p)}_{uu}$ and $T^{(p)}_{dd}$ are two transmission transfer matrices, and $R^{(p)}_{ud}$ and $R^{(p)}_{du}$ are two reflection transfer matrices. We assume that the incident light comes only from the top; then $u^{(0)} = 0$, and only $T^{(p)}_{dd}$ and $R^{(p)}_{ud}$ are of interest. This is called half-matrix recursion in [18], and the transmission and reflection matrices can be obtained recursively by the following steps. First we calculate $Q^{(p)}_E$ and $Q^{(p)}_H$,

$$\begin{aligned} Q^{(p)}_E &= (E^{(p+1)})^{-1} E^{(p)}, \\ Q^{(p)}_H &= (H^{(p+1)})^{-1} H^{(p)}, \end{aligned} \tag{3.51}$$

where $E^{(p)}$ and $H^{(p)}$ are obtained from the eigenvectors. For 2D RCWA TE polarization $E^{(p)} = E^{(p)}_z$ is the eigenvector matrix, and $H^{(p)} = H^{(p)}_y = -E^{(p)}_z \tilde{k}^{(p)}_x$, where $\tilde{k}^{(p)}_x$ is a diagonal matrix with the diagonal elements $\tilde{k}_{xm}$; and for TM polarization $H^{(p)} = H^{(p)}_z$ is the eigenvector matrix, and $E^{(p)} = E^{(p)}_y = \bar{\varepsilon}^{(p)} H^{(p)}_z \tilde{k}^{(p)}_x$. If a layer is homogeneous, then for this layer we have $\tilde{k}_{xn} = \sqrt{\varepsilon_r - \tilde{k}^2_{yn}}$, and for TE polarization the matrices are $E_z = I$ and $H_y = -\tilde{k}_x$, while for TM polarization the matrices are $H_z = I$ and $E_y = \tilde{k}_x/\varepsilon_r$. To make the recursion stable, the imaginary part of $\tilde{k}_{xm}$ should be positive so that the fields are always decaying. For 3D RCWA if we use Eq. (3.48) then $E^{(p)}$ is the eigenvector, and $H^{(p)} = \tilde{G}E^{(p)}\Gamma^{-1}$, where $\tilde{G}$ is defined in Eq. (3.47) and Γ is a diagonal matrix with diagonal elements $\tilde{\gamma}_l$. Next we define

$$t_1^{(p)} = \left(Q_E^{(p)} + Q_H^{(p)}\right)/2,$$
$$t_2^{(p)} = \begin{cases} \left(Q_E^{(p)} - Q_H^{(p)}\right)/2, & \text{if the eigenvectors are electric fields,} \\ \left(Q_H^{(p)} - Q_E^{(p)}\right)/2, & \text{if the eigenvectors are magnetic fields,} \end{cases} \tag{3.52}$$
$$\boldsymbol{\Omega}^{(p)} = \boldsymbol{\Phi}^{(p)} \boldsymbol{R}_{ud}^{(p-1)} \boldsymbol{\Phi}^{(p)},$$

where $\boldsymbol{\Phi}^{(p)} = [\exp(\mathrm{i}k_{xm}^{(p)} h_p)]$ is a diagonal matrix. Finally $\boldsymbol{T}_{dd}^{(p)}$ and $\boldsymbol{R}_{ud}^{(p)}$ are calculated as

$$\boldsymbol{R}_{ud}^{(p)} = \left[t_2^{(p)} + t_1^{(p)}\boldsymbol{\Omega}^{(p)}\right]\left[t_1^{(p)} + t_2^{(p)}\boldsymbol{\Omega}^{(p)}\right]^{-1},$$
$$\boldsymbol{T}_{dd}^{(p)} = \boldsymbol{T}_{dd}^{(p-1)}\boldsymbol{\Phi}^{(p)}\left[t_1^{(p)} + t_2^{(p)}\boldsymbol{\Omega}^{(p)}\right]^{-1}. \tag{3.53}$$

For the first iteration, we use a virtual layer (the zeroth layer) with zero thickness below the first interface (the layer between the interface $p = 0$ and $p = -1$). Then the initial values are $\boldsymbol{T}_{dd}^{(-1)} = \boldsymbol{I}$, $\boldsymbol{R}_{ud}^{(-1)} = 0$, $\boldsymbol{\Phi}^{(0)} = \boldsymbol{I}$, and $\boldsymbol{\Omega}^{(0)} = 0$; hence $\boldsymbol{T}_{dd}^{(0)} = (t_1^{(0)})^{-1}$ and $\boldsymbol{R}_{ud}^{(p)} = t_2^{(0)}\boldsymbol{T}_{dd}^{(0)}$.

If a layer (the pth layer) is homogeneous, the eigenvalues and eigenvectors can easily be calculated. For 2D RCWA TE polarization the eigenvalues are $\tilde{k}_{xm}^2 = n_p^2 - \tilde{k}_{yn}^2$, where n_p is the refractive index, and the eigenvectors are $\boldsymbol{E}^{(p)} = \boldsymbol{I}$ and $\boldsymbol{H}^{(p)} = -[\tilde{\boldsymbol{k}}_x]$. The eigenvalues in TM polarization are the same as for TE polarization, and the eigenvectors are $\boldsymbol{H}^{(p)} = \boldsymbol{I}$ and $\boldsymbol{E}^{(p)} = [\tilde{\boldsymbol{k}}_x]/n_p^2$. For 3D RCWA the eigenvalues are $\gamma^2 = n_p^2 - \alpha^2 - \beta^2$, and the eigenvectors are $\boldsymbol{E}^{(p)} = \boldsymbol{I}$ and

$$\boldsymbol{H}^{(p)} = \begin{bmatrix} \dfrac{\alpha\beta}{\gamma} & \dfrac{\beta^2}{\gamma} + \gamma \\ -\dfrac{\alpha^2}{\gamma} - \gamma & -\dfrac{\alpha\beta}{\gamma} \end{bmatrix}.$$

If the incident wave is designated by $\boldsymbol{d}^{(N+1)}$, then the transmitted fields are calculated as $\boldsymbol{d}^{(0)} = \boldsymbol{T}_{dd}^{(N)}\boldsymbol{d}^{(N+1)}$, and the reflected fields are calculated as $\boldsymbol{u}^{(N+1)} = \boldsymbol{R}_{ud}^{(N)}\boldsymbol{d}^{(N+1)}$. Note that if the eigenvectors are electric fields then the coefficients are for electric fields, whereas if the eigenvectors are magnetic fields then the coefficients are also for magnetic fields.

3.2.2 Modeling 2D and 3D plasmonic nanostructures with RCWA

Modeling 2D plasmonic gratings

Two-dimensional RCWA is capable of modeling plasmonic gratings accurately and efficiently; we show this experimentally with a gold grating illustrated in Fig. 3.3 [21]. In each unit cell, a gold strip was fabricated on a 15-nm-thick indium tin oxide (ITO)-coated glass substrate using electron-beam lithography. The designed dimensions of the sample are listed in Table 3.1. A field-emission scanning electron microscope

Table 3.1 The designed and fitted parameters of the sample

	Period (nm)	Thickness (nm)	Width (nm)	Loss factor
Design	400	50	200	1
Fitted	403	49	210	2.2

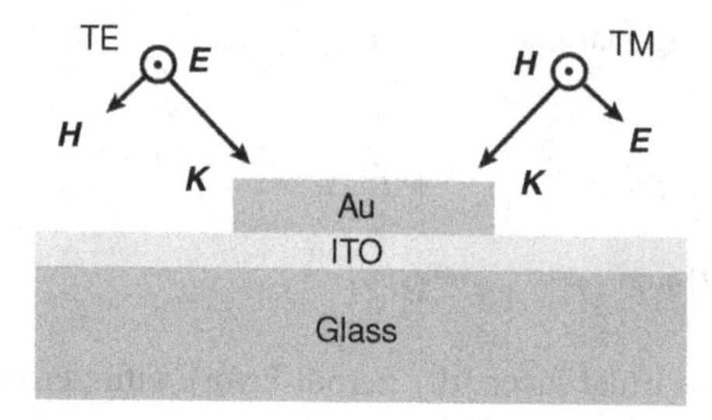

Figure 3.3 The unit cell of the structure used for 2D RCWA verification.

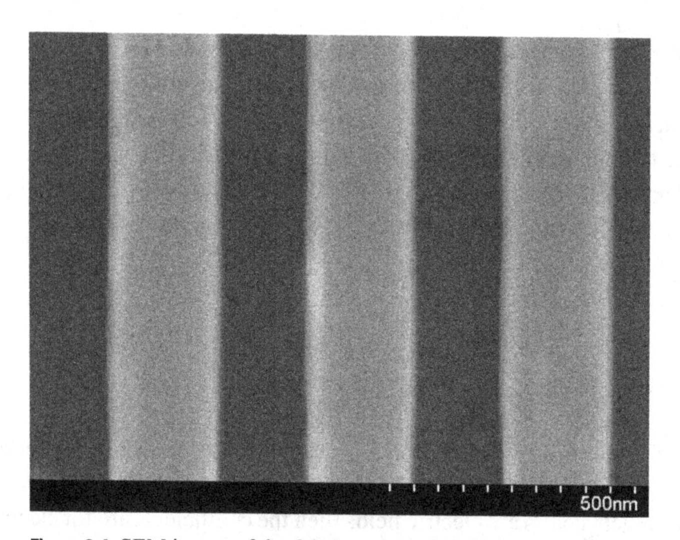

Figure 3.4 SEM image of the fabricated grating device.

(FESEM) image shows the top view of the fabricated device (Fig. 3.4). Owing to the nature of nanofabrication, the actual dimensions of the device could not be controlled exactly, and deviations from the designed values were inevitable. The device was then characterized using optical transmittance spectroscopy with the incidence angle varying from 0 to 30°. The transmittance spectra both for TE and for TM polarization were taken in the wavelength range from 450 nm to 850 nm (Fig. 3.5).

The transmittance of the device was simulated using 2D RCWA with 5 Fourier modes for TE and 15 Fourier modes for TM polarizations. The refractive indices of the glass substrate and the ITO film were obtained using spectroscopic ellipsometry.

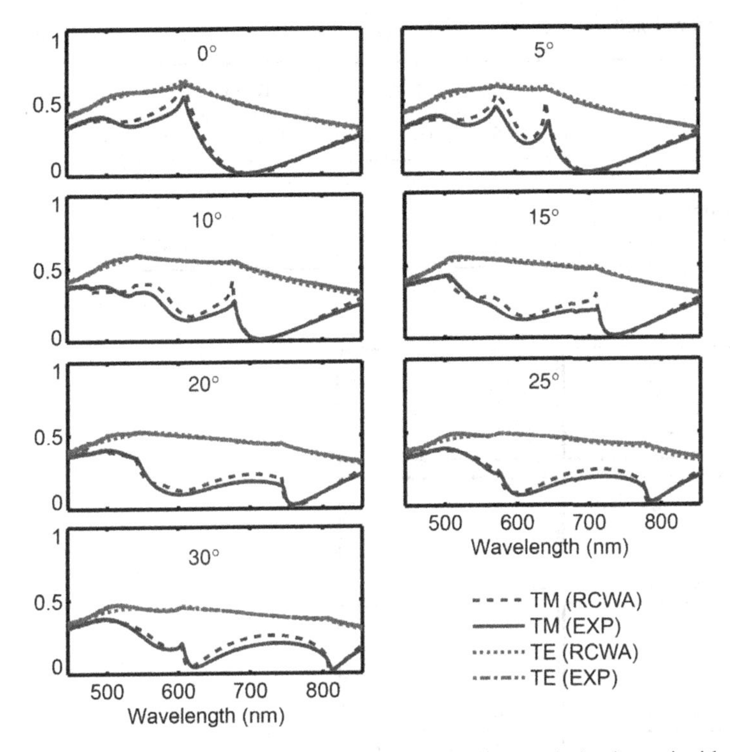

Figure 3.5 Transmittance spectra for both TE and TM polarization at incidence angles of 0, 5, 10, 20, 25, and 30°. The experimental and RCWA simulated spectra are both shown for comparison.

The refractive index of gold was taken from [22]. Since the actual device was different from the design, we fine-tuned the dimensions, including the period and the width and thickness of the gold strips, to match the experimental spectra. It has been observed that the loss in nano-structured metal could be higher than in bulk metal, and the higher loss can be described by a loss factor [23], therefore in our simulations a loss factor for gold was introduced as a fitting parameter. The fitted parameters are listed in Table 3.1. The simulated transmittance spectra using the fitted parameters are compared with experimental spectra in Fig. 3.5.

The spectral features in Fig. 3.5 were caused by a resonance in the TM polarization, such as the dip at 700 nm for 10° incidence, and by diffraction. The wavelengths at which the -1 and 1 transmission diffraction orders start can be calculated using $d(n_2 + \sin\theta_i)$ and $d(n_2 - \sin\theta_i)$, respectively, where d is the period, n_2 is the refractive index of the glass substrate, and θ_i is the incidence angle [24]. The wavelength at which the -1 reflection diffraction order starts can be calculated from $d(1 + \sin\theta_i)$. These wavelengths for different incidence angles are listed in Table 3.2. On comparing the list of wavelengths in Table 3.2 with the spectra in Fig. 3.5, we see that all the diffraction features within the wavelength range we studied can be identified in the transmittance spectra. Overall the simulated spectra matched the experimental spectra well, with all the important

Table 3.2 The wavelengths of different diffraction orders

	Angle of incidence (degrees)						
	0	5	10	15	20	25	30
Transmission: −1 (nm)	613	647	683	717	750	783	814
Transmission: 1 (nm)	613	577	543	508	475	442	411
Reflection: −1 (nm)	403	438	473	507	541	573	604

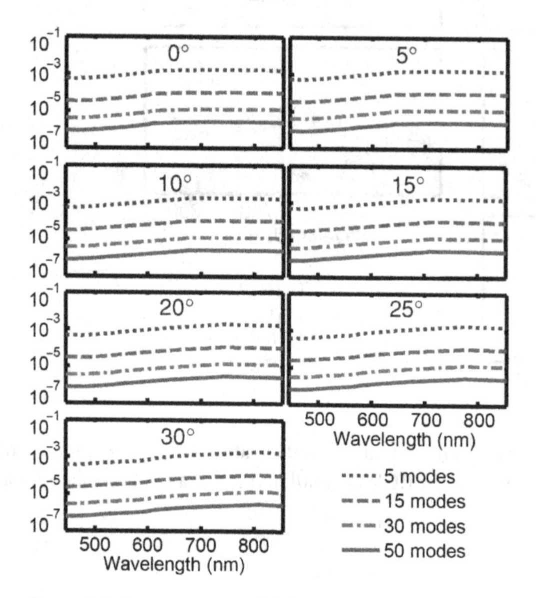

Figure 3.6 Convergence of 2D RCWA for TE polarization. The difference spectra between 5, 15, 30, and 50 Fourier modes and the reference are plotted. The reference spectra are simulated using 2D RCWA with 100 modes.

features reproduced, especially the plasmonic resonances originating from subwavelength metallic structures.

The mismatch between the experimental and numerical results is likely due to the nonuniformity of the sample: the width of the strips had small variations, and the metal was deposited by electron-beam evaporation, therefore the grain size was not uniform, both of which effects can lead to spectral feature broadening but were not accounted for in the RCWA simulations [23, 25]. From Fig. 3.5 we conclude that, for both polarizations and all incidence angles, the 2D RCWA simulations reproduced the experimental spectra with high fidelity.

The convergence rate of RCWA has been an important issue, especially for TM polarization [14, 15]. Figure 3.6 shows the convergence of 2D RCWA for TE polarization. We used spectra with 100 Fourier modes as a reference, and plotted the difference for spectra with 5, 15, 30, and 50 modes. With only five modes for TE polarization 2D RCWA gives quite accurate results, and the convergence is very uniform for all

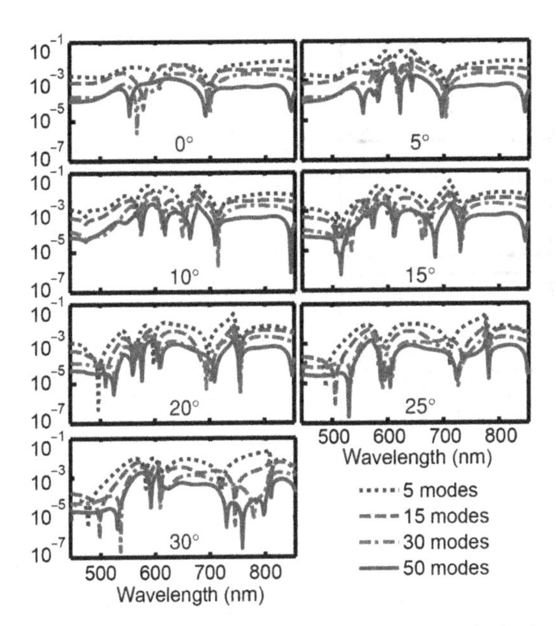

Figure 3.7 Convergence of 2D RCWA for TM polarization. The difference spectra between 5, 15, 30, and 50 Fourier modes and the reference are plotted. The reference spectra are simulated using 2D RCWA with 100 modes.

incidence angles and all wavelengths in the studied spectrum. The same plots for TM polarization are shown in Fig. 3.7, and clearly the convergence for TM polarization is more complicated. The averages of the difference spectra for different numbers of modes are shown in Fig. 3.8. It is evident that 2D RCWA simulations converge faster for TE polarization than for TM polarization. Since the tolerance of experimental measurements is typically larger than 1%, using 5 modes for TE polarization and 15 modes for TM polarization provides enough accuracy for the purpose of fitting experimental spectra. The higher number of modes required for TM polarization is due to the resonance in TM polarization. At resonance, the fields near the grating are strong and highly localized, therefore more modes are needed to resolve these fields. This is true for all incidence angles.

A 2D RCWA solver has been staged on NanoHub (https://nanohub.org/tools/sha2d/), so it is freely available on-line and is accessible from any standard web browser [26].

Modeling 3D nanoantennas

RCWA is also widely used to model 3D plasmonic nanostructures, such as nanoantennas and metamaterials [27, 28]. Optical nanoantennas have been of great interest due to their ability to support a highly efficient, localized surface plasmon resonance and produce significantly enhanced and highly confined electromagnetic fields. Such enhanced local fields have many applications, such as biosensors, near-field scanning optical microscopy (NSOM), and quantum optical information processing, enhanced Raman scattering as

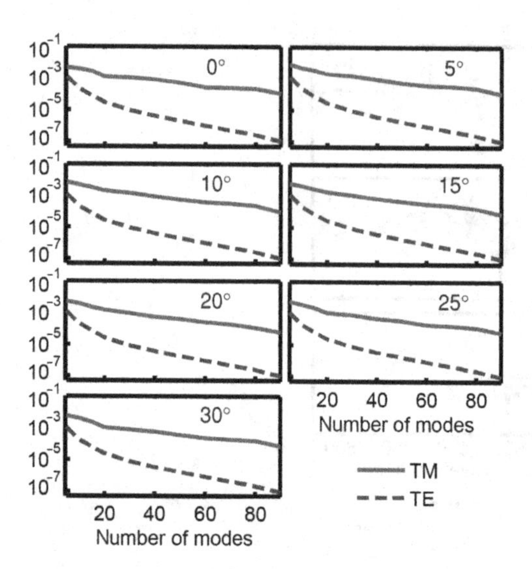

Figure 3.8 Average difference versus number of modes. The reference spectra are simulated using 2D RCWA with 100 modes. For each number of modes the difference is averaged over the spectrum.

well as other optical processes [29–34]. In this section we show that 3D RCWA is instrumental for the modeling and design of nanoantennas.

We use 3D RCWA to model a nanoantenna array whose unit cell is shown in Fig. 3.9(a) [27]. The array is composed of gold nanoparticles with elliptic cylinder shapes. Each unit cell has two nanoparticles with a gap between the two particles along their major axes. The periods along the major and minor axes are 400 nm and 200 nm, respectively.

The array was fabricated and then examined by field emission electron scanning microscopy (FESEM). Representative FESEM images are shown in Fig. 3.9(b). Although all unit cells in an array have the same design, each particle randomly deviates from the designed shape to a certain extent. The transmittance and reflectance spectra of these samples were measured in the visible range with incident light normal to the sample surface. The incident light was linearly polarized, and we used two polarizations: the principal polarization, for which the electric field is parallel to the major axis; and the secondary polarization, for which the electric field is normal to the major axis. The measured spectra are shown in Fig. 3.10 [27]. The sample shows a strong resonance for the principal polarization at 660 nm. For the secondary polarization only a weak resonance is visible.

We used both RCWA and FEM to model the nanoantenna array. In our modeling the dimensions of the major axis, the minor axis, and the gap in the models were tuned in order to match the simulated resonance wavelengths to the experimental results, and the resulting dimensions used in simulations were as follows: major axis 110 nm, minor axis 55 nm, and gap 17 nm. These values are all within the ranges of dimensions obtained from SEM measurements. The 40 nm thickness and the 400 nm and 200 nm periods along the major axis and minor axis, respectively, were taken from the initial design. The

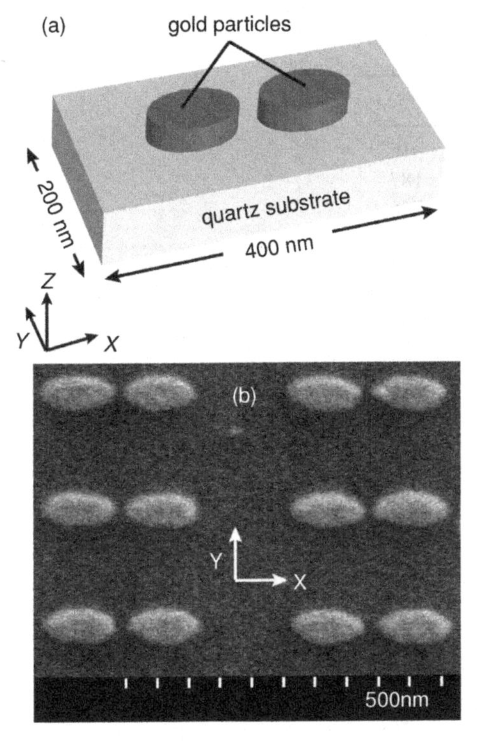

Figure 3.9 A unit cell of the nanoantenna array. (b) FESEM images of the nanoantenna array.

results are compared with the experimental spectra in Fig. 3.10. We can see that they all agree with one another quite well. The numbers of Fourier modes used in RCWA simulations are indicated by the legends, for example, the legend "RCWA 40 $\times$ 20 (R)" means that the reflectance was obtained using 40 modes for the x direction and 20 modes for the y direction. For both polarizations we increased the number of modes until the results did not significantly change, which indicated that convergence had been achieved. Because of the strong resonance in the primary polarization, we usually need a large number of modes to achieve convergence; and, because there is a small gap along the x direction, we need more modes for the x direction than for the y direction. For the secondary polarization fewer modes are needed in order to obtain accurate results. Generally both RCWA and FEM can achieve good accuracy in modeling nanoantenna arrays; however, FEM requires much less in the way of resources. Table 3.3 compares the memory and time resources required by FEM and RCWA. The tests were performed on a Linux cluster node, which had eight Intel Xeon 2.33-GHz cores and 16 GB of memory. It is clear that as the number of modes used increases the resources needed by RCWA increase very fast.

Although FEM requires less in the way of resources, RCWA has a distinct advantage: all diffraction modes are naturally separated. In FEM simulations the total fields are solved for, so, if there are higher-order diffractions under oblique light incidence, it is difficult to obtain the fields for each individual diffraction order. RCWA and FEM can

Table 3.3 Time and memory requirements for RCWA and FEM

	Memory (GB)	Time (s)
FEM (COMSOL)	2.5	10
RCWA (20 × 20)	0.5	187
RCWA (40 × 20)	2.6	1776
RCWA (50 × 25)	5	4345

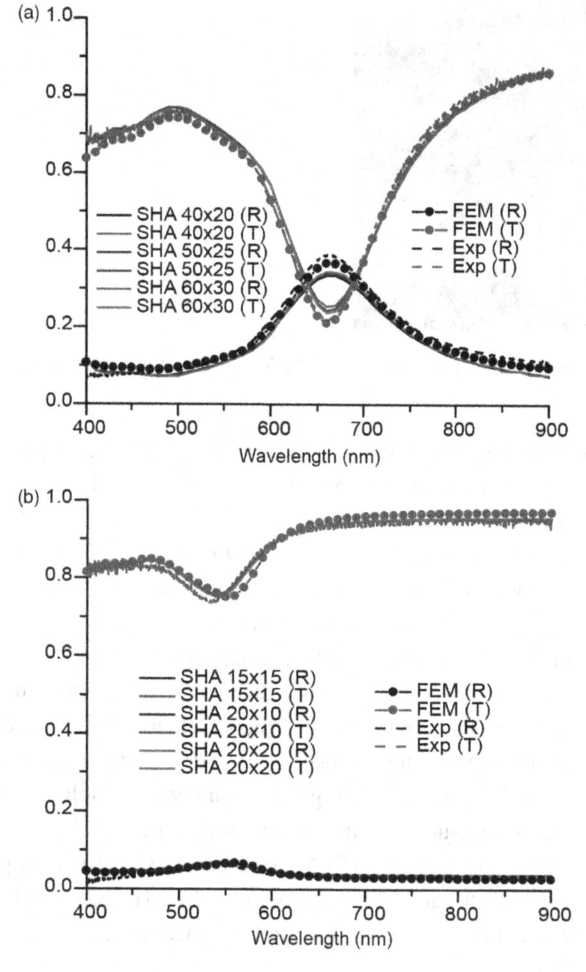

Figure 3.10 The reflectance (R) and transmittance (T) spectra obtained by FEM, experiments (Exp), and RCWA for (a) primary polarization and (b) secondary polarization.

also provide cross-references for each other, which is critical in numerical modeling, especially when other reference methods are not available. Therefore RCWA is an important tool in the study of optical nanoantennas.

3.3 A semi-analytical method for near-field coupling study

3.3.1 Superlens and subwavelength imaging

The concept of negative refractive index was first discussed by V. G. Veselago in 1968 [35], and was revived by Sir J. Pendry in 2000 [6]. Pendry introduced subwavelength near-field imaging (NFI), an important application of negative-index material (NIM), using a thin slab of NIM as a superlens. The general requirement for NIM functionality is $\mathrm{Re}(\varepsilon) < 0$ for permittivity and $\mathrm{Re}(\mu) < 0$ for permeability. For imaging with a p-polarized wave, only negative permittivity, $\mathrm{Re}(\varepsilon) < 0$, would be sufficient for NIM functionality, which can be satisfied by metals (e.g. silver, gold, and copper) [6]. The permittivity of a superlens must also match that of the surrounding dielectric, i.e. $\mathrm{Re}(\varepsilon) = -\mathrm{Re}(\varepsilon_{\mathrm{d}})$, where ε_{d} is the permittivity of the surrounding dielectric. Simulations have shown that subwavelength imaging can indeed theoretically be achieved by a superlens made of these metals [36–38]. Subsequent experimental results have confirmed the subwavelength-imaging capability of metallic-film lenses. The enhancement of the evanescent wave was also observed [39].

In most of the reported simulations, the objects to be imaged were hypothetical slits. These were rather different from the metal masks used in experiments, since mask–superlens coupling may play an important role in the overall result. However, the coupling cannot be studied with commonly used methods such as FEM and FDTD, since these methods always take into account the coupling. In this section we develop a rigorous full-wave semi-analytical frequency-domain method that can switch the coupling effect "on" and "off," to show the significance of the coupling effect. We use a circular cylindrical wire as a realistic imaging object and a planar metal film as a superlens. The classical scattering problem of a cylinder adjacent to a planar interface has been studied by many researchers [40–45]. We use the cylindrical-wave approach to study the near-field regime, and show that the near-field distributions are significantly different when the object–superlens coupling is switched on or off. Therefore it is imperative to take into account the coupling effect when designing a superlens imaging system.

3.3.2 Object–superlens coupling

Formulations

The geometry of the object–lens system is shown in Fig. 3.11 [46]. The superlens is a thin silver film of thickness d. An infinitely long thin wire (a circular cylinder) is placed in front of the lens with its axis parallel to the lens interface. The radius of the wire is a and the center-to-interface distance is $b\,(b > a)$. The superlens and the object are illuminated by a p-polarized plane wave with $\boldsymbol{H}_0 = \hat{\boldsymbol{z}} H_0$ $(H_0 = \mathrm{e}^{\mathrm{i}kx})$ and $\boldsymbol{E} = \hat{\boldsymbol{y}}(\mathrm{i}\omega\varepsilon)^{-1}\partial_x H_0$, where $\partial_x H_0 = \partial H_0/\partial x$ and ε is the complex permittivity of air.

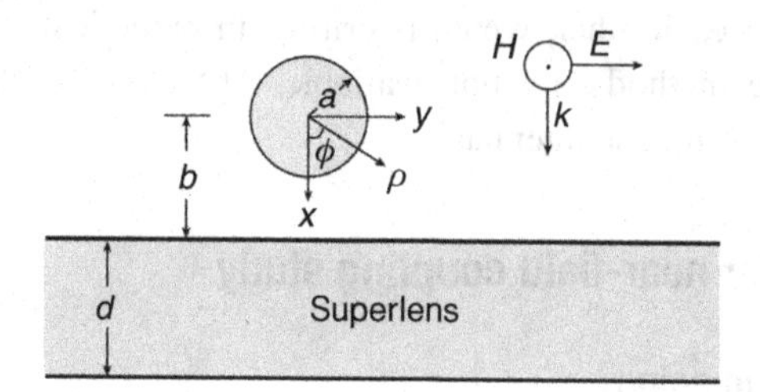

Figure 3.11 A schematic representation of a superlens-cylinder system.

First, we consider the cylinder as an isolated scatterer illuminated by incident light normal to the axis of the wire, with free-space wavelength λ and wavenumber $k = 2\pi/\lambda$. The temporal dependence is assumed to be $e^{-i\omega t}$. This trivial case can be treated using the following spectral equations [47]:

$$H_c = \sum_{m=0}^{\infty} c_c^m J_m(k\rho)\cos(m\phi), \quad \rho \le a_0,$$
$$H_s = \sum_{m=0}^{\infty} c_s^m H_m^{(1)}(k\rho)\cos(m\phi), \quad \rho \ge a_0, \tag{3.54}$$
$$H_i = \sum_{m=0}^{\infty} c_i^m J_m(n_c k\rho)\cos(m\phi), \quad \rho \le a_0,$$

where H_c is the incident field, H_s is the scattered field, H_i is the field inside the cylinder, and n_c is the complex refractive index of the cylinder. J_m and $H_m^{(1)}$ are the Bessel function and the Hankel function of the first kind. c_c^m, c_s^m and c_i^m are expansion coefficients. The tangential components of electric displacement are

$$D_c = \frac{k}{i\omega}\sum_{m=0}^{\infty} c_c^m J_m'(k\rho)\cos(m\phi), \quad \rho \le a_0,$$
$$D_s = \frac{k}{i\omega}\sum_{m=0}^{\infty} c_s^m H_m^{(1)\prime}(k\rho)\cos(m\phi), \quad \rho \ge a_0, \tag{3.55}$$
$$D_i = \frac{n_c k}{i\omega}\sum_{m=0}^{\infty} c_i^m J_m'(n_c k\rho)\cos(m\phi), \quad \rho \le a_0,$$

where $J_m' = dJ_m(x)/dx$ and $H_m^{(1)\prime}$ are the derivatives.

Magnetic field continuity together with the conservation of tangential electric field at the cylinder interface gives the standard boundary conditions, $H_c + H_s = H_i$ and $n_c^2(D_c + D_s) = D_i$, at the circumference of the cylinder. These boundary conditions give the following relations:

$$c_i^m = s_i^m c_c^m,$$
$$c_s^m = s^m c_c^m. \tag{3.56}$$

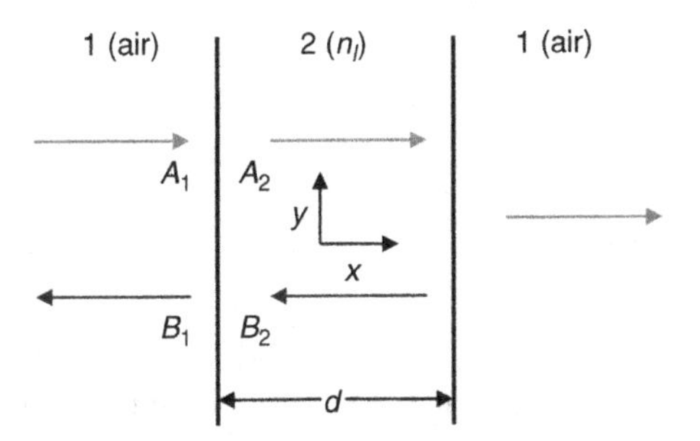

Figure 3.12 An air–thin-film–air system.

Here the transformation factors (s_{i}^m and s^m) are defined in the following manner:

$$
\begin{aligned}
s_{\mathrm{i}}^m &= \frac{2n_{\mathrm{c}}\mathrm{i}}{\pi k a_0} \frac{1}{n_{\mathrm{c}} J_m(n_{\mathrm{c}}ka) H_m^{(1)\prime}(ka) - J_m'(n_{\mathrm{c}}ka) H_m^{(1)}(ka)}, \\
s^m &= -\frac{n_{\mathrm{c}} J_m(n_{\mathrm{c}}ka) J_m'(ka) - J_m'(n_{\mathrm{c}}ka) J_m(ka)}{n_{\mathrm{c}} J_m(n_{\mathrm{c}}ka) H_m^{(1)\prime}(ka) - J_m'(n_{\mathrm{c}}ka) H_m^{(1)}(ka)}.
\end{aligned}
\tag{3.57}
$$

The above equation can be expressed in matrix form as

$$
\begin{aligned}
\boldsymbol{C}_{\mathrm{i}} &= \boldsymbol{S}_{\mathrm{i}} \boldsymbol{C}_{\mathrm{c}}, \\
\boldsymbol{C}_{\mathrm{s}} &= \boldsymbol{S} \boldsymbol{C}_{\mathrm{c}},
\end{aligned}
\tag{3.58}
$$

where

$$
\boldsymbol{S}_{\mathrm{i}} = \begin{pmatrix} s_{\mathrm{i}}^0 & 0 & 0 & \cdots \\ 0 & s_{\mathrm{i}}^1 & 0 & \cdots \\ 0 & 0 & s_{\mathrm{i}}^2 & \cdots \\ \vdots & \vdots & \vdots & \ddots \end{pmatrix}, \quad \boldsymbol{S} = \begin{pmatrix} s^0 & 0 & 0 & \cdots \\ 0 & s^1 & 0 & \cdots \\ 0 & 0 & s^2 & \cdots \\ \vdots & \vdots & \vdots & \ddots \end{pmatrix}. \tag{3.59}
$$

Next we consider a thin-film system as shown in Fig. 3.12. In this single-layer system, the incident wave comes from the left, A_1 and B_1 represent the electric fields of the forward- and backward-going waves right before the air/film interface, and A_2 and B_2 represent the electric fields of the forward- and backward-going waves right after the air/film interface. We define the following quantities: $\alpha = k_y/k$, $\beta = k_x/k = \sqrt{1-\alpha^2}$, $\alpha_2 = k_{2y}/k = \alpha$, and $\beta_2 = k_{2x}/k = \sqrt{n_l^2 - \alpha^2}$, where k is the wavevector in air, k_x and k_y are the respective components of the wavevector in air, and k_{2x} and k_{2y} are the respective components of the wavevector in the film. Therefore α, β, α_2, and β_2 are the normalized components of the wavevectors.

The reflection and transmission coefficients for electric fields can be expressed using the quantities

$$
r_{12} = \frac{\beta_2 - n_l^2\beta}{\beta_2 + n_l^2\beta}, \qquad t_{12} = \frac{2n_l\beta}{\beta_2 + n_l^2\beta}, \qquad t_{21} = \frac{2n_l\beta_2}{\beta_2 + n_l^2\beta},
$$

and $\phi = \beta_2 kd$ as the following equations [48]:

$$
\begin{aligned}
r &= \frac{r_{12} - r_{12}e^{2i\phi}}{1 - r_{12}^2 e^{2i\phi}}, \\
t &= \frac{t_{12}t_{21}e^{i\phi}}{1 - r_{12}^2 e^{2i\phi}}.
\end{aligned}
\tag{3.60}
$$

The results slightly differ from [48] because we use $e^{-i\omega t}$ instead of $e^{i\omega t}$ as the temporal dependence. Equations (3.60) are for the electric field only. For the magnetic field, we need to use

$$
\begin{aligned}
r &= -\frac{r_{12} - r_{12}e^{2i\phi}}{1 - r_{12}^2 e^{2i\phi}}, \\
t &= \frac{t_{12}t_{21}e^{i\phi}}{1 - r_{12}^2 e^{2i\phi}}.
\end{aligned}
\tag{3.61}
$$

Using A_1', B_1', A_2', and B_2' to denote the magnetic field associated with A_1, B_1, A_2, and B_2, at a point inside the film, the total magnetic field is

$$
H = \tau\left[e^{i\beta_2 kx} + r_{12}e^{i\beta_2 k(2d-x)}\right] A_1', \tag{3.62}
$$

where x is the distance between the interface and the point, and $\tau = n_l t_{12}/(1 - r_{12}^2 e^{2i\beta_2 kd})$.

The solutions for an isolated cylinder are then coupled with a thin-film system to solve the imaging problem. We assume that the incident wave is a plane wave, $H_0 = e^{ikx}$, and that the field scattered by the cylinder is H_s. Then both of them are reflected by the superlens, generating H_{0r} and H_{sr}, which are again scattered by the cylinder. H_{0r} can be easily calculated from $H_{0r} = r(0)e^{2ikb_0}e^{-ikx}$.

To find H_s and H_{sr}, note that H_0, H_{0r}, and H_{sr} are all incoming waves heading towards the cylinder, while H_s represents the outgoing waves. To use Eq. (3.58), all incoming waves must be expressed in terms of Bessel functions. H_0 and H_{0r} are usually in plane-wave form, so the plane waves must be expanded into Bessel functions. H_{sr} is generated by H_s, which itself is expressed as Hankel functions. H_s is then reflected and transmitted by the film, so, in order to use the reflection and transmission coefficients, which are for plane waves, the Hankel functions must be converted into plane waves. Those plane waves then must be transformed again into Bessel functions to use the boundary conditions along the cylinder surface.

Using the identity

$$
H_m^{(1)}(k\rho)e^{im\phi} = \frac{1}{\pi}\int_{-\infty}^{\infty} (\alpha - i\beta)^m \beta^{-1} e^{ik(\beta x + \alpha y)}\, d\alpha,
$$

where $(x, y) = (\rho\cos\phi, \rho\sin\phi)$ and $\beta = \sqrt{1-\alpha^2}$, we obtain

$$
H_m^{(1)}(k\rho)\cos(m\phi) = \frac{1}{\pi}\left[\int_0^{\infty} [(\alpha - i\beta)^m + (-1)^m(\alpha + i\beta)^m]\beta^{-1}e^{ik\beta x}\cos(k\alpha y)d\alpha\right].
\tag{3.63}
$$

This equation expands Hankel functions in terms of plane waves. Then the scattered field H_s is expressed as

$$H_s = \sum_{m=0}^{\infty} \frac{c_s^m}{2\pi} \int_{-\infty}^{\infty} [(\alpha - i\beta)^m + (-1)^m (\alpha + i\beta)^m] \beta^{-1} e^{ik(\beta x + \alpha y)}\, d\alpha. \tag{3.64}$$

At the air/film interface, the scattered field is reflected, and the resulting field is

$$H_{sr} = \sum_{m=0}^{\infty} \frac{c_s^m}{\pi} \int_0^{\infty} r(\alpha)[(\alpha - i\beta)^m + (-1)^m (\alpha + i\beta)^m] \beta^{-1} e^{ik\beta(2b_0 - x)} \cos(k\alpha y) d\alpha, \tag{3.65}$$

where b_0 is the distance between the interface and the cylinder center. Equation (3.65) can be rewritten as

$$H_{sr} = \sum_{m=0}^{\infty} \frac{c_s^m}{\pi} \int_0^{\infty} r(\alpha) f_m(\alpha) e^{2ik\beta b_0} e^{-ik\beta x} \cos(k\alpha y) d\alpha, \tag{3.66}$$

by defining $f_m(\alpha) = [(\alpha - i\beta)^m + (-1)^m(\alpha + i\beta)^m]\beta^{-1}$. Using another identity,

$$e^{-ik\rho(\beta\cos\phi + \alpha\sin\phi)} = \sum_{n=-\infty}^{\infty} (\alpha - i\beta)^n J_n(k\rho) e^{in\phi},$$

we can transform plane waves into Bessel functions. We write the identity as

$$e^{-ik\beta x}\cos(k\alpha y) = J_0(k\rho) + \sum_{n=1}^{\infty} [(\alpha - i\beta)^n + (-1)^n(\alpha + i\beta)^n] J_n(k\rho)\cos(n\phi), \tag{3.67}$$

where $(x, y) = (\rho\cos\phi, \rho\sin\phi)$ and $\beta = \sqrt{1 - \alpha^2}$. We define

$$g_n(\alpha) = \begin{cases} 1, & n = 0, \\ (\alpha - i\beta)^n + (-1)^n(\alpha + i\beta)^n, & n \neq 0, \end{cases} \tag{3.68}$$

and rewrite Eq. (3.67) as

$$e^{-ik\beta x}\cos(k\alpha y) = \sum_{n=0}^{\infty} g_n(\alpha) J_n(k\rho)\cos(n\phi). \tag{3.69}$$

On substituting Eq. (3.69) into (3.66), we have

$$H_{sr} = \sum_{n=0}^{\infty}\sum_{m=0}^{\infty} c_s^m J_n(k\rho)\cos(n\phi) \frac{1}{\pi} \int_0^{\infty} r(\alpha) f_m(\alpha) g_n(\alpha) e^{2ik\beta b_0}\, d\alpha, \tag{3.70}$$

which can be written in matrix form

$$H_{sr} = \boldsymbol{J}^{\mathrm{T}} \boldsymbol{Q} \boldsymbol{C}_s, \tag{3.71}$$

where

$$\boldsymbol{J} = \begin{bmatrix} J_0(k\rho) \\ J_1(k\rho)\cos\phi \\ J_2(k\rho)\cos(2\phi) \\ \vdots \end{bmatrix}, \quad \boldsymbol{C}_s = \begin{bmatrix} c_s^0 \\ c_s^1 \\ c_s^2 \\ \vdots \end{bmatrix}, \tag{3.72}$$

and

$$\boldsymbol{Q} = \begin{pmatrix} q_{00} & q_{01} & \cdots \\ q_{10} & q_{11} & \cdots \\ \vdots & \vdots & \ddots \end{pmatrix}, \quad q_{nm} = \frac{1}{\pi}\int_0^{\infty} r(\alpha) f_m(\alpha) g_n(\alpha) \mathrm{e}^{2\mathrm{i}k\beta b_0}\,\mathrm{d}\alpha. \tag{3.73}$$

Using the identities

$$\begin{aligned} \mathrm{e}^{\mathrm{i}kx} &= \sum_{n=-\infty}^{\infty} \mathrm{i}^n J_n(k\rho)\mathrm{e}^{\mathrm{i}n\phi}, \\ \mathrm{e}^{-\mathrm{i}kx} &= \sum_{n=-\infty}^{\infty} \mathrm{i}^{(-n)} J_n(k\rho)\mathrm{e}^{\mathrm{i}n\phi}, \end{aligned} \tag{3.74}$$

both H_0 and $H_{0\mathrm{r}}$ can be expressed in terms of $\boldsymbol{J}$:

$$\begin{aligned} H_0 &= \boldsymbol{J}^{\mathrm{T}}\boldsymbol{C}_0, \\ H_{0\mathrm{r}} &= r(0)\mathrm{e}^{2\mathrm{i}kb_0}\boldsymbol{J}^{\mathrm{T}}\boldsymbol{C}_{0\mathrm{r}}, \end{aligned} \tag{3.75}$$

where

$$c_0^n = \begin{cases} 1, & n = 0, \\ 2\mathrm{i}^n, & \text{otherwise}, \end{cases} \qquad c_{0\mathrm{r}}^n = \begin{cases} 1, & n = 0, \\ 2\mathrm{i}^{(-n)}, & \text{otherwise}. \end{cases}$$

Then the total incoming wave to the cylinder is $H_{\mathrm{c}} = \boldsymbol{J}^{\mathrm{T}}\boldsymbol{C}_{\mathrm{c}}$, where $\boldsymbol{C}_{\mathrm{c}} = \boldsymbol{C}_0 + r(0)\mathrm{e}^{2\mathrm{i}kb_0}\boldsymbol{C}_{0\mathrm{r}} + \boldsymbol{Q}\boldsymbol{C}_{\mathrm{s}}$. According to Eq. (3.58), the following equation holds:

$$\boldsymbol{C}_{\mathrm{s}} = \boldsymbol{S}\left[\boldsymbol{C}_0 + r(0)\mathrm{e}^{2\mathrm{i}kb_0}\boldsymbol{C}_{0\mathrm{r}} + \boldsymbol{Q}\boldsymbol{C}_{\mathrm{s}}\right]. \tag{3.76}$$

On rearranging the above equation we arrive at the matrix equation

$$(\boldsymbol{I} - \boldsymbol{S}\boldsymbol{Q})\,\boldsymbol{C}_{\mathrm{s}} = \boldsymbol{S}\left[\boldsymbol{C}_0 + r\,(0)\,\mathrm{e}^{2\mathrm{i}kb_0}\boldsymbol{C}_{0\mathrm{r}}\right]. \tag{3.77}$$

Solving Eq. (3.77) gives $\boldsymbol{C}_{\mathrm{s}}$, and hence H_{s} and H_{sr}.

Two wave components, H_0 and H_{s}, are transmitted through the superlens. The field after the superlens therefore can be calculated from

$$H_{\mathrm{t}} = t(0)\mathrm{e}^{\mathrm{i}k(b_0+x)} + \sum_{m=0}^{\infty}\frac{c_{\mathrm{s}}^m}{2\pi}\int_{-\infty}^{\infty} t(\alpha) f_m(\alpha)\mathrm{e}^{\mathrm{i}k\beta b_0}\mathrm{e}^{\mathrm{i}k(\beta x+\alpha y)}\,\mathrm{d}\alpha, \tag{3.78}$$

where x is the distance between the rear interface and the point at interest.

Before the superlens, the total field inside and outside the cylinder can be calculated separately. Inside the cylinder, the field is

$$H_\mathrm{i} = \sum_{m=0}^{\infty} c_\mathrm{i}^m J_m(n_\mathrm{c} k\rho)\cos(m\phi), \quad \rho \le a_0,$$
$$\boldsymbol{C}_\mathrm{i} = \boldsymbol{S}_\mathrm{i}\left[\mathbf{C}_0 + r(0)\mathrm{e}^{2\mathrm{i}kb_0}\mathbf{C}_{0\mathrm{r}} + \boldsymbol{Q}\boldsymbol{C}_\mathrm{s}\right]. \tag{3.79}$$

Outside the cylinder, the field is

$$H_\mathrm{b} = \mathrm{e}^{\mathrm{i}kx} + r(0)\mathrm{e}^{\mathrm{i}k(2b_0 - x)} + \sum_{m=0}^{\infty} c_\mathrm{s}^m H_m^{(1)}(k\rho)\cos(m\phi)$$
$$+ \sum_{m=0}^{\infty} \frac{c_\mathrm{s}^m}{\pi} \int_0^{\infty} r(\alpha) f_m(\alpha)\mathrm{e}^{2\mathrm{i}k\beta b_0}\mathrm{e}^{-\mathrm{i}k\beta x}\cos(k\alpha y)\mathrm{d}\alpha, \tag{3.80}$$

where the origin is at the cylinder center.

Within the superlens, the total field can be calculated as

$$H_\mathrm{f} = \mathrm{e}^{\mathrm{i}kb_0}\tau(0)\left[\mathrm{e}^{\mathrm{i}\beta_2(0)kx} + r_{12}(0)\mathrm{e}^{\mathrm{i}\beta_2(0)k(2d-x)}\right]$$
$$+ \sum_{m=0}^{\infty} \frac{c_\mathrm{s}^m}{2\pi} \int_{-\infty}^{\infty} \tau(\alpha) f_m(\alpha)\left[\mathrm{e}^{\mathrm{i}\beta_2 kx} + r_{12}\mathrm{e}^{\mathrm{i}\beta_2 k(2d-x)}\right]\mathrm{e}^{\mathrm{i}k\beta b_0}\mathrm{e}^{\mathrm{i}k\alpha y}\,\mathrm{d}\alpha, \tag{3.81}$$

where x is the distance between the interface and the point of interest.

The above discussion is based on the actual situation of the object–superlens system, which involves coupling. We also need to simulate the system without coupling, which is a hypothetical situation. The key difference between the cases with and without coupling is that without coupling the wave reflected by the superlens is not scattered by the cylinder, and therefore there is no feedback interaction between the reflected and scattered waves. In the case of there being no coupling, the incident wave is scattered by the cylinder first, creating an object of the cylinder. Then both the incident wave and the scattered wave are reflected and transmitted by the superlens, and the reflected waves then do not interact with the cylinder and simply propagate to the far field. Therefore the scattered fields and reflected/transmitted fields can be calculated sequentially. The scattered magnetic fields are calculated from Eq. (3.58) using $\boldsymbol{C}_\mathrm{c} = \boldsymbol{C}_0$. Then the scattered and transmitted fields are obtained by using Eq. (3.61).

Numerical results for near-field coupling

Here we show an example given by $a = 25$ nm, $b = 35$ nm, and $d = 20$ nm at a wavelength of 340 nm. The refractive index of silver is taken from [22]. The permittivity of silver at 340 nm is $-1.17 + 0.28\mathrm{i}$, which is close to the surface plasmon resonance condition. The electric and magnetic fields are plotted in Fig. 3.13(a) and (b), respectively, and they are clearly distinct [46]. After the lens there is a shadow region where the electric field is low, which is considered to be a near-field shadow of the cylinder. In comparison, the magnetic field in Fig. 3.13(b) does not exhibit this shadow feature.

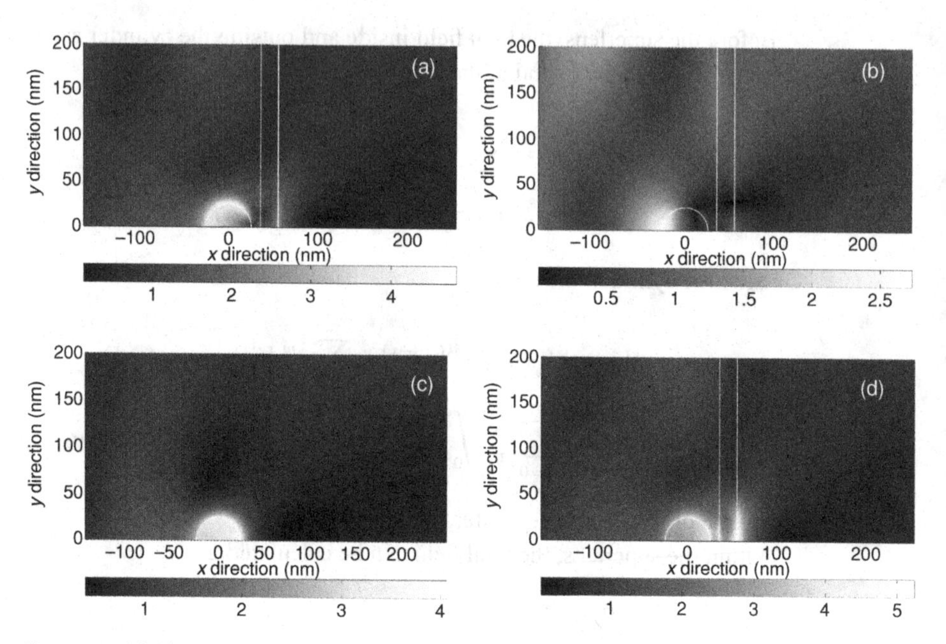

Figure 3.13 (a) The normalized electric field and (b) the normalized magnetic field of the wire–lens system. (c) The normalized electric field of a single cylinder. (d) The normalized electric field of the wire–lens system without coupling. (Only half of each field map need be shown due to symmetry.)

As shown in Fig. 3.13(c), a similar shadow (negative image) is present without the lens, and the lens clearly translates the image in the x direction. Other features in the field maps are part of the diffraction pattern caused by the cylinder. The field map without coupling is shown in Fig. 3.13(d). A comparison of Figs. 3.13(a) and (d) clearly indicates that the coupling between the wire and the lens plays an important role, modifying the field distribution and suppressing the enhancement of the electric field beyond the lens. On comparing the electric-field distributions around the cylinder in Figs. 3.13(a), (c), and (d), it can clearly be seen that the coupling not only modifies the image field, but also changes the object field. A gold cylinder with a silver superlens was also simulated, and the results were similar – the coupling effect significantly changes the field distribution.

Figure 3.13(a) also shows that the electric field can be enhanced beyond the lens with an enhancement factor of about two. For comparison, the electric field of the same wire without the lens is shown in Fig. 3.13(c). In this case, the best enhancement factor is about four close to the cylinder surface. This result demonstrates the possibility of transferring the field enhancement from the cylinder surface to the other side of the lens, albeit with a loss. The spectral behavior of the normalized fields right beyond the lens is plotted in Fig. 3.14 [46]. The highest electric field enhancement (by a factor of about 2.7) occurs at a wavelength of 340 nm. This wavelength is coincident with the surface plasmon resonance wavelength, and is where the superlens effect is predicted to occur [6].

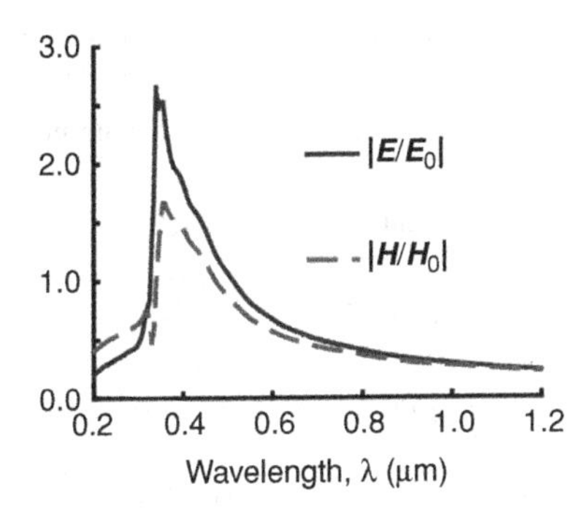

Figure 3.14 The normalized fields at the lens interface.

3.4 Summary

In summary, the imaging of a realistic object (a silver cylinder) is different from the imaging of hypothetical slits due to near-field coupling. The object was illuminated by a p-polarized plane wave. Without the lens, the object produces a near-field shadow (negative image) where the field intensity is lower than the background. This is not surprising since a slit allows light to pass through, while a cylinder blocks light. It has also been shown that a thin silver film translates the negative image of the object to the imaging plane. The lens is also capable of transferring the field enhancement resulting from the surface plasmon resonance of the silver cylinder to the image side of the lens. The coupling between the wire and the lens decreases the transferred enhancement.

The near-field coupling in an object–superlens system makes it necessary to study the system as whole, rather than treating the object and the superlens separately. The significance of the coupling effect has both positive and negative implications. On the one hand the coupling effect makes the performance of a superlens object-dependent, indicating that no universal superlens designs exist that work with all objects. On the other hand, the coupling effect makes it possible to design a superlens without the constraint of matching permittivities. Because metals usually have strong dispersion in the optical frequency range the previously reported superlens designs can work only at a few wavelengths due to the permittivity requirement $\mathrm{Re}(\varepsilon) = -\mathrm{Re}(\varepsilon_d)$. Once the permittivity requirement has been removed, it is possible to design a superlens in a broad spectral range with simple materials. Many devices and applications, such as nanoantenna superlenses [49–51], can greatly benefit from this finding.

References

[1] K. A. Willets and R. P. Van Duyne, "Localized surface plasmon resonance spectroscopy and sensing," *Ann. Rev. Phys. Chem.*, vol. 58, pp. 267–297, 2007.

[2] M. E. Stewart, C. R. Anderton, L. B. Thompson *et al.*, "Nanostructured plasmonic sensors," *Chem. Rev.*, vol. 108, pp. 494–521, 2008.

[3] E. Petryayeva and U. J. Krull, “Localized surface plasmon resonance: Nanostructures, bioassays and biosensing – A review,” *Anal. Chim. Acta.*, vol. 706, pp. 8–24, 2011.
[4] C. Hoppener and L. Novotny, “Exploiting the light–metal interaction for biomolecular sensing and imaging,” *Q. Rev. Biophys.*, vol. 45, pp. 209–255, 2012.
[5] J. Grandidier, G. C. D. Francs, S. Massenot *et al.*, “Gain-assisted propagation in a plasmonic waveguide at telecom wavelength,” *Nano Lett.*, vol. 9, pp. 2935–2939, 2009.
[6] J. B. Pendry, “Negative refraction makes a perfect lens,” *Phys. Rev. Lett.*, vol. 85, pp. 3966–3969, 2000.
[7] J. Jin, *The Finite Element Method in Electromagnetics*, 2nd ed. New York: Wiley-IEEE Press, 2002.
[8] A. Taflove and S. C. Hagness, *Computational Electrodynamics: The Finite-Difference Time-Domain Method*, 3rd ed. New York: Artech House, 2005.
[9] C. B. Burckhardt, “Diffraction of a plane wave at a sinusoidally stratified dielectric grating,” *J. Opt. Soc. Am.*, vol. 56, pp. 1502–1509, 1966.
[10] F. G. Kaspar, “Diffraction by thick, periodically stratified gratings with complex dielectric-constant,” *J. Opt. Soc. Am.*, vol. 63, pp. 37–45, 1973.
[11] K. Knop, “Rigorous diffraction theory for transmission phase gratings with deep rectangular grooves,” *J. Opt. Soc. Am.*, vol. 68, pp. 1206–1210, 1978.
[12] M. G. Moharam and T. K. Gaylord, “Rigorous coupled-wave analysis of planar-grating diffraction,” *J. Opt. Soc. Am.*, vol. 71, pp. 811–818, 1981.
[13] M. G. Moharam and T. K. Gaylord, “Rigorous coupled-wave analysis of grating diffraction – E-mode polarization and losses,” *J. Opt. Soc. Am.*, vol. 73, pp. 451–455, 1983.
[14] P. Lalanne and G. M. Morris, “Highly improved convergence of the coupled-wave method for TM polarization,” *J. Opt. Soc. Am. A – Opt. Image Sci. Vision*, vol. 13, pp. 779–784, 1996.
[15] G. Granet and B. Guizal, “Efficient implementation of the coupled-wave method for metallic lamellar gratings in TM polarization,” *J. Opt. Soc. Am. A – Opt. Image Sci. Vision*, vol. 13, pp. 1019–1023, 1996.
[16] L. F. Li, “Use of Fourier series in the analysis of discontinuous periodic structures,” *J. Opt. Soc. Am. A – Opt. Image Sci. Vision*, vol. 13, pp. 1870–1876, 1996.
[17] L. F. Li, “Formulation and comparison of two recursive matrix algorithms for modeling layered diffraction gratings,” *J. Opt. Soc. Am. A – Opt. Image Sci. Vision*, vol. 13, pp. 1024–1035, 1996.
[18] L. F. Li, “Note on the S-matrix propagation algorithm,” *J. Opt. Soc. Am. A – Opt. Image Sci. Vision*, vol. 20, pp. 655–660, 2003.
[19] E. Noponen and J. Turunen, “Eigenmode method for electromagnetic synthesis of diffractive elements with 3-dimensional profiles,” *J. Opt. Soc. Am. A – Opt. Image Sci. Vision*, vol. 11, pp. 2494–2502, 1994.
[20] L. F. Li, “New formulation of the Fourier modal method for crossed surface-relief gratings,” *J. Opt. Soc. Am. A – Opt. Image Sci. Vision*, vol. 14, pp. 2758–2767, 1997.
[21] Z. T. Liu, K. P. Chen, X. J. Ni *et al.*, “Experimental verification of two-dimensional spatial harmonic analysis at oblique light incidence,” *J. Opt. Soc. Am. B – Opt. Phy.*, vol. 27, pp. 2465–2470, 2010.
[22] P. B. Johnson and R. W. Christy, “Optical constants of noble metals,” *Phys. Rev. B*, vol. 6, pp. 4370–4379, 1972.
[23] V. P. Drachev, U. K. Chettiar, A. V. Kildishev *et al.*, “The Ag dielectric function in plasmonic metamaterials,” *Opt. Express*, vol. 16, pp. 1186–1195, 2008.

[24] E. G. Loewen and E. Popov, *Diffraction Gratings and Applications*. New York: Marcel Dekker, Inc., 1997.

[25] K. P. Chen, V. P. Drachev, J. D. Borneman, A. V. Kildishev, and V. M. Shalaev, "Drude relaxation rate in grained gold nanoantennas," *Nano Lett.*, vol. 10, pp. 916–922, 2010.

[26] X. Ni, Z. Liu, F. Gu *et al.*, *Photonics SHA-2D: Modeling of Single-Period Multilayer Optical Gratings and Metamaterials*, 2009. Available from: https://nanohub.org/tools/sha2d/.

[27] Z. Liu, A. Boltasseva, R. H. Pedersen *et al.*, "Plasmonic nanoantenna arrays for the visible," *Metamaterials*, vol. 2, pp. 45–51, 2008.

[28] S. Zhang, W. J. Fan, N. C. Panoiu *et al.*, "Experimental demonstration of near-infrared negative-index metamaterials," *Phys. Rev. Lett.*, vol. 95, 137404, 2005.

[29] J. P. Kottmann and O. J. F. Martin, "Plasmon resonant coupling in metallic nanowires," *Opt. Express*, vol. 8, pp. 655–663, 2001.

[30] S. A. Maier, M. L. Brongersma, P. G. Kik *et al.*, "Plasmonics – A route to nanoscale optical devices," *Adv. Mater.*, vol. 13, pp. 1501–1505, 2001.

[31] M. Moskovits, "Surface-enhanced Raman spectroscopy: A brief retrospective," *J. Raman Spectrosc.*, vol. 36, pp. 485–496, 2005.

[32] P. Muhlschlegel, H. J. Eisler, O. J. F. Martin, B. Hecht, and D. W. Pohl, "Resonant optical antennas," *Science*, vol. 308, pp. 1607–1609, 2005.

[33] E. Y. Poliakov, V. M. Shalaev, V. A. Markel, and R. Botet, "Enhanced Raman scattering from self-affine thin films," *Opt. Lett.*, vol. 21, pp. 1628–1630, 1996.

[34] W. Rechberger, A. Hohenau, A. Leitner *et al.*, "Optical properties of two interacting gold nanoparticles," *Opt. Commun.*, vol. 220, pp. 137–141, 2003.

[35] V. G. Veselago, "Electrodynamics of substances with simultaneously negative values of sigma and mu," *Sov. Phys. Usp.*, vol. 10, pp. 509–514, 1968.

[36] S. H. Jiang and R. Pike, "A full electromagnetic simulation study of near-field imaging using silver films," *New J. Phys.*, vol. 7, 169, 2005.

[37] S. A. Ramakrishna and J. B. Pendry, "The asymmetric lossy near-perfect lens," *J. Mod. Opt.*, vol. 49, pp. 1747–1762, 2002.

[38] S. A. Ramakrishna, J. B. Pendry, M. C. K. Wiltshire, and W. J. Stewart, "Imaging the near field," *J. Mod. Opt.*, vol. 50, pp. 1419–1430, 2003.

[39] Z. W. Liu, N. Fang, T. J. Yen, and X. Zhang, "Rapid growth of evanescent wave by a silver superlens," *Appl. Phys. Lett.*, vol. 83, pp. 5184–5186, 2003.

[40] R. Borghi, F. Gori, M. Santarsiero, F. Frezza, and G. Schettini, "Plane-wave scattering by a perfectly conducting circular cylinder near a plane surface: Cylindrical-wave approach," *J. Opt. Soc. Am. A – Opt. Image Sci. Vision*, vol. 13, pp. 483–493, 1996.

[41] R. Borghi, M. Santarsiero, F. Frezza, and G. Schettini, "Plane-wave scattering by a dielectric circular cylinder parallel to a general reflecting flat surface," *J. Opt. Soc. Am. A – Opt. Image Sci. Vision*, vol. 14, pp. 1500–1504, 1997.

[42] J. J. Bowman, T. B. A. Senior, and P. L. E. Uslenghi, *Electromagnetic and Acoustic Scattering by Simple Shapes*, revised ed. Amsterdam: North-Holland, 1969.

[43] A. Madrazo and M. Nietovesperinas, "Scattering of electromagnetic-waves from a cylinder in front of a conducting plane," *J. Opt. Soc. Am. A – Opt. Image Sci. Vision*, vol. 12, pp. 1298–1309, 1995.

[44] P. J. Valle, F. Gonzalez, and F. Moreno, "Electromagnetic-wave scattering from conducting cylindrical structures on flat substrates – Study by means of the extinction theorem," *Appl. Opt.*, vol. 33, pp. 512–523, 1994.

[45] P. J. Valle, F. Moreno, J. M. Saiz, and F. Gonzalez, "Near-field scattering from subwavelength metallic protuberances on conducting flat substrates," *Phys. Rev. B*, vol. 51, pp. 13681–13690, 1995.

[46] Z. T. Liu, V. M. Shalaev, and A. V. Kildishev, "Coupling effect in a near-field object–superlens system," *Appl. Phys. A – Mater. Sci. Processing*, vol. 107, pp. 83–88, 2012.

[47] C. F. Bohren and D. R. Huffman, *Absorption and Scattering of Light by Small Particles*. New York: John Wiley & Sons, 1983.

[48] P. Yeh, *Optical Waves in Layered Media*. New York: John Wiley & Sons, 1988.

[49] Z. T. Liu, M. D. Thoreson, A. V. Kildishev, and V. M. Shalaev, "Translation of nanoantenna hot spots by a metal–dielectric composite superlens," *Appl. Phys. Lett.*, vol. 95, 033114, 2009.

[50] M. D. Thoreson, R. B. Nielsen, P. R. West *et al.*, "Studies of plasmonic hot-spot translation by a metal–dielectric layered superlens," in *Metamaterials: Fundamentals and Applications IV*, A. D. Boardman, N. Engheta, M. A. Noginov, and N. I. Zheludev, Eds. Bellingham, WA: SPIE – International Society for Optical Engineering, pp. 80931J1–80931J20, 2011.

[51] Z. T. Liu, E. P. Li, V. M. Shalaev, and A. V. Kildishev, "Near field enhancement in silver nanoantenna–superlens systems," *Appl. Phys. Lett.*, vol. 101, pp. 021109–021111, 2012.

4 Time-domain simulation for plasmonic devices

In this chapter, the finite-difference time-domain method (FDTD) is developed and implemented for the modeling and simulation of passive and active plasmonic devices. For the simulation of passive devices, the Lorentz–Drude (LD) dispersive model is incorporated into the time-dependent Maxwell equations. For the simulation of active plasmonics, a hybrid approach, which combines the multilevel multi-electron quantum model (to simulate the solid state part of a structure) and the LD dispersive model (to simulate the metallic part of the structure), is used. In addition, the multilevel multi-electron quantum mode (solid-state model) is modified to simulate the semiconductor plasmonics. For numerical results, the methodologies developed here are applied to simulate nanoparticles, metal–semiconductor–metal (MSM) waveguides, microcavity resonators, spasers, and surface plasmon polariton (SPP) extraction from spaser. To enhance the simulation speed, graphics processing units (GPUs) are used for the computation, and, as an example, an application of a passive plasmonic device is examined.

4.1 Introduction

The diffraction limit was a challenge in the miniaturization of photonics devices, which restricted the minimum size of a component to being equivalent to $\lambda/2$. The new emerging field of plasmonics has recently made it possible to overcome the diffraction limit of photonic devices. In plasmonics, the wave propagates at the interface of a metal and dielectric, and remains bounded. This feature allows the miniaturization of photonics devices below the diffraction limit. Some plasmonic structures, which guide and manipulate the electromagnetic signals, have been presented in the literature [1–7]. The interesting plasmonic structures include nanoantennas, lenses, resonators, sensors, and waveguides. Most of the work in this area has been done on passive plasmonics [1–3]. In recent years, active plasmonics has attracted much attention owing to the capability of light-wave manipulation [4–7]. At the same time, due to size and RC time delay limits in the field of electronics, the complementary metal–oxide–semiconductor (CMOS) technology is also facing challenges, and reaching its limits. It is difficult to abandon the CMOS technology due its numerous applications, cheap manufacturing process, and mature fabrication techniques; however, the limitations can be overcome by interfacing plasmonics with electronics. Active plasmonics is believed to be a perfect candidate for this purpose, because the interface between these two technologies (which use similar semiconductor materials) is easier to realize than in passive plasmonics.

For simulation of active plasmonic devices, solid-state materials are required in addition to metallic materials. Therefore, to simulate solid-state materials, the multilevel multi-electron quantum system approach is incorporated into Maxwell's equations. The electron dynamics in the solid-state material is managed by the Pauli exclusion principle, state filling, and dynamical Fermi–Dirac thermalization, and the approach is also applicable to the modeling of molecular or atomic media [8]. The solid-state approach has successfully been applied for the simulation of active photonics devices such as lasers and optical switches [9–11]. For the simulation of the metallic part of a plasmonic structure, the LD model [12, 13] is incorporated into Maxwell's equations. The LD model deals with free electrons and bounded electrons in metals. Both a multilevel multi-electron quantum approach and an LD approach are incorporated into Maxwell's equations, and the resulting equations are used to simulate active plasmonic devices. For numerical analysis, the FDTD method [14] is applied to the resulting equations, because the method has potential to model complex dynamical media and their physics. The active plasmonics methodology has been applied to simulate different devices in [15–17]. A review of some other numerical techniques to couple carrier dynamics with full wave dynamics is presented in [18], in which a stochastic ensemble Monte Carlo (EMC) approach is applied to simulate carrier transport, while the FDTD method is applied to simulate Maxwell's equations. A recent approach that is conceptually similar to that presented in [18] to simulate nano-devices has been discussed in Ref. [15, 19]. Here coupled Schrödinger and Maxwell equations are used to simulate the coupled carrier and full wave dynamics. The method presented in this chapter has two main advantages compared with the methods presented in [15, 18, 19]. The first advantage is that the proposed method can model intraband and interband electron transitions, i.e. transition from one energy level to another, transition from the valence band to the conduction band, and vice versa. This provides a realistic modeling of the electron's energy in the bulk medium. The second advantage is that the proposed methods can deal with stimulated emission. These reasons demonstrated that the proposed method is more promising and realistic. In addition, the active-media approach is also modified to simulate all semiconductor plasmonic devices [20]. The proposed approaches are also implemented on single [21] and multiple GPUs for efficient simulation.

Section 4.2 provides details about the formulation of the passive and active plasmonics, and how the FDTD method is applied in simulation. In Section 4.3, simulations of different devices such as MSM waveguides and microcavities, surface plasmon polariton (SPP) generation, and SPP generation in semiconductor devices are studied. In Section 4.4, we discuss how a GPU can be used to simulate plasmonic devices and present a comparison with a CPU.

4.2 Formulation

In opto-electronic systems, electrons and photons are two important components. Photon absorption causes a transition of the electron from a lower energy state to a higher energy state. In the inverse process, when an electron moves from the conduction band to the valence band, it emits a photon. Therefore, the interaction between an electron and a photon is a key factor in active photonics devices; the same principle is adopted here for the modeling and simulation of active plasmonics devices. The LD model is used to deal with the intraband (Drude model) and interband (Lorentz model) effects of a metal. The solid-state model is used to deal with transient intraband and interband electron dynamics, and the carrier thermal equilibrium process for direct-band-gap semiconductors. The solid-state model is formulated to automatically incorporate the band filling and nonlinear optical effects associated with electron and hole carrier dynamics. Although the formulation developed to simulate active plasmonics devices is complicated, it covers all the necessary physical effects efficiently. The formulation given in Section 4.2.1 is used to simulate passive plasmonic devices. The solid-state model in Section 4.2.2 is used to simulate semiconductor devices. Later, both models are hybridized to simulate active plasmonics. The proposed methodology consists of two parts, one for metals and the other for semiconductors. The time-domain Maxwell equations (1.1) and (1.2) can be used to formulate the proposed approach. Equations (1.1) and (1.2) are written in the CGS system, but in this chapter we use the SI system. Therefore, Maxwell's equations can be written in the SI system as

$$\nabla \times \boldsymbol{H} = \varepsilon_0 \varepsilon_\mathrm{r} \frac{\partial \boldsymbol{E}}{\partial t}, \tag{4.1}$$

$$\nabla \times \boldsymbol{E} = -\mu_0 \mu_\mathrm{r} \frac{\partial \boldsymbol{H}}{\partial t}. \tag{4.2}$$

These equations can be manipulated for solid-state and metallic dispersive models. For the metallic model we use the frequency-dependent permittivity in Eq. (4.1) (details are given in Section 4.2.1), and for the solid-state model we add the polarization term in Eq. (4.1) (details are given in Section 4.2.2).

4.2.1 A model for metals

At near-infrared and optical frequencies, the relative permittivity ε_r of numerous metals becomes frequency-dependent, $\varepsilon_\mathrm{r}(\omega)$, and can be described by different dispersive models depending on the nature of the application [22]. In this chapter the LD model is considered, which is written as

$$\varepsilon_\mathrm{r}(\omega) = \varepsilon_\infty + \frac{\omega_\mathrm{pD}^2}{\mathrm{i}^2\omega^2 + \mathrm{i}\Gamma_\mathrm{D}\omega} + \frac{\Delta\varepsilon_\mathrm{L}\,\omega_\mathrm{pL}^2}{\mathrm{i}^2\omega^2 + \mathrm{i}\omega\Gamma_\mathrm{L} + \omega_\mathrm{L}^2}, \tag{4.3}$$

where ω_pD is the plasma frequency and Γ_D is the damping coefficient for the Drude model. $\Delta\varepsilon_\mathrm{L}$ is a weighting factor, ω_pL is the resonance frequency, and Γ_L is the spectral width for the Lorentz model. For the Lorentz model, different numbers of oscillators can be considered. However, for simplicity, we considered only one oscillator and ω_L is its oscillation frequency. After putting Eq. (4.3) into Eq. (4.1) and into frequency-dependent form, we get

$$\nabla \times \boldsymbol{H} = \mathrm{i}\varepsilon_0\varepsilon_\infty\omega\boldsymbol{E} + \frac{\omega_{\mathrm{pD}}^2}{\mathrm{i}^2\omega^2 + \mathrm{i}\Gamma_{\mathrm{D}}\omega}\mathrm{i}\omega\boldsymbol{E} + \frac{\Delta\varepsilon_{\mathrm{L}}\,\omega_{\mathrm{pL}}^2}{\mathrm{i}^2\omega^2 + \mathrm{i}\omega\Gamma_{\mathrm{L}} + \omega_{\mathrm{L}}^2}\mathrm{i}\omega\boldsymbol{E}. \tag{4.4}$$

To put this equation into the time domain, different approaches such as recursive convolution (RC), piecewise linear recursive convolution (PLRC), use of the z-transform, and use of the auxiliary differential equation (ADE) can be used. The PLRC and ADE approaches are more common due to their high accuracy and efficiency. PLRC is an integral approach and needs convolution, whereas the ADE is a differential approach and seems to fit with FDTD due to its differential nature. Therefore, the ADE approach is applied to simplify Eq. (4.4). In Eq. (4.4), the terms with subscript D represent the Drude model and those with subscript L indicate the Lorentz model. First simplify the Drude model:

$$\mathrm{i}\omega\varepsilon_0\frac{\omega_{\mathrm{pD}}^2}{\mathrm{i}^2\omega^2 + \mathrm{i}\Gamma_{\mathrm{D}}\omega}\boldsymbol{E} = \boldsymbol{J}_{\mathrm{D}}, \tag{4.5}$$

$$\mathrm{i}\omega\omega_{\mathrm{pD}}^2\varepsilon_0\boldsymbol{E} = \boldsymbol{J}_{\mathrm{D}}(\mathrm{i}^2\omega^2 + \mathrm{i}\Gamma_{\mathrm{D}}\omega). \tag{4.6}$$

After cancelling out $\mathrm{i}\omega$ from both sides of Eq. (4.6) and converting into differential form, we get

$$\omega_{\mathrm{pD}}^2\varepsilon_0\boldsymbol{E} = \frac{\partial \boldsymbol{J}_{\mathrm{D}}}{\partial t} + \boldsymbol{J}_{\mathrm{D}}\Gamma_{\mathrm{D}}. \tag{4.7}$$

The Lorentz term is given as

$$\frac{\Delta\varepsilon_{\mathrm{L}}\,\omega_{\mathrm{pL}}^2}{\mathrm{i}^2\omega^2 + \mathrm{i}\omega\Gamma_{\mathrm{L}} + \omega_{\mathrm{L}}^2}\boldsymbol{E} = \boldsymbol{Q}_{\mathrm{L}}. \tag{4.8}$$

After some mathematical manipulation Eq. (4.8) is written as

$$\Delta\varepsilon_{\mathrm{L}}\,\omega_{\mathrm{pL}}^2\boldsymbol{E} = \frac{\partial^2 \boldsymbol{Q}_{\mathrm{L}}}{\partial t^2} + \Gamma_{\mathrm{L}}\frac{\partial \boldsymbol{Q}_{\mathrm{L}}}{\partial t} + \omega_{\mathrm{L}}^2\boldsymbol{Q}_{\mathrm{L}}. \tag{4.9}$$

By putting Eqs. (4.5) and (4.8) into Eq. (4.4), and converting to the time domain, we obtain

$$\nabla \times \boldsymbol{H} = \varepsilon_0\varepsilon_\infty\frac{\partial \boldsymbol{E}}{\partial t} + \boldsymbol{J}_{\mathrm{D}} + \frac{\varepsilon_0\,\partial \boldsymbol{Q}_{\mathrm{L}}}{\partial t}, \tag{4.10}$$

where Eq. (4.10) is Maxwell's equation with both Drude, $\boldsymbol{J}_{\mathrm{D}}$, and Lorentz, $\varepsilon_0\,\partial\boldsymbol{Q}_{\mathrm{L}}/\partial t$, terms. Equation (4.7) is for the Drude model and Eq. (4.9) is for the Lorentz model in the time domain, and both are obtained from Eq. (4.4). Equation (4.2) in scalar form is written as

$$\frac{\partial H_x}{\partial t} = \frac{1}{\mu_0 \mu_r}\left(\frac{\partial E_y}{\partial z} - \frac{\partial E_z}{\partial y}\right), \tag{4.11}$$

$$\frac{\partial H_y}{\partial t} = \frac{1}{\mu_0 \mu_r}\left(\frac{\partial E_z}{\partial x} - \frac{\partial E_x}{\partial z}\right), \tag{4.12}$$

$$\frac{\partial H_z}{\partial t} = \frac{1}{\mu_0 \mu_r}\left(\frac{\partial E_x}{\partial y} - \frac{\partial E_y}{\partial x}\right). \tag{4.13}$$

Equation (4.10) in scalar form with both Drude and Lorentz terms is written as

$$\frac{\partial E_x}{\partial t} = \frac{1}{\varepsilon_0 \varepsilon_\infty}\left(\frac{\partial H_z}{\partial y} - \frac{\partial H_y}{\partial z}\right) - \frac{1}{\varepsilon_0 \varepsilon_\infty} J_{\mathrm{D}x} - \frac{1}{\varepsilon_\infty}\frac{\partial Q_{\mathrm{L}x}}{\partial t}, \tag{4.14}$$

$$\frac{\partial E_y}{\partial t} = \frac{1}{\varepsilon_0 \varepsilon_\infty}\left(\frac{\partial H_x}{\partial z} - \frac{\partial H_z}{\partial x}\right) - \frac{1}{\varepsilon_0 \varepsilon_\infty} J_{\mathrm{D}y} - \frac{1}{\varepsilon_\infty}\frac{\partial Q_{\mathrm{L}y}}{\partial t}, \tag{4.15}$$

$$\frac{\partial E_z}{\partial t} = \frac{1}{\varepsilon_0 \varepsilon_\infty}\left(\frac{\partial H_y}{\partial x} - \frac{\partial H_x}{\partial y}\right) - \frac{1}{\varepsilon_0 \varepsilon_\infty} J_{\mathrm{D}z} - \frac{1}{\varepsilon_\infty}\frac{\partial Q_{\mathrm{L}z}}{\partial t}. \tag{4.16}$$

By using the FDTD method Eqs. (4.11)–(4.16) are discretized as

$$\begin{aligned} H_x^{n+1/2}\left(i, j+\frac{1}{2}, k+\frac{1}{2}\right) &= H_x^{n-1/2}\left(i, j+\frac{1}{2}, k+\frac{1}{2}\right) + \frac{\Delta t}{\mu} \\ &\times \left[\frac{E_y^n\left(i, j+\frac{1}{2}, k+1\right) - E_y^n\left(i, j+\frac{1}{2}, k\right)}{\Delta z}\right. \\ &\left. - \frac{E_z^n\left(i, j+1, k+\frac{1}{2}\right) - E_z^n\left(i, j, k+\frac{1}{2}\right)}{\Delta y}\right], \end{aligned} \tag{4.17}$$

$$\begin{aligned} H_y^{n+1/2}\left(i+\frac{1}{2}, j, k+\frac{1}{2}\right) &= H_y^{n-1/2}\left(i+\frac{1}{2}, j, k+\frac{1}{2}\right) + \frac{\Delta t}{\mu} \\ &\times \left[\frac{E_z^n\left(i+1, j, k+\frac{1}{2}\right) - E_z^n\left(i, j, k+\frac{1}{2}\right)}{\Delta x}\right. \\ &\left. - \frac{E_x^n\left(i+\frac{1}{2}, j, k+1\right) - E_x^n\left(i+\frac{1}{2}, j, k\right)}{\Delta z}\right], \end{aligned} \tag{4.18}$$

$$\begin{aligned} H_z^{n+1/2}\left(i+\frac{1}{2}, j+\frac{1}{2}, k\right) &= H_z^{n-1/2}\left(i+\frac{1}{2}, j+\frac{1}{2}, k\right) + \frac{\Delta t}{\mu} \\ &\times \left[\frac{E_x^n\left(i+\frac{1}{2}, j+1, k\right) - E_x^n\left(i+\frac{1}{2}, j, k\right)}{\Delta y}\right. \\ &\left. - \frac{E_y^n\left(i+1, j+\frac{1}{2}, k\right) - E_y^n\left(i, j+\frac{1}{2}, k\right)}{\Delta x}\right]. \end{aligned} \tag{4.19}$$

As an example, after simplifying Eqs. (4.7) and (4.9) and then by inserting them into Eqs. (4.14)–(4.16), we get the following equations. The electric field in the x direction is written as

$$\begin{aligned}
E_x^{n+1}\left(i+\frac{1}{2},j,k\right) &= \frac{1}{\Omega_x}E_x^n\left(i+\frac{1}{2},j,k\right)+\frac{\Delta t}{\Omega_x\varepsilon_0\varepsilon_\infty} \\
&\quad\times\left[\frac{H_z^{n+1/2}\left(i+\frac{1}{2},j+\frac{1}{2},k\right)-H_z^{n+1/2}\left(i+\frac{1}{2},j-\frac{1}{2},k\right)}{\Delta y}\right. \\
&\quad\left.-\frac{H_y^{n+1/2}\left(i+\frac{1}{2},j,k+\frac{1}{2}\right)-H_y^{n+1/2}\left(i+\frac{1}{2},j,k-\frac{1}{2}\right)}{\Delta z}\right] \\
&\quad-\frac{\Delta t}{2\Omega_x\varepsilon_0\varepsilon_\infty}\left[\alpha_x J_{\mathrm{D}x}^n\left(i+\frac{1}{2},j,k\right)+\beta_x E_x^n\left(i+\frac{1}{2},j,k\right)\right. \\
&\quad\left.+J_{\mathrm{D}x}^n\left(i+\frac{1}{2},j,k\right)\right] \\
&\quad-\frac{1}{\Omega_x\varepsilon_\infty}\left[\varsigma_x E_x^n\left(i+\frac{1}{2},j,k\right)+\tau_x Q_{\mathrm{L}x}^n\left(i+\frac{1}{2},j,k\right)\right. \\
&\quad\left.-\rho_x Q_{\mathrm{L}x}^{n-1}\left(i+\frac{1}{2},j,k\right)-Q_{\mathrm{L}x}^n\left(i+\frac{1}{2},j,k\right)\right],
\end{aligned} \tag{4.20}$$

whereas the Drude and Lorentz equations are discretized as

$$\begin{aligned}
J_{\mathrm{D}x}^{n+1}\left(i+\frac{1}{2},j,k\right) &= \alpha_x J_{\mathrm{D}x}^n\left(i+\frac{1}{2},j,k\right) \\
&\quad+\beta_x\left[E_x^{n+1}\left(i+\frac{1}{2},j,k\right)+E_x^n\left(i+\frac{1}{2},j,k\right)\right],
\end{aligned} \tag{4.21}$$

$$\begin{aligned}
Q_{\mathrm{L}x}^{n+1}\left(i+\frac{1}{2},j,k\right) &= \varsigma_x\left[E_x^{n+1}\left(i+\frac{1}{2},j,k\right)+E_x^n\left(i+\frac{1}{2},j,k\right)\right] \\
&\quad+\tau_x Q_{\mathrm{L}x}^n\left(i+\frac{1}{2},j,k\right)-\rho_x Q_{\mathrm{L}x}^{n-1}\left(i+\frac{1}{2},j,k\right),
\end{aligned} \tag{4.22}$$

where

$$\Omega_x=\frac{\varepsilon_0\varsigma_x}{\varepsilon_0\varepsilon_\infty}+1+\frac{\Delta t\,\beta_x}{2\varepsilon_0\varepsilon_\infty},\qquad \alpha_x=\frac{1-\Delta t\,\Gamma_{\mathrm{D}}/2}{1+\Delta t\Gamma_{\mathrm{D}}/2},\qquad \beta_x=\frac{\Delta t\,\omega_{\mathrm{pD}}^2\varepsilon_0/2}{1+\Delta t\Gamma_{\mathrm{D}}/2},$$

$$\varsigma_x=\frac{\Delta t^2\,\Delta\varepsilon_{\mathrm{L}}\omega_{\mathrm{pL}}^2/2}{1+\Delta t\,\Gamma_{\mathrm{L}}+(\Delta t^2/2)\omega_{\mathrm{L}}^2}\qquad \tau_x=\frac{2+\Delta t\,\Gamma_{\mathrm{L}}-(\Delta t^2/2)\omega_{\mathrm{L}}^2}{1+\Delta t\,\Gamma_{\mathrm{L}}+(\Delta t^2/2)\omega_{\mathrm{L}}^2},$$

$$\rho_x=\frac{1}{1+\Delta t\,\Gamma_{\mathrm{L}}+(\Delta t^2/2)\,\omega_{\mathrm{L}}^2}.$$

The electric field in the y direction is given as

$$
\begin{aligned}
E_y^{n+1}\left(i, j+\frac{1}{2}, k\right) = {} & \frac{1}{\Omega_y} E_y^n\left(i, j+\frac{1}{2}, k\right) + \frac{\Delta t}{\Omega_y \varepsilon_0 \varepsilon_\infty} \\
& \times \left[\frac{H_x^{n+1/2}\left(i, j+\frac{1}{2}, k+\frac{1}{2}\right) - H_x^{n+1/2}\left(i, j+\frac{1}{2}, k-\frac{1}{2}\right)}{\Delta z}\right. \\
& \left. - \frac{H_z^{n+1/2}\left(i+\frac{1}{2}, j+\frac{1}{2}, k\right) - H_z^{n+1/2}\left(i-\frac{1}{2}, j+\frac{1}{2}, k\right)}{\Delta x}\right] \\
& - \frac{\Delta t}{2\Omega_y \varepsilon_0 \varepsilon_\infty}\left[\alpha_y J_{\mathrm{D}y}^n\left(i, j+\frac{1}{2}, k\right) + \beta_y E_y^n\left(i, j+\frac{1}{2}, k\right)\right. \\
& \left. + J_{\mathrm{D}y}^n\left(i, j+\frac{1}{2}, k\right)\right] \\
& - \frac{1}{\Omega_y \varepsilon_\infty}\left[\varsigma_y E_y^n\left(i, j+\frac{1}{2}, k\right) + \tau_y Q_{\mathrm{L}y}^n\left(i, j+\frac{1}{2}, k\right)\right. \\
& \left. - \rho_y Q_{Ly}^{n-1}\left(i, j+\frac{1}{2}, k\right) - Q_{Ly}^n\left(i, j+\frac{1}{2}, k\right)\right],
\end{aligned}
\tag{4.23}
$$

whereas the Drude and Lorentz equations are discretized as

$$
\begin{aligned}
J_{\mathrm{D}y}^{n+1}\left(i, j+\frac{1}{2}, k\right) = {} & \alpha_y J_{\mathrm{D}y}^n\left(i, j+\frac{1}{2}, k\right) \\
& + \beta_y\left[E_y^{n+1}\left(i, j+\frac{1}{2}, k\right) + E_y^n\left(i, j+\frac{1}{2}, k\right)\right],
\end{aligned}
\tag{4.24}
$$

$$
\begin{aligned}
Q_y^{n+1}\left(i, j+\frac{1}{2}, k\right) = {} & \varsigma_y\left[E_y^{n+1}\left(i, j+\frac{1}{2}, k\right) + E_y^n\left(i, j+\frac{1}{2}, k\right)\right] \\
& + \tau_y Q_y^n\left(i, j+\frac{1}{2}, k\right) - \rho_y Q_y^{n-1}\left(i, j+\frac{1}{2}, k\right),
\end{aligned}
\tag{4.25}
$$

where

$$
\Omega_y = \frac{\varepsilon_0 \varsigma_y}{\varepsilon_0 \varepsilon_\infty} + 1 + \frac{\Delta t\, \beta_y}{2\varepsilon_0 \varepsilon_\infty}, \qquad \alpha_y = \frac{1 - \Delta t\, \Gamma_{\mathrm{D}}/2}{1 + \Delta t\, \Gamma_{\mathrm{D}}/2}, \qquad \beta_y = \frac{\Delta t\, \omega_{\mathrm{pD}}^2 \varepsilon_0/2}{1 + \Delta t\, \Gamma_{\mathrm{D}}/2},
$$

$$
\varsigma_y = \frac{\Delta t^2\, \Delta\varepsilon_{\mathrm{L}}\, \omega_{\mathrm{pL}}^2/2}{1 + \Delta t\, \Gamma_{\mathrm{L}} + (\Delta t^2/2)\omega_{\mathrm{L}}^2}, \qquad \tau_y = \frac{2 + \Delta t\, \Gamma_{\mathrm{L}} - (\Delta t^2/2)\omega_{\mathrm{L}}^2}{1 + \Delta t\, \Gamma_{\mathrm{L}} + (\Delta t^2/2)\omega_{\mathrm{L}}^2},
$$

$$
\rho_y = \frac{1}{1 + \Delta t\, \Gamma_{\mathrm{L}} + (\Delta t^2/2)\omega_{\mathrm{L}}^2}.
$$

The electric field in the z direction is given as

$$
\begin{aligned}
E_z^{n+1}\left(i,j,k+\frac{1}{2}\right) &= \frac{1}{\Omega_z}E_z^n\left(i,j,k+\frac{1}{2}\right) + \frac{\Delta t}{\Omega_z\varepsilon_0\varepsilon_\infty} \\
&\times \left[\frac{H_y^{n+1/2}\left(i+\frac{1}{2},j,k+\frac{1}{2}\right) - H_y^{n+1/2}\left(i-\frac{1}{2},j,k+\frac{1}{2}\right)}{\Delta x}\right. \\
&\left. - \frac{H_x^{n+1/2}\left(i,j+\frac{1}{2},k+\frac{1}{2}\right) - H_x^{n+1/2}\left(i,j-\frac{1}{2},k+\frac{1}{2}\right)}{\Delta y}\right] \\
&- \frac{\Delta t}{2\Omega_z\varepsilon_0\varepsilon_\infty}\left[\alpha_z J_{\mathrm{D}z}^n\left(i,j,k+\frac{1}{2}\right) + \beta_z E_z^n\left(i,j,k+\frac{1}{2}\right)\right. \\
&\left. + J_{\mathrm{D}z}^n\left(i,j,k+\frac{1}{2}\right)\right] \\
&- \frac{1}{\Omega_z\varepsilon_\infty}\left[\varsigma_z E_z^n\left(i,j,k+\frac{1}{2}\right) + \tau_z Q_{\mathrm{L}z}^n\left(i,j,k+\frac{1}{2}\right)\right. \\
&\left. - \rho_z Q_{\mathrm{L}z}^{n-1}\left(i,j,k+\frac{1}{2}\right) - Q_{\mathrm{L}z}^n\left(i,j,k+\frac{1}{2}\right)\right],
\end{aligned}
\tag{4.26}
$$

whereas the Drude and Lorentz equations are discretized as

$$
\begin{aligned}
J_z^{n+1}\left(i,j,k+\frac{1}{2}\right) &= \alpha_z J_z^n\left(i,j,k+\frac{1}{2}\right) \\
&+ \beta_z\left[E_z^{n+1}\left(i,j,k+\frac{1}{2}\right) + E_z^n\left(i,j,k+\frac{1}{2}\right)\right],
\end{aligned}
\tag{4.27}
$$

$$
\begin{aligned}
P_z^{n+1}\left(i,j,k+\frac{1}{2}\right) &= \varsigma_z\left[E_z^{n+1}\left(i,j,k+\frac{1}{2}\right) + E_z^n\left(i,j,k+\frac{1}{2}\right)\right] \\
&+ \tau_z P_z^n\left(i,j,k+\frac{1}{2}\right) - \rho_z P_z^{n-1}\left(i,j,k+\frac{1}{2}\right),
\end{aligned}
\tag{4.28}
$$

where

$$
\Omega_z = \frac{\varepsilon_0\varsigma_z}{\varepsilon_0\varepsilon_\infty} + 1 + \frac{\Delta t\,\beta_z}{2\varepsilon_0\varepsilon_\infty}, \qquad \alpha_z = \frac{1-\Delta t\,\Gamma_{\mathrm{D}}/2}{1+\Delta t\,\Gamma_{\mathrm{D}}/2}, \qquad \beta_z = \frac{\Delta t\,\omega_{\mathrm{pD}}^2\varepsilon_0/2}{1+\Delta t\,\Gamma_{\mathrm{D}}/2},
$$

$$
\varsigma_z = \frac{\Delta t^2\,\Delta\varepsilon_{\mathrm{L}}\,\omega_{\mathrm{pL}}^2/2}{1+\Delta t\,\Gamma_{\mathrm{L}} + (\Delta t^2/2)\omega_{\mathrm{L}}^2}, \qquad \tau_z = \frac{2+\Delta t\,\Gamma_{\mathrm{L}} - (\Delta t^2/2)\omega_{\mathrm{L}}^2}{1+\Delta t\,\Gamma_{\mathrm{L}} + (\Delta t^2/2)\omega_{\mathrm{L}}^2},
$$

$$
\rho_z = \frac{1}{1+\Delta t\,\Gamma_{\mathrm{L}} + (\Delta t^2/2)\omega_{\mathrm{L}}^2}.
$$

Equations (4.17)–(4.28) can be used for the simulation of dispersive materials such as gold, silver etc. The structures which can be simulated with this model are referred to as passive plasmonic devices. An example of this model (passive plasmonic) is simulated in Section 4.4. For the simulation of active plasmonic devices, there is a need for a model that can include active effects. For this purpose a solid-state model is developed and explained in the following subsection.

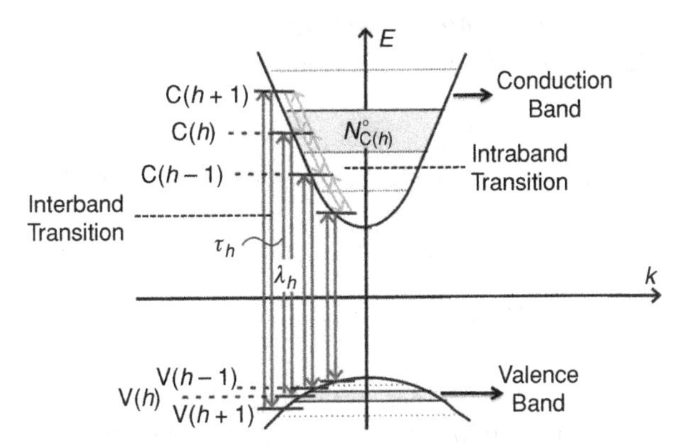

Figure 4.1 The discretization of conduction and valence bands for a multilevel multi-electron model of a direct-band-gap semiconductor for simulation with the FDTD method.

4.2.2 A model for solid-state materials

Equation (4.1) can be modified to

$$\nabla \times \boldsymbol{H} = \varepsilon_0 n^2 \frac{\partial \boldsymbol{E}}{\partial t} - \frac{\partial \boldsymbol{P}}{\partial t}, \tag{4.29}$$

whereas Eq. (4.2) remains unchanged. In Eq. (4.29), n is the refractive index and $\boldsymbol{P}$ is the macroscopic polarization density, which represents the total dipole moment per unit volume. It is expressed as

$$\boldsymbol{P}(r, t) = U_{\mathrm{m}}(t) \sum_h N_{\mathrm{dip},h}(r), \tag{4.30}$$

where $U_{\mathrm{m}}(t)$ is the atomic dipole moment, $N_{\mathrm{dip},h}(r)$ is the dipole volume density for energy level h within energy width (δE). $N_{\mathrm{dip},h}(r)$ is specified by the number of dipoles (N_0) divided by unit volume (δV) and r represents the x, y, and z directions. In a solid-state (semiconductor) medium, the electron dynamics is modeled by discretizing the conduction and valence band into different energy levels as shown in Fig. 4.1. The lines drawn between different levels of conduction and valence bands show interband transitions, whereas the lines drawn in between different energy levels of the conduction band represent intraband transitions, and similarly for the case of the valence band. The valence band levels are labeled V(h), while the conduction-band levels are labeled C(h). The occupation probability with respect to time for each level in the conduction and valence band, and the effect of carrier densities for different level pairs, have been studied in [8]. It was observed that the electrons relax to equilibrium much slower in the conduction band than in the valence band. This is because holes have a bigger effective mass than that of electrons.

Different energy levels can be considered both for the conduction band and for the valence band. From these levels, we can calculate intraband and interband transition times. These levels are also helpful for integrating Fermi–Dirac thermalization dynamics in the semiconductor, and will be discussed later in this section.

Intraband and interband transition equations

The expression for interband transition and spontaneous decay from energy level C(h) to V(h), is given as

$$\Delta N_{(h)}|^n_{i+1/2,j,k} = \frac{\omega_{ah}}{\hbar}\left(A_x|^n_{i+1/2,j,k}P_{hx}|^n_{i+1/2,j,k} + A_y|^n_{i+1/2,j,k}P_{hy}|^n_{i+1/2,j,k} + A_z|^n_{i+1/2,j,k}P_{hz}|^n_{i+1/2,j,k}\right) + \frac{N_{\mathrm{C}(h)}|^n_{i+1/2,j,k}(1 - N_{\mathrm{V}(h)}|^n_{i+1/2,j,k}/N^{\circ}_{\mathrm{V}(h)}|_{i+1/2,j,k})}{\tau_h}, \quad (4.31)$$

where $\Delta N_{(h)}$ is the number of electrons per unit volume transferred from conduction-band level C(h) to valence-band level V(h), ω_{ah} is the interband transition frequency, and τ_h is the interband transition time. Similarly the relations for intraband transition for the conduction and valence bands are also derived and are given below.

Intraband transition for the conduction band from energy level C(h) to energy level C($h - 1$) is given by

$$\Delta N_{\mathrm{C}(h,h-1)}|^n_{i+1/2,j,k} = \frac{N_{\mathrm{C}(h)}|^n_{i+1/2,j,k}\left(1 - N_{\mathrm{C}(h-1)}|^n_{i+1/2,j,k}/N^{\circ}_{\mathrm{C}(h-1)}|_{i+1/2,j,k}\right)}{\tau_{\mathrm{C}(h,h-1)}} - \frac{N_{\mathrm{C}(h-1)}|^n_{i+1/2,j,k}\left(1 - N_{\mathrm{C}(h)}|^n_{i+1/2,j,k}/N^{\circ}_{\mathrm{C}(h)}|_{i+1/2,j,k}\right)}{\tau_{\mathrm{C}(h-1,h)}}, \quad (4.32)$$

where $\Delta N_{\mathrm{C}(h,h-1)}$ is the number of electrons per unit volume transferred from level C(h) to level C($h - 1$), and τ_{C} is the intraband transition time in the conduction band.

Intraband transition for the valence band from energy level V(h) to energy level V($h - 1$) is given by

$$\Delta N_{\mathrm{V}(h,h-1)}|^n_{i+1/2,j,k} = \frac{N_{\mathrm{V}(h)}|^n_{i+1/2,j,k}\left(1 - N_{\mathrm{V}(h-1)}|^n_{i+1/2,j,k}/N^{\circ}_{\mathrm{V}(h-1)}|_{i+1/2,j,k}\right)}{\tau_{\mathrm{V}(h,h-1)}} - \frac{N_{\mathrm{V}(h-1)}|^n_{i+1/2,j,k}\left(1 - N_{\mathrm{V}(h)}|^n_{i+1/2,j,k}/N^{\circ}_{\mathrm{V}(h)}|_{i+1/2,j,k}\right)}{\tau_{\mathrm{V}(h-1,h)}}, \quad (4.33)$$

where $\Delta N_{\mathrm{V}(h,h-1)}$ is the number of electrons per unit volume transferred from level V(h) to level V($h - 1$) and τ_{V} is the intraband transition time in the valence band.

In Eqs. (4.31)–(4.33), $N^{\circ}_{\mathrm{V}(h)}$ and $N^{\circ}_{\mathrm{C}(h)}$ are the initial values of the volume densities of the energy states at level h in the valence and conduction band, respectively, while $N_{\mathrm{V}(h)}$ and $N_{\mathrm{C}(h)}$ are the values at a later time.

The polarization equation

After some mathematical derivations, Eq. (4.30) for polarization can be written as

$$\frac{\mathrm{d}^2 P_{hk}(r,t)}{\mathrm{d}t^2} + \Gamma_h \frac{\mathrm{d}P_{hk}(r,t)}{\mathrm{d}t} + \frac{\omega^2_{ah}}{h^2}\left[h^2 + 4|U_{kh}|^2 A^2_k(r,t)\right] P_{hk}(r,t) = \frac{2\omega_{ah}}{h}|U_{kh}|^2\left[\frac{N_{\mathrm{dip}-h}(r)}{N^{\circ}_{\mathrm{V}h}(r)}N_{\mathrm{V}h}(r,t) - \frac{N_{\mathrm{dip}-h}(r)}{N^{\circ}_{\mathrm{C}h}(r)}N_{\mathrm{C}h}(r,t)\right] E_k(r,t), \quad (4.34)$$

where $k = x, y$, and z, Γ_h represents the de-phasing rate at the hth energy level after excitation, and A_k is the vector potential, $|U_{kh}|^2 = 3\pi\hbar\varepsilon_0 c^3/(\omega_{ah}^3 \tau_h)$, where $\hbar = h/(2\pi)$, h is Planck's constant, c is the speed of light, and ε_0 is the permittivity of free space.

Equation (4.34) is further simplified and discretized. In the x direction, it is written as

$$\begin{aligned} P_{hx}|_{i+1/2,j,k}^{n+1} &= \frac{4 - 2\,\Delta t^2\left(\omega_{ah}^2 + 4(\omega_{ah}^2/\hbar^2)|U_h|^2 A_x^2|_{i+1/2,j,k}^n\right)}{2 + \Delta t\,\gamma_h} P_{hx}|_{i+1/2,j,k}^n \\ &+ \frac{\Delta t\,\gamma_h - 2}{2 + \Delta t\,\gamma_h} P_{hx}|_{i+1/2,j,k}^{n-1} - \frac{4\,\Delta t^2\,\omega_{ah}}{\hbar(2 + \Delta t\,\gamma_h)}|U_{kh}|^2 \\ &\times \left[N_{C(h)}|_{i+1/2,j,k}^n - N_{V(h)}|_{i+1/2,j,k}^n\right] E_x|_{i+1/2,j,k}^n . \end{aligned} \tag{4.35}$$

Similar polarization equations can be written for the y and z directions also.

Rate equations

The rate equations for the conduction and valence bands are written as

$$\begin{aligned} N_{C(h)}|_{i+1/2,j,k}^{n+1} = N_{C(h)}|_{i+1/2,j,k}^{n-1} &+ 2\,\Delta t\,\left(-\Delta N_{(h)}|_{i+1/2,j,k}^n - \Delta N_{(h,h-1)}|_{i+1/2,j,k}^n\right. \\ &\left. + \Delta N_{(h+1,h)}|_{i+1/2,j,k}^n + W_{\text{pump}}\right), \end{aligned} \tag{4.36}$$

$$\begin{aligned} N_{V(h)}|_{i+1/2,j,k}^{n+1} = N_{V(h)}|_{i+1/2,j,k}^{n-1} &+ 2\,\Delta t\,\left(-\Delta N_{(h)}|_{i+1/2,j,k}^n - \Delta N_{(h,h-1)}|_{i+1/2,j,k}^n\right. \\ &\left. + \Delta N_{(h+1,h)}|_{i+1/2,j,k}^n - W_{\text{pump}}\right). \end{aligned} \tag{4.37}$$

In Eqs. (4.36) and (4.37), the term W_{pump} represents electrical pumping. The Fermi–Dirac thermalization dynamics in the semiconductor is obtained by taking the ratio between upward and downward intraband transitions for two neighboring energy levels. As soon as the intraband transition between these two energy levels reaches steady state, the number of electrons per unit volume transferred from one energy level to the other level reduces to zero (i.e. $\Delta N_{V(h,h-1)} = 0$). The relation between the two energy levels can be obtained by use of the intraband transition-rate equations. As an example, Eq. (4.33) for the valence band can be written as

$$\frac{\tau_{V(h-1,h)}}{\tau_{V(h,h-1)}} = \frac{N_{V(h-1)}^{\circ}(r)}{N_{V(h)}^{\circ}(r)} \exp\left(\frac{E_{V(h)} - E_{V(h-1)}}{k_B T}\right), \tag{4.38}$$

where $\tau_{V(h,h-1)}$ is the downward transition rate and $\tau_{V(h-1,h)}$ is the upward transition rate between levels $V(h)$ and $V(h-1)$ in the valence band, k_B is the Boltzmann constant, T represents temperature, and $E_{V(h)}$ and $E_{V(h-1)}$ are the energies at levels $V(h)$ and $V(h-1)$, respectively. A similar relation can be obtained for the conduction band. The approach adopted for intraband thermalization can be applied to interband transition also, but the contribution is negligible due to a large energy gap. Therefore, the thermalization effect is considered only for intraband transitions.

After incorporation of the above effects into the polarization equation and then of the resulting equation into Maxwell's equation, the discretized electric field equations for the FDTD approach are written as

$$E_x\big|_{i+1/2,j,k}^{n+1} = E_x\big|_{i+1/2,j,k}^{n} + \frac{\Delta t}{\varepsilon\,\Delta y}\left(H_z\big|_{i+1/2,j+1/2,k}^{n+1/2} - H_z\big|_{i+1/2,j-1/2,k}^{n+1/2}\right) + \frac{\Delta t}{\varepsilon\,\Delta z}\left(H_z\big|_{i+1/2,j,k+1/2}^{n+1/2} - H_z\big|_{i+1/2,j,k-1/2}^{n+1/2}\right) - \frac{1}{\varepsilon}\sum_{h=1}^{M}\left(P_{h,x}\big|_{i+1/2,j,k}^{n+1} - P_{h,x}\big|_{i+1/2,j,k}^{n}\right), \tag{4.39}$$

$$E_y\big|_{i,j+1/2,k}^{n+1} = E_y\big|_{i,j+1/2,k}^{n} + \frac{\Delta t}{\varepsilon\,\Delta z}\left(H_x\big|_{i,j+1/2,k+1/2}^{n+1/2} - H_x\big|_{i,j+1/2,k-1/2}^{n+1/2}\right) - \frac{\Delta t}{\varepsilon\,\Delta x}\left(H_z\big|_{i,j+1/2,k+1/2}^{n+1/2} - H_z\big|_{i,j+1/2,k-1/2}^{n+1/2}\right) - \frac{1}{\varepsilon}\sum_{h=1}^{M}\left(P_{h,y}\big|_{i,j+1/2,k}^{n+1} - P_{h,y}\big|_{i,j+1/2,k}^{n}\right), \tag{4.40}$$

$$E_z\big|_{i,j,k+1/2}^{n+1} = E_z\big|_{i,j,k+1/2}^{n} + \frac{\Delta t}{\varepsilon\,\Delta x}\left(H_y\big|_{i+1/2,j,k+1/2}^{n+1/2} - H_y\big|_{i-1/2,j,k+1/2}^{n+1/2}\right) - \frac{\Delta t}{\varepsilon\,\Delta y}\left(H_x\big|_{i,j+1/2,k+1/2}^{n+1/2} - H_x\big|_{i,j-1/2,k+1/2}^{n+1/2}\right) - \frac{1}{\varepsilon}\sum_{h=1}^{M}\left(P_{h,z}\big|_{i,j,k+1/2}^{n+1} - P_{h,z}\big|_{i,j,k+1/2}^{n}\right). \tag{4.41}$$

For further illustration, the differences in relative permittivities for both semiconductor and metal cases, and their effects on the electric field density, are discussed below.

In general, the electric field density can be written as

$$\boldsymbol{D} = \varepsilon_0\varepsilon_r\boldsymbol{E}, \tag{4.42a}$$

$$\boldsymbol{D} = \varepsilon_0(1+\chi)\boldsymbol{E}, \tag{4.42b}$$

$$\boldsymbol{D} = \varepsilon_0\boldsymbol{E} + \varepsilon_0\chi\boldsymbol{E}, \tag{4.42c}$$

$$\boldsymbol{p} = \varepsilon_0\chi\,\boldsymbol{E}, \tag{4.42d}$$

$$\boldsymbol{D} = \varepsilon_0 E + \boldsymbol{p}, \tag{4.42e}$$

where $\boldsymbol{p}$ is the polarization density and is linked to the susceptibility of the material. Here for a solid-state material we add a term $\boldsymbol{p}$, which is dependent on the dipole volume density, pumping density, and atomic dipole moment:

$$\boldsymbol{D} = \varepsilon_0\boldsymbol{E} + \varepsilon_0\chi\boldsymbol{E} + \boldsymbol{P}, \quad 1+\chi = \varepsilon_r, \tag{4.43}$$

$$\boldsymbol{D} = \varepsilon_0\varepsilon_r\boldsymbol{E} + \boldsymbol{P}. \tag{4.44}$$

After putting Eq. (4.44) into Eq. (4.1), we get Eq. (4.29).

For the LD model Eq. (4.42) becomes frequency-dependent and is written as

$$D = \varepsilon_0 \varepsilon_r(\omega) E, \quad \varepsilon_r(\omega) = \varepsilon_\infty + \varepsilon_D + \varepsilon_L,$$

$$D = \varepsilon_0(\varepsilon_\infty + \varepsilon_D + \varepsilon_L) E, \tag{4.45a}$$

where ε_∞ is a constant value, whereas ε_D and ε_L are complex values. Here ε_D is the frequency-dependent permittivity from the Drude model and ε_L is the frequency-dependent permittivity from the Lorentz model:

$$D = \varepsilon_0(\varepsilon_\infty + \varepsilon_D + \varepsilon_L) E, \tag{4.45b}$$

$$D = \varepsilon_0(1 + \chi) E, \quad \chi = \varepsilon_D + \varepsilon_L,$$

$$D = \varepsilon_0 \varepsilon_\infty E + \varepsilon_0 \chi E, \tag{4.45c}$$

$$P = \varepsilon_0 \chi E, \tag{4.45d}$$

$$D = \varepsilon_0 \varepsilon_\infty E + P. \tag{4.46}$$

It can be observed that the permittivity-dependent electric field density Eqs. (4.44) and (4.46), for semiconductors and metals, respectively, have similar relations. Both equations are dependent on polarization. In the case of a semiconductor, it is obtained by the electron dynamics between the conduction and valence bands, depending on the Pauli exclusion principle, the state-filling effect, and Fermi–Dirac thermalization. In the case of a metal, the relative permittivity is dependent on the resonance frequency and relaxation time of free and bound electrons (the Drude–Lorentz model). The formulation solves a similar problem but in two different ways. This is because in metals conduction and valence bands overlap, whereas in the case of semiconductors, they are separated due to the band gap in between them. This has resulted in different simulation approaches. Although the semiconductor part of the proposed method contains essential carrier dynamics, more features such as noise due to spontaneous emission, carrier diffusion, and carrier heating and cooling effects could be added to the solid-state material model. For the LD model, a greater number of oscillators can be included, depending on the range and accuracy requirements in the optical spectrum. For simplicity, we used one oscillator, which is enough for the applications studied. In Sections 4.2.3 and 4.2.4 different applications are simulated with the developed active model.

4.2.3 Simulation of an MSM waveguide and a microcavity

For numerical results, a metal–semiconductor–metal (MSM) waveguide and a microcavity are simulated. The MSM structure is shown in Fig. 4.2. It consists of a gallium arsenide (GaAs) slab, which is sandwiched between two parallel gold (Au) plates. In the structure, the length of the semiconductor slab is 4 μm. Its thickness is 50 nm, while the width is 100 nm. The thickness of each gold plate is 100 nm, whereas the length and width are the same as those of the semiconductor slab. For the semiconductor, the effective masses of electrons and holes in the conduction and valence bands are $0.047 m_e$ and $0.36 m_e$, respectively, where m_e is the mass of a free electron. The refractive index of the semiconductor layer is 3.54. The carrier density is 3×10^{22} m^{-3}. Ten

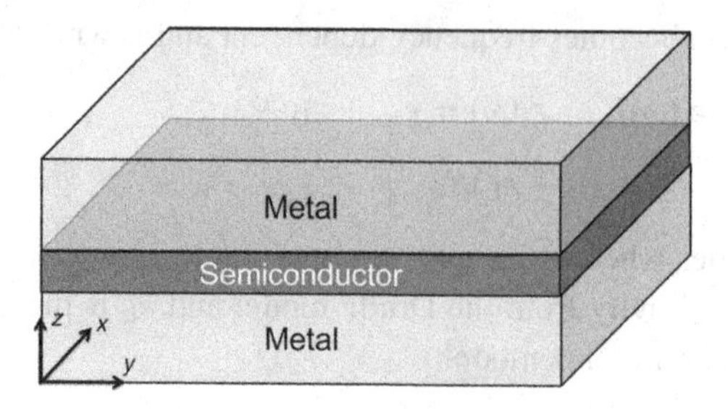

Figure 4.2 A semiconductor slab sandwiched between two parallel gold plates.

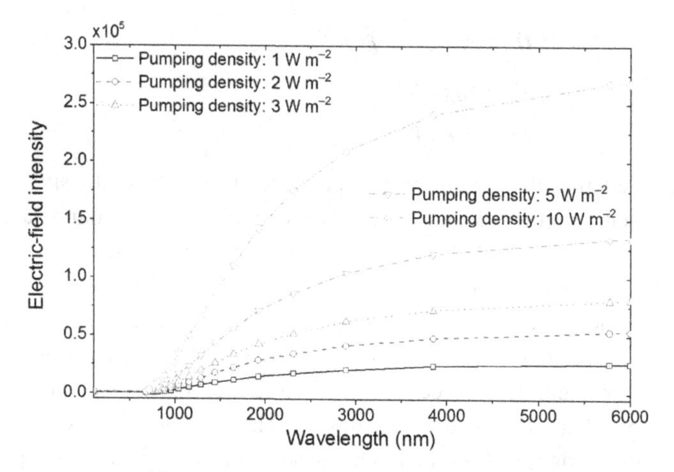

Figure 4.3 Field intensity with respect to wavelength, at different pumping densities.

energy levels are used both for the conduction band and for the valence band. The transition-time parameters are taken from Ref. [22]. The parameters for the Lorentz–Drude model are $\omega_{pD} = 2\pi \times 1903.41 \times 10^{12}$ rad s^{-1}, $\Gamma_D = 2\pi \times 12.81 \times 10^{12}$ rad s^{-1}, $\Gamma_L = 2\pi \times 58.27 \times 10^{12}$ rad s^{-1}, $\varepsilon_\infty = 1$, and $\Delta\varepsilon_L = 0.024$, and are taken from Ref. [12]. A beam of light with Gaussian profile and a wavelength of 800 nm is injected as a source at the center of the semiconductor slab. The field propagates equally towards both ends of the slab.

The electric-field intensity increases with increasing pumping density, and vice versa. Owing to the plasmonic effect, the field at the interface of a metal and a semiconductor is enhanced at optical frequencies. This field enhancement creates more electron–hole pairs in the semiconductor, which increases the propagation distance of the field.

An approach employed to enhance the propagation distance by incorporating quantum dots into the dielectric medium is studied in [5] for splitters and interferometers. Figure 4.3 illustrates the electric-field intensity with respect to wavelength at different pumping densities. It is observed that the cutoff wavelength is around 800 nm. This shows that at higher wavelengths, depending on the pumping field density, there is no further change in the field intensity and curves almost become horizontal; in other words, the semiconductor medium saturates and allows smooth and maximum field transmission. Figure 4.3 also shows that at higher pumping densities there is a shift in

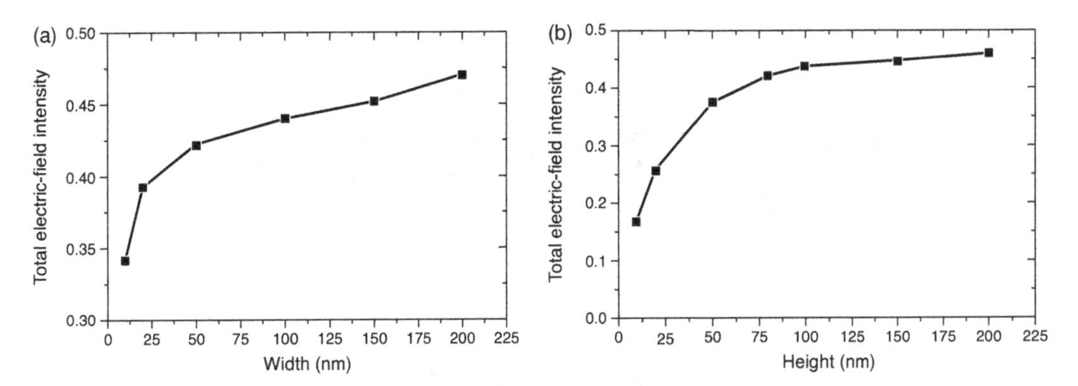

Figure 4.4 The total electric-field intensity for the MSM waveguide for different widths and heights of slab: (a) the electric-field intensity with respect to the width of the slab and (b) the electric-field intensity with respect to the height of the slab.

the cutoff wavelength. The reason is that at higher pumping density the carrier level inside the semiconductor increases due to the band-filling effect. As a result, there is a change in the refractive index, causing a shift in the wavelength.

For further analysis, the width and height of the semiconductor slab are varied. Figure 4.4 illustrates the effect of different widths and heights of the semiconductor slab on the propagating field that is observed at a distance of 1 μm away from the source. The pumping density for this analysis is 1 W m^{-2}. Figure 4.4(a) shows that the total electric-field intensity increases with increasing slab width, while the height is kept constant (100 nm). For example, when the slab width is 200 nm, the loss in the field intensity at a distance of 1 μm from the source is 3% (the actual signal is 0.5, because the source is equally transmitting in both directions, so the field intensity at the observation point is 0.47). It is observed that, as the width increases, the rate of increase in field intensity becomes smaller. A further increase in width may cause other higher modes to be excited and propagate. This may affect the field intensity at the observation point and, as a result, the propagation distance. Figure 4.4(b) illustrates the total electric-field intensity for different heights of the slab, while the width is kept constant (100 nm).

The total electric-field intensity increases with increasing height. For example, when the slab height is 100 nm, the signal loss is 3.91%, after traveling the same distance as in Fig. 4.4(a). When the slab height is increased to more than 100 nm, the increase in electric-field intensity is not significant even with a further increase in height. Therefore 100 nm is an appropriate thickness for the application.

In the second example, a microcavity with radius $R = 700$ nm and thickness $t =$ 196 nm is simulated using the solid-state model. The layout of the cavity is shown in Fig. 4.5(a). Figure 4.5(b) shows the field distribution in a conventional semiconductor resonator, and depicts a symmetrical field distribution pattern along the E_z component. Later, we added gold nanoparticles in the semiconductor microcavity, and used an active simulation approach with a different orientation of the nanoparticles. In Fig. 4.5(c), two cylindrical gold nanoparticles are embedded into the microcavity. Each nanoparticle has a radius of 140 nm and a thickness of 224 nm. It can be seen that the electric field

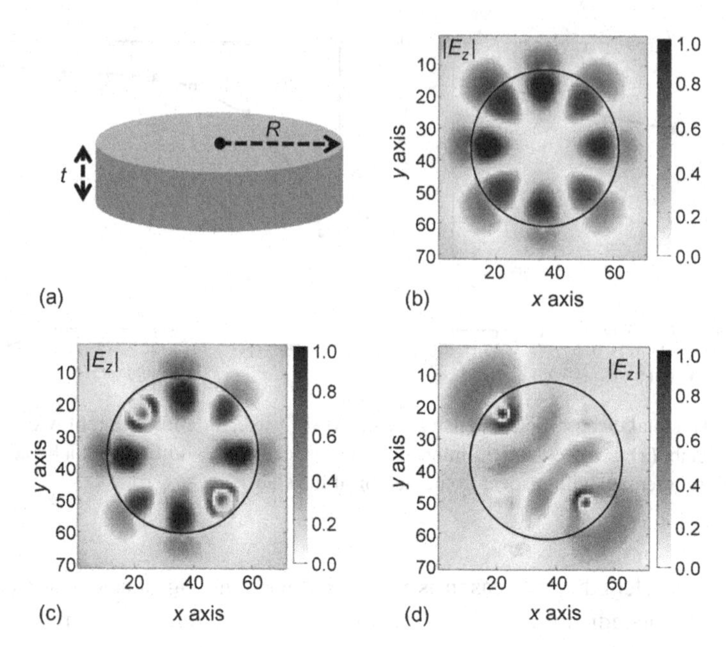

Figure 4.5 Structure and field patterns for a microcavity: (a) three-dimensional view, (b) electric field distribution in the conventional microcavity, (c) electric field distribution in a microcavity with two cylindrically shaped gold nanoparticles, and (d) electric field distribution with four cylindrical gold nanoparticles embedded inside the microcavity.

around nanoparticles is confined, and the symmetry of the field distribution is broken. Figure 4.5(d) shows the field distribution with four cylindrically shaped nanoparticles embedded inside the microcavity, with field oscillation along two nanoparticles only and the field suppressed in other directions. This arrangement demonstrates a dipole antenna-like behavior of the microcavity. Figure 4.6 shows that, by embedding the gold nanoparticles in the microcavity, the resonance frequency can be shifted. The resonance wavelength with two nanoparticles is 640.2 nm, while it is 661.20 nm with four nanoparticles, and without nanoparticles it is 642.8 nm. The results show that, by using different combinations of gold nanoparticles inside the semiconductor microcavity, it is possible to manipulate the resonance frequency to the desired value.

4.2.4 SPP generation using an MSM microdisk

Usually a surface plasmon source is obtained by using a light source from free space, which is then coupled to plasmonic devices. However, it is known that the surface plasmon mode momentum, $\hbar\mathbf{k}_S$, is greater than the free-space photon momentum $\hbar\mathbf{k}_0$ at the same frequency. Therefore, a change of momentum is required in order to couple light from free space to surface plasmons at a metal–dielectric interface. Three main techniques are commonly used to achieve this coupling: using prisms, scattering, and gratings. Monochromatic lasers are commonly used light sources for these experiments. It is beneficial to have a surface plasmon source similar to lasers that can be coupled directly to waveguides.

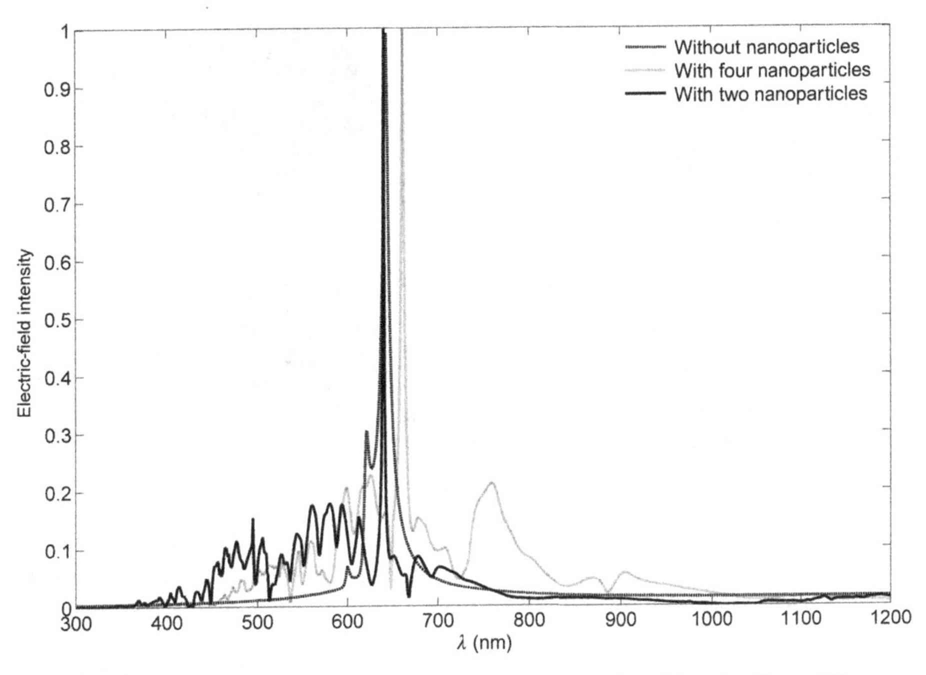

Figure 4.6 Resonance wavelengths for a semiconductor microcavity with and without different combinations of gold nanoparticles.

In 2003, Bergman and Stockman showed that stimulated amplification of a plasmon mode can be achieved [23]. They calculated the gain of the stimulated amplification emission when a metal structure was embedded inside a gain medium. In this way, the local field was excited using surface plasmon amplification by stimulated emission of radiation (spaser). Akimov *et al.* showed that the light from a quantum dot can be coupled to the surface plasmon of the metallic nanowires [24]. Koller *et al.* successfully demonstrated a surface plasmon emitter that is based on organic light-emitting diodes and is useful for organic devices [25]. Walters *et al.* proposed a silicon-based electrical source for an SPP [26]. The surface plasmon was excited by using an impact ionization process due to the high electric field of the metal layers, and the surface plasmon was detected by scattering out the wave using nanoslits.

In this section an active device that generates an SPP is presented. The active device can be coupled directly to plasmonic devices, and it eliminates the complexity of coupling loss from free space to plasmonic devices. The active device for the surface plasmon source is based on a microdisk laser, and the structure is shown in Fig. 4.7. Light is reflected internally as it travels around the microdisk, where its modes allow only certain frequencies to survive and the circular shape acts as a resonator. It is well known that light waves in a microdisk travel in a whispering-gallery-mode (WGM) fashion. The main advantage of this structure is that a cavity structure is utilized to enhance the SPP intensity, and that its cross-section is similar to that of a metal–insulator–metal (MIM) or MSM plasmonic waveguide (as shown in Figs. 4.7(c) and (d)). This enables one to

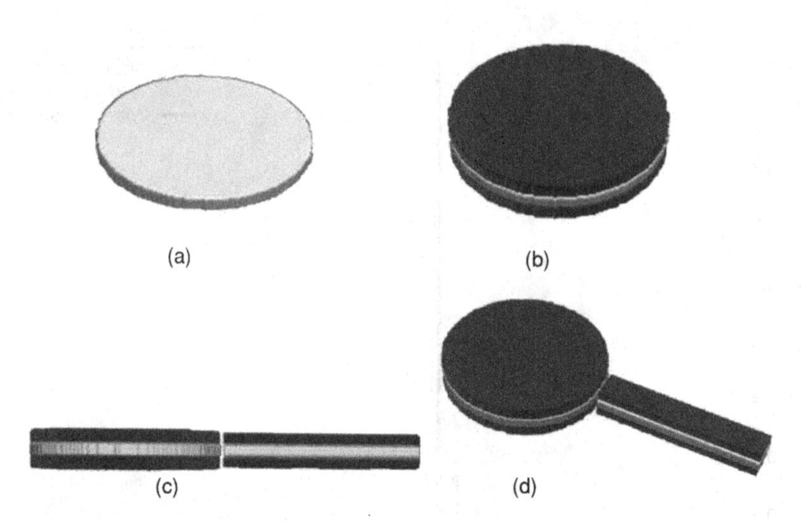

Figure 4.7 Dark regions indicate metals and light regions indicate dielectric or semiconductor. (a) A conventional microdisk laser. (b) A plasmonic microdisk with metal on the top and bottom of the gain medium. (c) A side view of the proposed structure, a plasmonic microdisk coupled to a plasmonic waveguide. (d) A bird's-eye view of the proposed structure.

efficiently couple the SPP from the plasmonic microdisk to the MIM/MSM plasmonic waveguide.

The plasmonic microdisk structure is studied by using the full three-dimensional (3D) solid-state and LD dispersive models [16] (elaborated in the previous subsections). Most of the previous studies were not able to take into account the electron dynamics in the semiconductor media and consider the semiconductor layer as just a simple dielectric. A 3D simulation is required since the light wave would normally travel in the x–y plane through a WGM while the metal confinement is in the z direction. The hybridized surface plasmon mode can be extracted efficiently to the MIM or MSM waveguide. The high extraction efficiency is mainly caused by the gap between the plasmonic microdisk and the plasmonic waveguide. The effect of varying gap distance between the plasmonic microdisk and the waveguide is also studied. A microdisk of radius 1 μm and thickness 120 nm is considered, as shown in Figs. 4.7(a)–(d). The microdisk is made of a III–V semiconductor material with a band-gap wavelength of 1.55 μm and a refractive index of 3.5. The microdisk is sandwiched between two metal layers (gold) with the same radius and thickness. Next to the plasmonic microdisk, as an example, one places a metal–insulator–metal (MIM) waveguide with a nano-sized gap. For simplicity, air is used as insulator in the MIM waveguide. The thicknesses of all layers in the MIM waveguide are the same as for the layers in the plasmonic microdisk. The width of the waveguide is 360 nm, while the length is 2 μm. The fabrication of such a structure is feasible with current technology. A similar structure of a disk sandwiched between metal layers has previously been fabricated by using physical vapor deposition and a focused ion beam to etch the disk [27], and resonant modes of $Ag/SiO_2/Ag$ disks have been studied. Another experimental design of a plasmonic resonator has also been reported in [28], in which a laser design based on half-circle disk resonators is covered with metal.

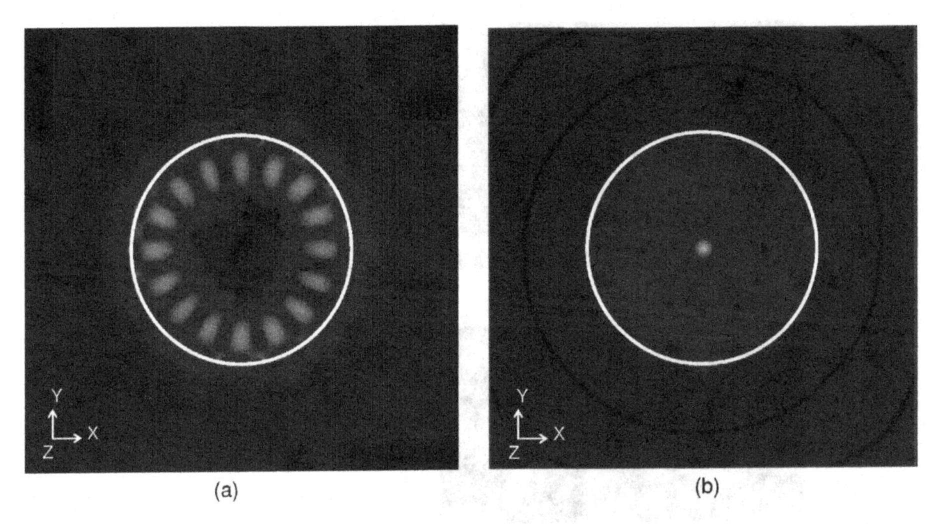

Figure 4.8 Near-field patterns of (a) a conventional microdisk and (b) a plasmonic microdisk. The total electric field is on a logarithmic scale. The field distribution is taken at the center plane of the semiconductor material. The white circle denotes the microdisk edges.

To model the semiconductor medium accurately with the solid-state model, it is recommended that one should use more energy levels in the conduction and valence band for electron dynamics. In the multilevel multi-electron model, the conduction- and valence-band states are divided into several groups. In this example, five energy levels are taken into account within each band, and are enough for the application. The energy-level separation is $\Delta\lambda = 25\,\text{nm}$, and the pumping rate into the microdisk is 3×10^9 electrons per second. This methodology allows us to take into account the electron transitions for stimulated emissions in the system. Therefore, this model is able to simulate lasing effects in a microdisk structure. To model the gold part of the structure, the Lorentz–Drude dispersive model is used, and the parameters are the same as in Section 4.2.1.

The microdisk is first considered without the waveguide. The simulation is run until the field inside the microdisk reaches a steady state. The total electric field is measured and captured at the middle plane of the semiconductor materials. This allows us to observe the electric field pattern inside the microdisk. The electric field pattern of a plasmonic microdisk is different from that of a conventional microdisk laser. The comparison of total electric field distributions in a conventional microdisk laser and a plasmonic microdisk is shown in Fig. 4.8. The white circle denotes the microdisk boundary. The WGM of the conventional microdisk laser is observed in Fig. 4.8(a). On the other hand, Fig. 4.8(b) shows that the total electric field pattern is changed significantly. The WGM is hybridized with the SPP mode inside the plasmonic microdisk. Calculation shows that the direction of the dominant field changes from the radial direction (TE mode) in a conventional microdisk to the z (normal to the disk plane) direction in a plasmonic microdisk. In Refs. [29, 30], a similar structure has been studied, but the work was limited to the loss and the Q factor of the device.

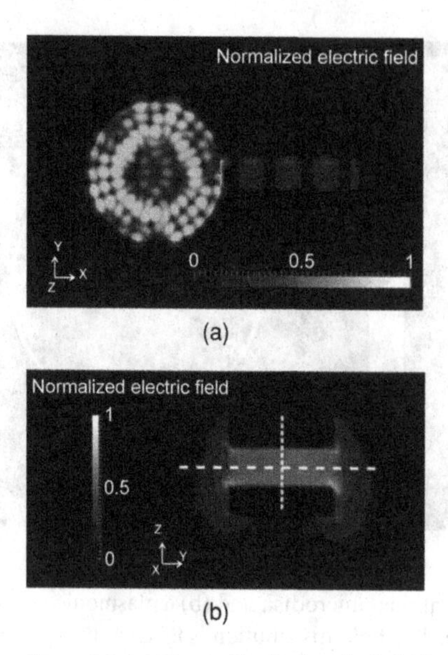

Figure 4.9 (a) Normalized electric field in the x–y plane. (b) Normalized electric field inside the plasmonic waveguide in the y–z plane.

To investigate this hybridization further, the electric field across the MSM layer is measured. The electromagnetic field for a plasmonic microdisk is no longer a WGM but rather is hybridized with the SPP mode. A surface plasmon mode would have a maximum intensity at the interface of semiconductor and metal, and the field decays exponentially inside the metal.

The hybridized mode confines the field nearer to the interface between the dielectric and the metal. The mode observed here seems to be a hybridization between the WGM and the SPP. A pure SPP would have the maximum field intensity at the interface. Such a mode is observed inside the waveguide after the hybrid mode has been coupled to the waveguide. Another observation is that attaching metal layers results in a significant enhancement of E_z. At a wavelength of 1.47 μm, the intensity enhancement is by a factor of about 20 000. Similar results have been reported by Perahia *et al.* [31] with a single-metal-layer microdisk laser studied experimentally. They have shown that the WGMs start to hybridize with the SPP modes as the metal layer's area increases. In our case, it is observed that the SPP waves at the top and bottom layers coupled together.

For SPP mode extraction, a metal–insulator–metal (MIM) plasmonic waveguide is placed near the microdisk. The mode is extracted to the waveguide through the nanometer gap. Figure 4.9(a) shows the total electric field distribution when the SPP wave couples to the waveguide. The field distribution is taken at the middle plane, which corresponds to the center of the semiconductor layer inside the microdisk or the center of the insulator inside the waveguide. The field intensity inside the plasmonic microdisk is higher than the intensity inside the waveguide, as expected. The microcavity builds up the field intensity and competes with the losses until it reaches steady state. At steady state, we observe that the field inside the waveguide propagates from the left end of

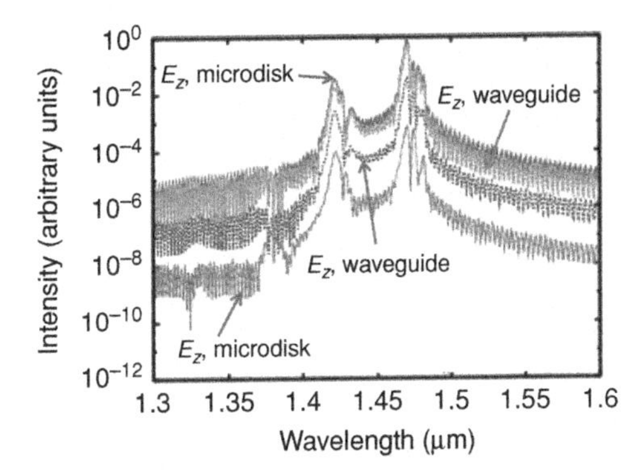

Figure 4.10 The normalized electric-field intensity as a function of wavelength both for the microdisk and for the MIM waveguide. Note that E_z is the dominant mode.

the waveguide down to the far end. Figure 4.9(b) shows the electric field inside the waveguide in the y–z plane.

Figure 4.10 shows the fast Fourier transform of the electric field both for the plasmonic microdisk and for the plasmonic waveguide on a logarithmic scale. The fields are measured at the center plane of the semiconductor inside the microdisk and at the center plane of the dielectric layer inside the waveguide. The observation point inside the microdisk is located about 20 nm from the gap adjacent to the waveguide. On the other hand, the probe for the field of the waveguide is located about 20 nm from the gap inside the waveguide. It can be observed that the z component of the electric field inside the waveguide follows that of the plasmonic microdisk for the wavelengths of interest from 1.3 μm to 1.6 μm.

The coupling efficiency is calculated by taking the Fourier transform of the square of the electric field, which gives the energy. Energy is computed at both the microdisk and the waveguide at the centre of the semiconductor layer. It was calculated that the intensity at the waveguide is about 60% of the intensity inside the plasmonic microdisk. This field-extraction result is considered efficient for a photonic integrated-circuit system. As a rough comparison with a previous work on conventional microdisk lasers, the light-extraction efficiency of an elliptical microdisk laser was successfully increased from 9.2% to about 30% by using an external magnetic field [9].

In addition, the effect of varying the gap distance between the microdisk and the waveguides on the extraction efficiency is also studied. When varying the gap distance, the other parameters are kept constant. The gap distance between the plasmonic microdisk and the waveguide is varied from 20 nm to 120 nm, with an interval of 20 nm as shown in Fig. 4.11. The plot indicates that the extraction efficiency deteriorates as the gap distance increases. It is expected that the field enhancement between two metal structures increases as the gap distance decreases, and vice versa. A similar phenomenon is observed here. As the gap distance between the microdisk and the waveguide decreases, the field enhancement in the gap increases. As a result, a higher extraction efficiency is obtained.

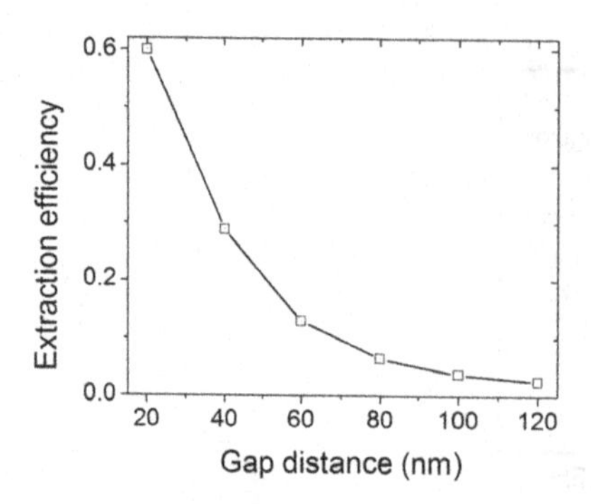

Figure 4.11 The extraction efficiency reaches 60% for a gap distance of 20 nm, and the efficiency decreases with increasing gap distance.

4.3 Surface plasmon generation in semiconductor devices

Many plasmonic devices such as multiplexers, plasmonic lasers, and waveguide couplers using subwavelength field confinement have emerged. However, this confinement has high optical loss, and the short lifetime of the SPPs limits its usefulness. It would be interesting to investigate other possible ways in which high optical loss in metallic materials can be overcome by embedding them in low-loss materials. Therefore, it is important to understand the reason for using metals for SPPs. Theoretically, SPPs are excited on a metal surface due to the presence of a large amount of electrons. This large amount of electrons accumulates as an electron plasma, which interacts with incident photons from a light source, forming SPPs. This means that, if there are enough electrons to create the plasma cloud, SPPs could possibly be generated. This understanding provides the motivation to explore the possibility of excitation of SPPs in semiconductor devices and microcavities with lower radiation loss. The surface mode which is being excited in such semiconductor devices would be referred to as a surface plasmon (SP) mode in order to differentiate it from the usual excitation in a metal/dielectric interface. As an example, a case of a semiconductor microcavity is studied.

To obtain and control the excitation of the SP mode, and to manipulate the electron path in the semiconductor microcavity, an external magnetic field is applied. The advantages of this technique are as follows:

(1) It provides different degrees of freedom to control the excitation of the SP mode in the microcavity. By controlling the strength of the external magnetic field and the pumping rate into the microcavity, the SP mode can be switched "on" or "off" logically.
(2) This is a unique and direct technique for the excitation of the SP mode in a semiconductor microcavity laser. The laser source generated in the microcavity by WGM interacts directly with the electron plasma accumulated on the surface of the microcavity to excite the SP mode. This reduces coupling loss, increases compactness, and reduces the system size required in order to excite the SP mode. This excitation

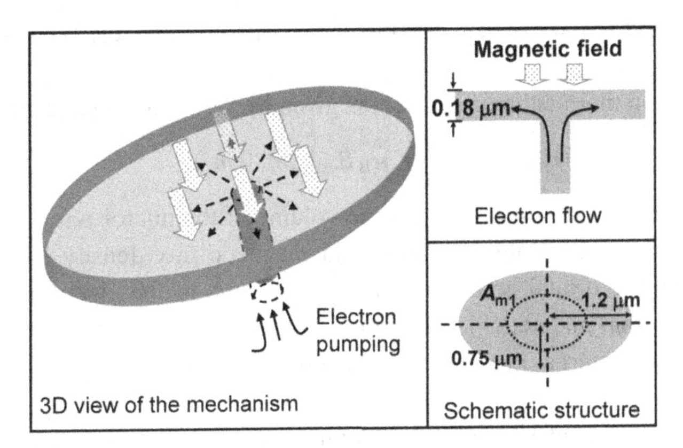

Figure 4.12 (a) A schematic 3D view of the microcavity under study, showing the confinement of the external magnetic field in a small area and also the direction in which electrons move after emerging from the supporting pedestal. (b) A side view of the layout that also shows the electron flow. (c) A top view of the elliptical microcavity, where the dotted elliptical area inside the main ellipse shows the external magnetic field confinement area A_{m1}, and the dimensions of the elliptical microcavity.

method is fundamentally different from conventional methods that use gratings or glass prisms to excite the SPPs.

(3) The fabrication process would be cheaper and simpler because of the availability of already well-matured processing and fabrication technologies for semiconductor materials. The proposed semiconductor microcavity is made of GaAs, which is commonly used to fabricate semiconductor microdisk lasers.

A 3D view of the microcavity under study is shown in Fig. 4.12(a). It is an elliptical microcavity supported by a pedestal. The pedestal acts as a supporting pillar. It also acts as a medium to transfer electrons from an external pumping source to the microcavity in order to achieve population inversion for lasing. Figure 4.12(b) shows the electron flow and a side view of the structure. The refractive index and thickness of semiconductor microcavity is 3.45 and 180 nm, respectively. The dimensions of the semi-major and semi-minor axes of the elliptical microcavity are 1.2 μm and 0.75 μm, respectively, as shown in Fig. 4.12(c).

The solid-state model in Section 4.2.2 is applied to simulate the structure. Twenty energy levels in the conduction and valence bands with an energy-level separation of $\Delta\lambda = 25$ nm are used in the model. The discrete transition energies are given by the equation $E_i = 0.8 - 0.129i$, where i refers to the ith energy level and $i = 1, 2, 3, \ldots, 20$. The atomic parameters such as effective masses of electrons and holes, band levels, and intraband and interband transition times take the values given in Refs. [8, 32]. During simulation, the microcavity is electrically pumped at the rate of 8×10^9 s^{-1}. A short optical pulse with Gaussian profile and centered at $\lambda = 1.55$ μm is injected into the microcavity to initiate lasing at temperature $T = 350$ K. The elliptical microcavity forms the WGM and the active semiconductor material provides the laser beam with electron pumping.

The magnetization momentum coefficient is given by the following equation:

$$M_{\mathrm{orb/spin}} \propto N_0 V \mu_{\mathrm{B}} B_{\mathrm{extz}} (m_l + m_s), \tag{4.47}$$

where N_0 is the electron density and V is the volume of the microcavity. The parameters μ_{B} and B_{extz} are the Bohr magneton and magnetic flux density. The symbols m_l and m_s denote the orbital and spin quantum numbers of the electrons under the influence of an external magnetic field. For metallic structures or semiconductors with pumping current, the external magnetic field will have more influence on the electrons' properties.

The energy levels will change with the application of an external magnetic field. The energy level in the elliptical microcavity is now described by the Landau principle in terms of the Landau energy level. The equation for the new energy level under the application of an external magnetic field can be found in many references (for example Ref. 9). The final electric and magnetic field equations in the z direction are given as

$$\begin{aligned} H_z|^{n+1/2}_{i+1/2,j+1/2,k} &= H_z|^{n-1/2}_{i+1/2,j+1/2,k} + \frac{\Delta t}{\mu_0\,\Delta y}\left(E_x|^{n}_{i+1/2,j+1,k} - E_x|^{n}_{i+1/2,j,k}\right) \\ &\quad - \frac{\Delta t}{\mu_0\,\Delta x}\left(E_y|^{n}_{i+1,j+1/2,k} - E_y|^{n}_{i,j+1/2,k}\right) \\ &\quad - \frac{1}{\mu_0}\sum^{m_l,m_s}\left(M_z|^{n+1/2}_{i+1/2,j+1/2,k} - M_z|^{n-1/2}_{i+1/2,j+1/2,k}\right), \end{aligned} \tag{4.48a}$$

$$\begin{aligned} E_x|^{n+1}_{i+1/2,j,k} &= E_x|^{n}_{i+1/2,j,k} + \frac{\Delta t}{\varepsilon\,\Delta y}\left(H_z|^{n+1/2}_{i+1/2,j+1/2,k} - H_z|^{n+1/2}_{i+1/2,j-1/2,k}\right) \\ &\quad - \frac{\Delta t}{\varepsilon\,\Delta z}\left(H_y|^{n+1/2}_{i+1/2,j,k+1/2} - H_y|^{n+1/2}_{i+1/2,j,k-1/2}\right) \\ &\quad - \frac{J}{qN_0}\left(\frac{B_{\mathrm{extz}}|^{n+1}_{i+1/2,j,k} - B_{\mathrm{extz}}|^{n}_{i+1/2,j,k}}{2}\right) \\ &\quad - \frac{1}{\varepsilon_0}\sum_{i=1}^{N_L}\left(P_{i,x}|^{n+1}_{i+1/2,j,k} - P_{i,x}|^{n}_{i+1/2,j,k}\right), \end{aligned} \tag{4.48b}$$

where M is the coefficient of magnetization momentum contributed by the presence of spin and orbital momentum of the electrons in the presence of the external field.

The electron-transport mechanism is included in the solid-state model and is applied to simulate the microcavity. In the absence of an external magnetic field, electrons travel through the supporting pedestal and diffuse outward from the centre of the microcavity as shown in Figs. 4.12(a) and (b). The microcavity is considered thin enough for one to neglect the vertical diffusion of electrons. Hence, only the planar diffusion of electrons in the x–y plane is considered. From the center of the elliptical microcavity electrons spread outward in a radial pattern (Fig. 4.12(a)). The diffusion current flow can be expressed in terms of a linear combination in the x–y plane, and is given as [33–35]

$$J_{\mathrm{total}} = q\left[\frac{D_{\mathrm{n}}\,\mathrm{d}n - D_{\mathrm{p}}\,\mathrm{d}p}{\mathrm{d}x} + \frac{D_{\mathrm{n}}\,\mathrm{d}n - D_{\mathrm{p}}\,\mathrm{d}p}{\mathrm{d}y} + \left(\frac{\mathrm{d}n^2}{\mathrm{d}x\,\mathrm{d}y}D_{\mathrm{n}} - D_{\mathrm{p}}\right)\right]. \tag{4.49}$$

This equation shows that the radial diffusion of the electrons from the center is dependent on the diffusion constants of holes, D_{p}, and electrons, D_{n}. The total current flow is also dependent on the electron density, n, and hole density, p. For electron pumping, the presence of holes can be neglected because the density of electrons would be much greater than that of holes.

During the simulation of a semiconductor plasmonic structure (Fig. 4.12), an external magnetic field is applied to a small elliptical confined area on the elliptical microcavity. The external field is applied after achieving a steady-state situation in terms of the simulation time (stable lasing mode). The applied magnetic field produces Lorentz torque and "pushes" electrons to the edges of the microcavity. A strong electron plasma is built up at the interface of the microcavity and air (after accumulation of more electrons). The light source interacts with the electron plasma at the boundary of the microcavity and excites the surface plasmon mode. This set-up provides both SP and laser modes, and may be a perfect source for nanophotonics chips that have both photonic and plasmonic components.

The use of nonmetallic materials such as conducting polymers and semiconductors for the excitations of SPPs has been reported in [36, 37]. That was achieved by changing their molecular properties or by using grating structures. These techniques provide alternate materials to excite SPPs, but do not give control over the excitation process. Moreover, high loss would be expected due to the use of chemical or grating structures for excitation. In order to control the excitation of SPPs, an external force would be required.

Figure 4.13(a) shows the electric field distribution (E_z) pattern in the microcavity at steady state. It shows the conventional WGM of the elliptical microcavity. With electron pumping, the WGM accumulates enough field intensity to become a laser mode. Figure 4.13(b) indicates that increasing the electron pumping rate introduces more electrons into the microcavity. The introduction of additional electrons by increasing the pumping rate increases the number of collision between electrons. However, this also results in loss of the WGM in the microcavity, and hence the laser mode is not properly formed in the elliptical microcavity. It is observed that some surface and internal modes are present. This is because higher electron density allows more electrons to reside on the edge of the microcavity, and this forms some modes on the interface. It is important to note that this is not a clear surface plasmon mode since the electrons are still undergoing random collisions with each other in the microcavity.

Since the diffusion rate of the electrons is very slow due to multiple collisions with the semiconductor lattice and other electrons, an external magnetic field is added to the system in order to increase the speed of movement of electrons. The addition of a magnetic field produces Lorentz torque [38, 39], which forces the electrons to move faster and in a specific direction. The intermolecular collision and slow diffusion

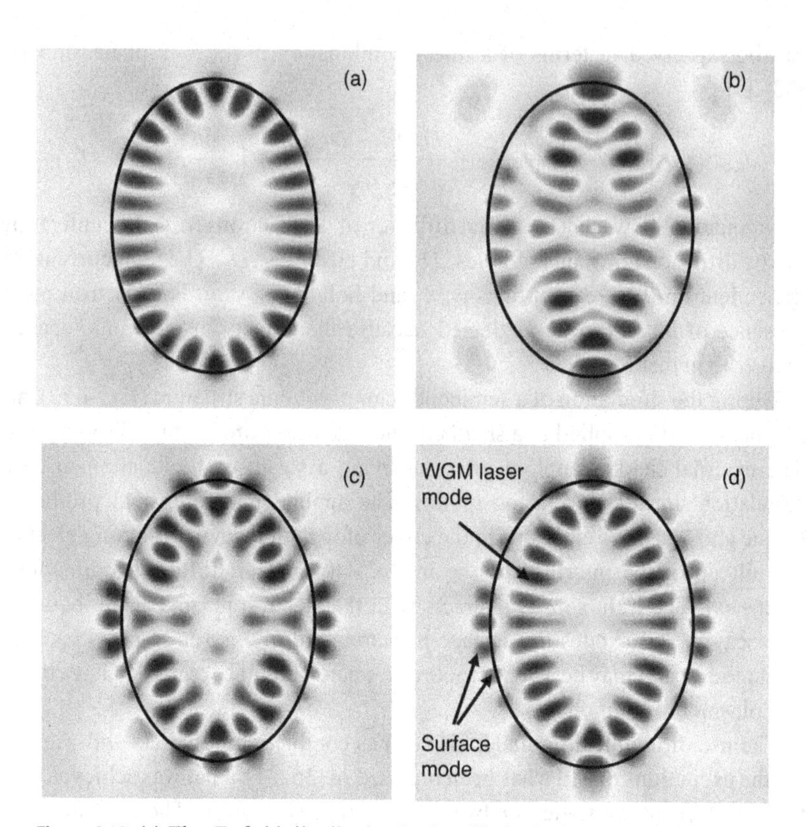

Figure 4.13 (a) The E_z field distribution in the elliptical microcavity with a pumping current rate of 6 kA s^{-1} without external magnetic field. This is the conventional WGM in the microcavity. (b) The E_z field distribution when the pumping rate increases to 15 kA s^{-1}. The distortion in the field distribution is due to the large amount of inter-electron collisions. (c) The E_z field distribution with an external magnetic field of 200 mT at 15 kA s^{-1}. (d) The E_z field distribution with an external magnetic field of 700 mT. The WGM laser and SPP modes are seen in the field diagram.

speed of electrons can be neglected since the magnitude of Lorentz torque is much stronger. The concept of an external magnetic field acting on electrons is analogous to the Hall effect [33–35]. Using the Fleming left-hand rule, the electrons will drift in the direction perpendicular to the electron flow and applied magnetic field. The change in the system of interaction between electrons and photons under the influence of a magnetic field has been reported in [9]. Under the influence of an external magnetic field, the energy level and energy interaction also change, and can be described by the Landau principle [40, 41]. However, the other conditions such as the parabolic band structure, optical source wavelength, and simulation temperature remain the same.

In [9], an external magnetic field is applied to the whole elliptical microcavity to create a large torque and to "push" the electrons to one side of the microcavity. This produces a redistribution of electrons and thus results in a strong electric field on one side of the elliptical microcavity. In the current application, the external magnetic field is applied in a small confined area labeled A_{ml} in Fig. 4.12(c). Owing to the external magnetic field, electrons are able to experience a torque perpendicular to the microcavity boundary. The resultant torque pushes the electrons equally to the elliptical boundary of the whole elliptical microcavity because of the confined external magnetic field.

The field distribution in the elliptical microcavity after applying a magnetic field of 200 mT is shown in Fig. 4.13(c). This indicates that the field distribution between the laser mode and the surface mode is not clearly defined. This is so because the magnetic field is still too weak to produce a strong torque on the electrons and "push" them to the surface of the microcavity.

When an external magnetic field of strength 700 mT is applied, two distinctive modes are observed in the field distribution, as shown in Fig. 4.13(d). One of them corresponds to the laser mode, and the other is the surface mode. To prove that the surface mode is an SP mode and not the conventional Sommerfeld–Zenneck [39, 40] surface mode, three conditions have to be met.

(1) The real part of the effective permittivity must be negative.
(2) The plasmon dispersion must be close to the light-line to ensure that the photons are coupling to the surface to form a surface plasmon.
(3) There must be a sufficient amount of electrons to interact with the light photons to form a surface plasmon.

These three conditions are satisfied with metallic materials. However, it has been found that semiconductor materials can also satisfy these conditions.

It is observed that the electron density is much higher at the boundary than in the inner part of the microcavity. The electrons were pushed to the boundaries of the microcavity. This accumulation of a large electron density forms electron plasmas on the boundary of the elliptical microcavity. At the same time, the laser mode in the microcavity also reappears. The effects studied in this section with different strengths of external magnetic field [42, 43] are of interest, and modeling and simulation of an elliptical microcavity is ideal with a large number of electrons. However, one potential problem that can arise with a high electron density in practical devices is that they can suffer from device breakdown or even meltdown due to overwhelming electron density at the edges of an elliptical microcavity [44, 45]. To avoid this situation, a pumping rate below 20 kA s^{-1} is recommended. For pumping rates lower than 20 kA s^{-1}, the SP mode can still be produced in semiconductor active devices, as demonstrated in Figs. 4.13(c) and (d).

4.4 Implementation of the LD model on a GPU

The popularity of the FDTD method is attributed to its simplicity, accuracy, robustness, capability to treat nonlinear behavior naturally, and capability to provide a real-time visualization response. The method has been applied to many areas, such as electromagnetics, electrodynamics, photonics, radiofrequency/microwaves, biosensors, plasmonics, and nanotechnology. With all these flexibilities and capabilities, the FDTD method requires very long simulation times for electrically large and fine-mesh-requiring structures. To enhance the simulation speed, in addition to different parallel processing techniques, recently two hardware accelerator approaches have been proposed: (i) field-programmable gate arrays (FPGAs) [46] and (ii) a graphics processing unit (GPU) [47].

In this chapter, the GPU approach is adopted. This approach is favored over the use of FPGAs due to the lower cost and its prevalent availability in mainstream computers. With easy access to this resource, it is relatively easy to test and implement different numerical techniques. The initial implementation of a GPU for numerical computation was tedious and time-consuming, primarily because the initial design of a GPU was only for graphics applications. In 2006, the CUDA technology was introduced by Nvidia, which made it easier to learn how to program and utilize GPUs. This new concept supports FDTD-type algorithms, which have natural characteristics of parallelization allowing them to run faster and accurately. The GPU-accelerated FDTD method has been applied to different applications. In [48], the two-dimensional (2D) FDTD method is implemented on a GPU for dispersive media using a single-pole Debye model with a piecewise linear recursive convolution (PLRC) method for microwave applications. In [49], the 3D FDTD method is implemented on a GPU for low- and mid-frequency acoustics applications. In [50], 2D FDTD using the Drude model is implemented on a GPU for double-negative (DNG) materials. An optimization of the FDTD method for computation on heterogeneous and GPU clusters is presented in [51]. Similarly to the FDTD method, the finite-difference frequency-domain (FDFD) method is also implemented on a GPU for electromagnetic scattering applications [52].

In this chapter, the FDTD method is implemented on a GPU for plasmonics applications. As an example, a nanosphere is studied. Numerical results obtained by using a central processing unit (CPU) are compared with those obtained by using a GPU. In addition to the GPU, we developed the same algorithm on MATLAB and C++, which execute on modern CPUs. We then ran representative numerical experiments for each version and compared the simulation speed performance throughput across all the implementations. We also ran additional tests to compare the accuracy of the CPU vs. GPU implementations so as to ensure that accuracy is not compromised. These developed codes are based on the FDTD method and can be utilized for the development of plasmonics applications.

At optical frequencies, numerous metals have a dispersive nature. In order to model them accurately, different dispersive models have been incorporated into Maxwell's equations. Most of the available dispersive models are in the frequency domain, so, to make them consistent with time-domain methods, different approaches such as recursive convolution (RC), piecewise linear recursive convolution (PLRC), use of the z-transform, and use of the auxiliary differential equation (ADE) [14] are used. PLRC and use of the ADE are the most commonly used approaches due to their accuracy and efficiency. PLRC is an integral approach, and numerical convolution is needed for its implementation, whereas use of the ADE is a differential approach. We use the Lorentz–Drude dispersive model, which has already been explained well in Section 4.2.1, and how to implement it on a GPU is explained in the next subsection and is also elaborated upon in [20].

4.4.1 GPU implementation

The 3D FDTD method is implemented on a GPU for plasmonics applications. For the simulation of plasmonic devices, the Lorentz–Drude (LD) dispersive model is incorporated into Maxwell's equations, while the ADE technique is applied to the LD model. Numerical simulation experiments based on typical domain sizes as well as a plasmonics environment demonstrate that our implementation of the FDTD method on a GPU offers significant speed-up compared with traditional CPU implementations.

The current generation of GPU has hundreds of processing cores. These processing cores are grouped into multiprocessors. Each multiprocessor contains 8–32 processing cores. For example, the Nvidia Tesla C2050 has 14 multiprocessors, each of 32 processing cores. In other words, this GPU has 448 processing cores. In order for a GPU to be utilized efficiently, thousands of processing threads have to be executed. A huge number of threads is needed to mitigate the effect of threads being stalled due to memory-access latency.

These threads are created and organized at two levels. For each multiprocessor, a block of threads, ranging from 1 to 1024, is created and executed. It will make sense to have more threads than the number of processing cores in the multiprocessor, so that each processing core has at least one thread to execute and another thread to switch to when the current thread is stalled. At the next level, such blocks are created to execute on all the multiprocessors. Similarly, it will be advantageous to have more blocks than the number of multiprocessors.

The other factor that affects the utilization of the GPU is related to memory access. A GPU is packaged on a board with its own memory, known as device memory. For example, the Nvidia Tesla C2050 has 3 gigabytes of device memory. The provision of device memory allows the GPU to do computations without the intervention of the CPU. However, in order to maximize the memory-access throughput, it is important to maximize coalescing. Coalescing is a mechanism to reduce the number of memory-access transactions. For example, if a warp of 32 threads is requested for a sequential set of memory locations, this request can be coalesced into one memory-access transaction.

A small section, 64 kilobytes on the Nvidia Tesla C2050, of the device memory can be classified as constant memory. The constraint for constant memory is that it allows only read access. However, the advantage is that there is an 8-kilobyte cache on the Nvidia Tesla C2050 per multiprocessor to reduce the memory-access latency. One has 48 kilobytes of on-chip memory per multiprocessor on the Nvidia Tesla C2050, which is known as the shared memory. As mentioned earlier, there is a block of threads that is executed on each multiprocessor. However, each thread has its own memory space, which is not accessible from another thread. Therefore, in order to solve this problem, the shared memory is used to allow the threads to access a common pool of memory on each multiprocessor. Since the shared memory is on the same chip as the GPU, it is as fast as the register.

Now we will look at the code segments that are used to update the H_x and E_x fields as an example. At the same time, the use of memory, coalescing, constant memory, and shared memory collectively to improve the overall performance of the GPU

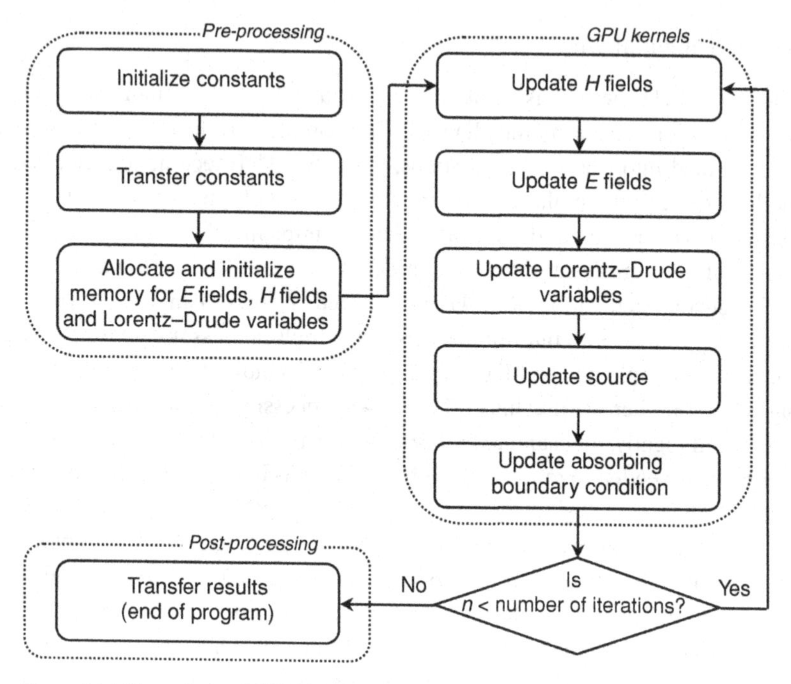

Figure 4.14 Flow chart of GPU implementation.

implementation is explained. Figure 4.14, shows the flow chart of the method, which consists of three parts: pre-processing, GPU kernel execution, and post-processing. Pre-processing, post-processing, and the conditional check are executed on the CPU, while the kernels are on the GPU.

Each thread updates the H_x equation on a unique $(i, j+\frac{1}{2}, k+\frac{1}{2})$ location. There is a loop in this thread to iterate through the range of k values without varying the i and j values. According to Eq. (4.17), in order to update $H_x(i, j+\frac{1}{2}, k+\frac{1}{2})$, there is one variable that is reused in the loop, another variable that is shared between two neighboring threads, and two constants that remain invariant when updating H_x for all the i, j, and k values. These two constants are stored in the constant memory, where there is cache to reduce the memory latency. The first two dependences are $E_y(i, j+\frac{1}{2}, k)$ and $E_y(i, j+\frac{1}{2}, k+1)$. Since each thread is iterating through the range of k values, $E_y(i, j+\frac{1}{2}, k+1)$ is also used in the next iteration. Therefore, it is beneficial to save the value in the register to be used in the next iteration.

The other two dependences are $E_z(i, j, k+\frac{1}{2})$ and $E_z(i, j+1, k+\frac{1}{2})$. Assuming that there is a block of 128 threads and accessing $E_z(i, j, k+\frac{1}{2})$ and $E_z(i, j+1, k+\frac{1}{2})$ as shown in Fig. 4.15, it is clear that neighboring threads are sharing the memory access. For example, thread 0 and thread 32, as highlighted in Fig. 4.15, are accessing the same E_z. Therefore, shared memory should be used to share the memory access between neighboring threads. Figure 4.15 also shows that the warps of 32 threads are accessing contiguous memory locations. Therefore, the memory-access request is coalesced into

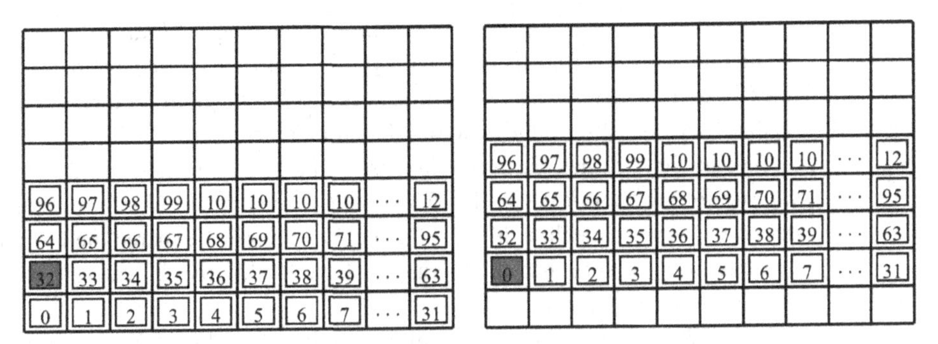

Figure 4.15 The display on the left shows the memory-access pattern when accessing $E_z(i, j, k + \frac{1}{2})$ and that on the right shows the memory-access pattern when accessing $E_z(i, j + 1, k + \frac{1}{2})$.

```
 1: __global__ void updateHx(double *hx, double *ey,
 2:                              double *ez, int
*material)
 3: {
 4:   double prevEy, currEy;
 5:   extern __shared__ double s_Ez[];
 6:
 7:   // Compute the increment and offset
 8:
 9:   prevEy = ey[index];
10:
11:   for (k=0; k<kEnd; k++)
12:   {
13:     currEy = ey[index+zIncr];
14:
15:     s_Ez[sIndex] = ez[index];
16:
17:     if (index < 32)
18:       s_Ez[sIndex+sIncr] = ez[index+yIncr];
19:
20:     __syncthreads();
21:
22:     hx[index] += c_chz[material[index]] *
23:                       (currEy - prevEy) -
24:                      c_chy[material[index]] *
25:                       (s_Ez[sIndex+offset] -
26:                        s_Ez[sIndex]);
27:
28:     index += zIncr;
29:     prevEy = currEy;
30:
31:     __syncthreads();
32:   }
```

Figure 4.16 The code segment for updating the H_x field component.

one memory-access transaction, and this reduces the memory-access latency. Figure 4.16 shows the code segment for updating the H_x field component.

On line 1, the "_global_" keyword is used to describe a function, also known as a kernel, that is executed on the GPU.

On line 5, the "_shared_" keyword is used to declare a variable in shared memory. The "extern" keyword is used to indicate that the size is specified at run-time when this function is executed.

On line 9, it loads the current $E_y(i, j + \frac{1}{2}, k)$, and it loads the next $E_y(i, j + \frac{1}{2}, k + 1)$ on line 13. It then stores the $E_y(i, j + \frac{1}{2}, k + 1)$ for the next iteration on line 29.

Line 11 describes a loop that iterates through the range of k values.

On line 15, each thread loads its E_z value into the shared memory. Subsequent lines 17 and 18 check whether the thread is the first block of 32 threads, then it loads the last block of 32 E_z values into the shared memory.

On line 20, one has the _syncthreads( ) function. This function acts as a barrier, waiting for all the threads in the block to reach this point before proceeding. This is necessary because the threads are using the E_z values in the shared memory. Therefore, one has to make sure that all the threads have loaded the E_z values into the shared memory before proceeding.

Lines 22–26 describe the computation to update the value of H_x. In this computation, the variables c_chy and c_chz are needed. Since these variables are read-only, they are stored in the constant memory. With the use of a constant memory cache, this helps to reduce the memory-access latency.

Finally, there is another _syncthreads( ) function on line 31. This is to ensure that all the threads in the block have finished their computation before proceeding. This is necessary because new values of E_z will be loaded into the shared memory in the next iteration. Figure 4.17 shows the code segment for updating the E_x field component and the Lorentz–Drude variables.

4.4.2 Applications

This section consists of two subsections, on numerical results and on the performance of various programming models and platforms for the FDTD method. The first is about the plasmonics application, and the accuracy of the method with and without GPU, while the second is about the performance efficiency of the GPU compared with the other programming models and platforms for the approach.

Numerical results

For numerical validation and assessing the accuracy of the method with and without the GPU for plasmonics applications, gold nanospheres of different radii are considered. Although different media around the sphere can be used, for simplicity in our application the surrounding medium is air, and a gold nanosphere with radius R is shown in Fig. 4.18. To truncate the free space around the nanosphere, Mur's absorbing boundary condition is used. The same cell size is considered in all three directions, i.e. $\Delta x = \Delta y = \Delta z = 1$ nm. The parameters used for the LD model are the same as given in [16], i.e. $\varepsilon_\infty = 5.9673$, $\omega_{\mathrm{PD}}/(2\pi) = 2113.6$ THz, $\Gamma_{\mathrm{D}}/(2\pi) = 15.92$ THz, $\omega_{\mathrm{PL}}/(2\pi) = 650.07$ THz, $\Gamma_{\mathrm{L}}/(2\pi) = 104.86$ THz, and $\Delta\varepsilon_{\mathrm{L}} = 1.09$.

The simulation was run until it reached a steady state. The electric field distribution inside and outside a 25-nm gold nanosphere in free space is depicted in Fig. 4.19. In Fig. 4.19(a), the field is obtained with the GPU, while in Fig. 4.19(b) it is obtained with the CPU. The electric field distributions in these two cases are similar. Figure 4.20 shows the electric-field intensity with respect to wavelength for different radii of nanospheres calculated with and without the GPU. With the changes in radii, there is a change in the resonance wavelength, but the simulation results with and without the GPU are in very

```
 1: __global__ void updateEx(double *oldEx, double
*newEx,
 2:                                 double *hy, double
*hz,
 3:                                 double *jx, double
*oldLx,
 4:                                 double *newLx, int
*material)
 5: {
 6:    double prevHy, currHy;
 7:    extern __shared__ double s_Hz[];
 8:
 9:    // Compute the increment and offset
10:
11:    prevHy = hy[index-zIncr];
12:
13:    for (k=1; k<kEnd; k++)
14:    {
15:      currHy = hy[index];
16:
17:      s_Hz[sIndex] = hz[index];
18:
19:      if (index < 32)
20:        s_Hz[sIndex-sIncr] = hz[index-yIncr];
21:
22:      __syncthreads();
23:
24:      newEx[index] = c_ca[material[index]] *
25:                        oldEx[index] +
26:                     c_cay[material[index]] *
27:                        (s_Hz[sIndex] -
28:                         s_Hz[sIndex-offset]) -
29:                     c_caz[material[index]] *
30:                        (currHy - prevHy) -
31:                     c_cad[material[index]] *
32:                        jx[index] -
33:                     c_caL1[material[index]] *
34:                        oldLx[index] +
35:                     c_caL2[material[index]] *
36:                        newLx[index];
37:
38:      jx[index] = c_alfa * jx[index] +
39:                  c_beta * (oldEx[index] +
40:                            newEx[index]);
41:
42:      newLx[index] = c_tau * oldLx[index] -
43:                     c_rho * newLx[index] +
44:                     c_eta * (oldEx[index] +
45:                              newEx[index]);
46:
47:      index += zIncr;
48:      prevHy = currHy;
49:
50:      __syncthreads();
51:    }
52: }
```

Figure 4.17 The code segment for updating the E_x field and Lorentz–Drude variables.

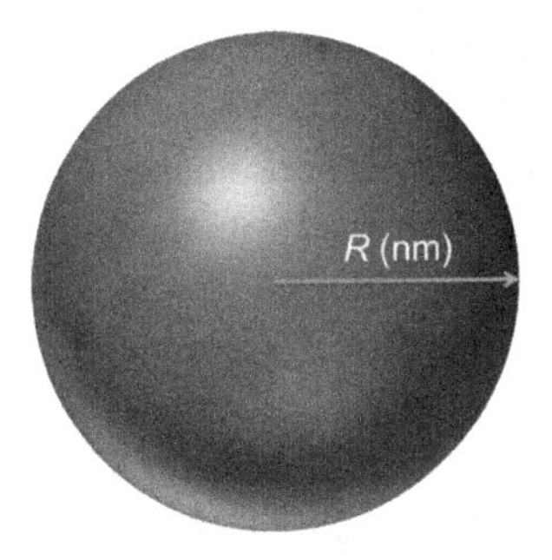

Figure 4.18 The gold nanosphere structure.

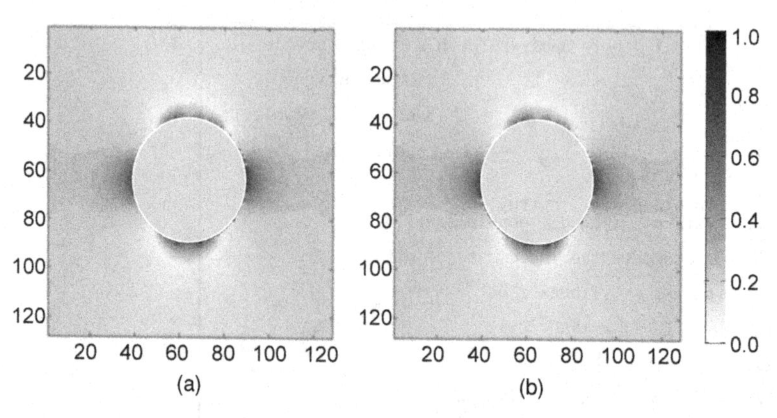

Figure 4.19 Electric field distributions inside and outside a 25-nm gold nanosphere in free space simulated (a) with a GPU and (b) without a GPU.

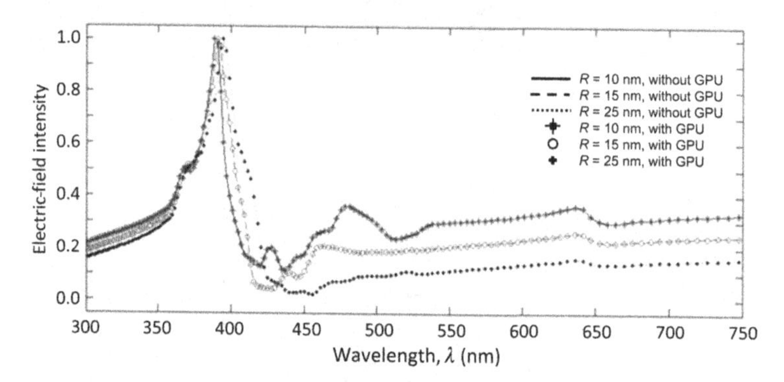

Figure 4.20 The electric-field intensity with respect to wavelength for different sizes of the nanosphere with and without a GPU.

good agreement. However, a significant improvement in simulation speed is observed with the GPU, as shown in Figs. 4.21 and 4.22, and Tables 4.1 and 4.2. For example, with a domain size of $128 \times 128 \times 128$ and 60 000 simulation iterations, the GPU took 10.87 min, whereas MATLAB took 49.64 h.

Various programming models and platforms

For further analysis and benchmarks, we developed the FDTD code by using different programming models and platforms, and compared their performance with that of the GPU. A significant improvement in performance is observed with the GPU, while there is no effect on the accuracy of the results. The GPU card used was an Nvidia "Fermi" C2050, and the computer hardware used in the numerical experiments was an Intel Core 2 Quad 3.2-GHz workstation with 4 GB RAM. The software and compilers used were MATLAB version R2008a, GCC 4.4.3, and Intel Parallel Studio XE (v. 12.0). Two different domain sizes are tested, i.e. $64 \times 64 \times 64$ and $128 \times 128 \times 128$.

Figure 4.21 shows the performance bar graphs of the GPU, GCC-O3, and ICC-O3 with respect to the number of simulation iterations, while the domain size is $64 \times 64 \times 64$.

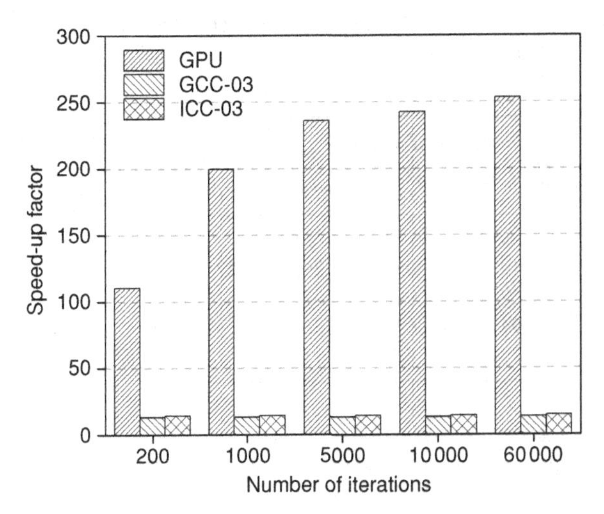

Figure 4.21 The speed-up factor of the GPU and C++ versions using different compilers compared with the original MATLAB version. The GPU version running on the Nvidia Fermi C2050 outperforms the MATLAB version by a factor of as much as 253. The domain size is $64 \times 64 \times 64$.

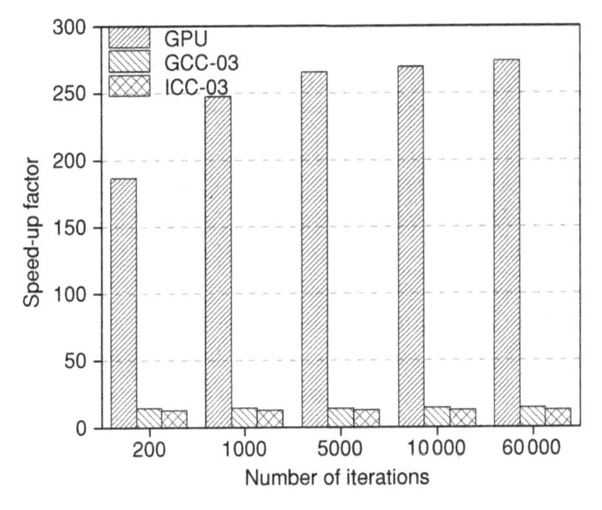

Figure 4.22 The speed-up factors of the GPU and C++ versions using different compilers compared with the original MATLAB version. The GPU version running on the Nvidia Fermi C2050 outperforms the MATLAB version by a factor of as much as 274 for 60 000 iterations. The domain size is $128 \times 128 \times 128$.

For this performance comparison, MATLAB code is used as a reference. Figure 4.21 shows that the GPU implementation outperforms the MATLAB version by as much as a factor of 253. The GPU outperforms the serial C++ versions by a factor of about 19.

Figure 4.22 shows the performance bar graphs for the GPU, GCC-O3, and ICC-O3 with respect to the number of simulation time steps for a larger domain size, i.e.

Table 4.1 The time taken (seconds) to complete the specified number of simulation steps using different programming platform/languages (the domain size is 64 × 64 × 64)

	Number of iterations				
	200	1000	5000	10 000	60 000
	Time (seconds)				
MATLAB	74.94	373.49	1858.05	3729.29	22 910.77
GCC-O3	5.31	26.27	132.09	262.5	1576.85
ICC-O3	5.88	28.94	143.65	286.9	1722.99
GPU	0.68	1.87	7.88	15.39	90.47

Table 4.2 The time taken (seconds) to complete the specified number of simulation steps using different programming platform/languages (the domain size is 128 × 128 × 128)

	Number of iterations				
	200	1000	5000	10 000	60 000
	Time (seconds)				
MATLAB	604.72	2950.38	14 688.61	29 515.71	178 716.48
GCC-O3	42.74	208.22	1056.27	2068.84	12 412.52
ICC-O3	47.02	230.79	1151.97	2296.57	13 774
GPU	3.24	11.93	55.36	109.66	652.2

128 × 128 × 128. The MATLAB code is used as a reference for this comparison. As shown in Fig. 4.22, the GPU implementation outperforms the MATLAB version by a factor of as much as 274. The GPU outperforms the serial C++ versions by a factor of about 21.

These results show that, if the simulation domain is larger, the GPU performance improves compared with the other options presented. From Figs. 4.21 and 4.22, it is observed that the GPU reduces the simulation time more for a larger domain than it does for a smaller domain. It is also observed that, when the number of simulation iterations increases, it becomes beneficial to run the FDTD program on a GPU platform.

4.5 Summary

In this chapter, different material models are incorporated into Maxwell's equations. The FDTD method is used to simulate different aspects of plasmonic devices. First, an approach for the simulation of active plasmonic devices is presented. It combines the solid-state model and a dispersive material model to simulate active plasmonic devices. The solid-state model includes a multilevel multi-electron quantum approach for the simulation of the semiconductor part, and the dispersive model employ the Lorentz–Drude model to simulate the metallic part of active plasmonic devices. The combined methodology is applied to two examples: an MSM waveguide and a microcavity resonator. The combined approach is also applied to simulate a plasmonic structure that acts as a source for generating SPP waves. The SPP extraction efficiency from the source is almost 60%. The extraction efficiency drops as the gap between the plasmonic microdisk and the waveguide increases. Furthermore, it is shown that the plasmonic waveguide perturbs the field inside the plasmonic microdisk. With this device, the complexity of light extraction to a plasmonic device is simplified and the plasmonic source can be easily integrated on-chip with other photonic devices. An integrated plasmonic source is essential for the development of plasmonic sensors, transistors, switches, and many other devices. It is believed that the development of plasmonic sources would pave the way to enhancing the applications of nanophotonics.

In addition, the excitation of an SP mode in a semiconductor material is discussed. Pumping of the electrons into the elliptical microcavity first generates the lasing mode. Subsequently, applying an external magnetic field in a small confined region "pushes" electrons towards the boundary of the elliptical microcavity. High electron densities are built up at the boundaries of the microcavity, and the laser light directly excites the electron plasma to form the SP mode. This phenomenon shows that excitation of an SP mode by electron pumping is possible in a pure semiconductor microcavity. This can be useful in the development of low-loss SP devices, which will enhance the role of plasmonics in integrated circuits. In addition, direct coupling of an internally generated laser mode to excite the SP mode reduces coupling, providing a smaller coupling system and useful harnassing of light energy in structures of small volume. The potential applications of this system include digital switches, logic gates and flip–flops with an external magnetic field, biohazards detection, and imaging spectroscopy. It is believed that the proposed approach will pave the way for new applications in the field of active plasmonics, which is still in its infancy.

In the last part of the chapter, we implemented a 3D FDTD method on a GPU for plasmonics simulation. The method is based on the LD dispersive model, which is incorporated into Maxwell's equations, while the ADE technique is applied to the LD model. Through the extensive correctness tests, it has been shown clearly that our GPU-implemented algorithm provides results that are accurate and similar to those obtained using the conventional version without a GPU. The main highlight of the GPU implementation is the outstanding speed-up in the simulation time required for large-domain structures. The performance speed-up that our CUDA-based FDTD method offers is by a factor of up to 274 relative to the MATLAB version on the CPU. It is demonstrated that the larger the structure domain, the greater the speed improvement which can be achieved. This brings a new dawn to the computational era, allowing the modeling and simulation of sophisticated and complex circuits in a reasonable time frame.

References

[1] S. A. Maier, *Plasmonics: Fundamentals and Applications*. Berlin: Springer-Verlag, 2007.
[2] M. L. Brongersma and P. G. Kik, *Surface Plasmon Nanophotonics*. Dordrecht: Springer, 2007.

[3] I. Ahmed, C. E. Png, E. P. Li, and R. Vahldieck, "Electromagnetic propagation in a novel Ag nanoparticle based plasmonic structure," *Opt. Express,* vol. 17, pp. 337–345, 2009.
[4] K. F. MacDonald, Z. L. Samson, M. I. Stockman, and N. I. Zheludev, "Ultrafast active plasmonics," *Nature Photonics*, vol. 3, pp. 55–58, 2009.
[5] A. V. Krasavin and A. V. Zayats, "Three-dimensional numerical modeling of photonics integration with dielectric loaded SPP waveguides," *Phys. Rev. B.*, vol. 78, 045425 (8 pp.), 2008.
[6] J. A. Dionne, K. Diest, L. A. Sweatlock, and H. A. Atwater, "Plas-MOStor: A metal–oxide–Si field effect plasmonic modulator," *Nano Lett.*, vol. 9, pp. 897–902, 2009.
[7] M. T. Hill, Y.-S. Oei, B. Smalbrugge *et al.*, "Lasing in metal–insulator–metal sub-wavelength plasmonic waveguides," *Opt. Express*, vol. 17, pp. 11107–11112, 2009.
[8] Y. Huang and S. T. Ho, "Computational model of solid state, molecular, or atomic media for FDTD simulation based on a multi-level multi-electron system governed by Pauli exclusion and Fermi–Dirac thermalization with application to semiconductor photonics," *Opt. Express*, vol. 14, pp. 3569–3587, 2006.
[9] E. H. Khoo, I. Ahmed and E. P. Li, "Enhancement of light energy extraction from elliptical microcavity using external magnetic field for switching applications," *Appl. Phys. Lett.*, vol. 95, 121104–121106, 2009.
[10] E. H. Khoo, S. T. Ho, I. Ahmed, E. P. Li, and Y. Huang, "Light energy extraction from the minor surface arc of an electrically pumped elliptical microcavity laser," *IEEE J. Quant. Electron.*, vol. 46, pp. 128–136, 2010.
[11] Y. Huang and S. T. Ho, "Simulation of electrically-pumped nanophotonic lasers using dynamical semiconductor medium FDTD method," in *2nd IEEE International INEC*, pp. 202–205, 2008.
[12] D. Rakic, A. B. Djurisic, J. M. Elazar, and M. L. Majewski, "Optical properties of metallic films for vertical-cavity optoelectronic devices," *Appl. Optics*, vol. 37, pp. 5271–5283, 1998.
[13] I. Ahmed, E. P. Li, and E. H. Khoo, "Interactions between magnetic and non-magnetic materials for plasmonics," in *International Conference on Materials and Advanced Technologies*, Singapore, 2009.
[14] A. Taflove, *Computational Electrodynamics*, Norwood, MA: Artech House, 2005.
[15] I. Ahmed, E. H. Khoo, E. P. Li, and R. Mittra, "A hybrid approach for solving coupled Maxwell and Schrödinger equations arising in the simulation of nano-devices," *IEEE Antennas Wireless Propagation Lett.*, vol. 9, pp. 914–917, 2010.
[16] I. Ahmed, E. Khoo, O. Kurniawan, and E. Li, "Modeling and simulation of active plasmonics with the FDTD method by using solid state and Lorentz–Drude dispersive model," *J. Opt. Soc. Am. B*, vol. 28, pp. 352–359, 2011.
[17] O. Kurniawan, I. Ahmed, and E. P. Li, "Development of a plasmonics source based on nano-antenna concept for nano-photonics applications," *IEEE Photonics J.*, vol. 3, pp. 344–352, 2011.
[18] K. J. Willis, J. S. Ayubi-Moak, S. C. Hagness, and I. Knezevic, "Global modeling of carrier-field dynamics in semiconductor using EMC-FDTD," *J. Comput. Electron.*, vol. 8, pp. 153–171, 2009.
[19] I. Ahmed and E. P. Li, "Simulation of plasmonics nanodevices using coupled Maxwell and Schrödinger equations using the FDTD method," *J. Adv. Electromagn.*, vol. 1, pp. 76–83, 2012.
[20] E. H. Khoo, I. Ahmed, and E. P. Li, "Investigation of the light energy extraction efficiency using surface modes in electrically pumped semiconductor microcavity," *Proc. SPIE*, vol. 7764B, doi:10.1117/12.860618, 2010.

[21] K. H. Lee, I. Ahmed, R. S. M. Goh *et al.*, "Implementation of the FDTD method based on Lorentz–Drude model on GPU for plasmonics applications," *Prog. Electromagn. Res.*, vol. 116, pp. 441–456, 2011.

[22] I. Ahmed, and E. P. Li, "Time domain simulation of dispersive materials from microwave to optical frequencies," in *IEEE International Conference on Emerging Technologies*, 2011.

[23] D. J. Bergman and M. I. Stockman, "Surface plasmon amplification by stimulated emission of radiation: Quantum generation of coherent surface plasmons in nanosystems," *Phys. Rev. Lett.*, vol. 90, 027402, 2003.

[24] A. V. Akimov, A. Mukherjee, C. L. Yu *et al.*, "Generation of single optical plasmons in metallic nanowires coupled to quantum dots," *Nature*, vol. 450, pp. 402–406, 2007.

[25] D. Koller, A. Hohenau, H. Ditlbacher *et al.*, "Organic plasmon-emitting diode," *Nature Photonics*, vol. 2, pp. 684–687, 2008.

[26] R. J. Walters, R. V. A. van Loon, I. Brunets, J. Schmitz, and A. Polman, "A silicon-based electrical source of surface plasmon polaritons," *Nature Mater.*, vol. 9, pp. 21–25, 2010.

[27] M. Kuttge, F. J. García de Abajo, and A. Polman, "Ultrasmall mode volume plasmonic nanodisk resonators," *Nano Lett.*, vol. 10, pp. 1537–1541, 2010.

[28] C. Walther, G. Scalari, M. I. Amanti, M. Beck, and J. Faist, "Microcavity laser oscillating in a circuit-based resonator," *Science*, vol. 327, pp. 1495–1497, 2010.

[29] Y. H. Chen and L. J. Guo, "Analysis of surface plasmon guided sub-wavelength microdisk cavity," *Proc. IEEE LEOS*, pp. 320–321, 2008.

[30] M. Kim, P. Ku, and J. Guo, "Surface plasmon enabled sub-wavelength nano-cavity laser," *Proc. IEEE LEOS*, pp. 252–253, 2007.

[31] R. Perahia, T. P. M. Alegre, A. H. Safavi-Naeini, and O. Painter, "Surface-plasmon mode hybridization in subwavelength microdisk lasers," *Appl. Phys. Lett.*, vol. 95, 201114 (3 pp.), 2009.

[32] R. K. Chang and A. J. Campillo, *Optical Processes in Microcavities*. Singapore: World Scientific, 1996.

[33] E. Hall, "On a new action of the magnet on electric currents," *Am. J. Math.*, vol. 2, 287–292, 1879.

[34] S. Kasap, *Hall Effect in Semiconductors*, e-booklet, http://kasap3.usask.ca/samples/HallEffectSemicon.pdf, 1990–2001.

[35] N. A. Sinitsyn, "Semiclassical theories of the anomalous Hall effect," *J. Phys.: Condens. Matter*, vol. 20, 023201, 2008.

[36] K. Okamoto, A. Scherer, and Y. Kawakami, "Surface plasmon enhanced light emission from semiconductor materials," *Phys. Stat. Sol. C*, vol. 5, pp. 2822–2824 , 2008.

[37] J. Y. Lee, J. Z. Xue, W. J. Park, and A. Mickelson, "Surface plasmon polariton waveguides in nonlinear optical polymer, " *ACS Symp. Series*, vol. 1069, pp. 67–83, 2010.

[38] J. D. Jackson, *Classical Electrodynamics*. New York: John Wiley & Sons, 1998.

[39] O. Darrigol, *Electrodynamics from Ampère to Einstein*. Oxford: Oxford University Press, 2000.

[40] L. D. Landau and E. M. Lifshitz, *Quantum Mechanics: Nonrelativistic Theory*. Oxford: Pergamon Press, 1977.

[41] H. Aoki and T. Ando, "Critical localization in two-dimensional Landau quantization," *Phys. Rev. Lett.*, vol. 54, pp. 831–834, 1985.

[42] G. Goubau, "Surface waves and their applications to transmission lines," *J. Appl. Phys.*, vol. 21, pp. 1119–1128, 1950.

[43] A. Sommerfeld, "Propagation of waves in wireless telegraphy," *Ann. Phys.*, vol. 81, pp. 1367–1153, 1926.

[44] S. Cho and N. M. Jokerst, "A polymer microdisk photonic sensor integrated onto silicon," *IEEE Photonics Tech. Lett.*, vol. 18, pp. 2096–2098, 2006.

[45] B. E. Little, J. S. Foresi, G. Steinmeyer *et al.*, "Ultra-compact Si–SiO_2 microring resonator optical channel dropping filter," *IEEE Photonics Tech. Lett.* vol. 10, pp. 549–551, 1998.

[46] C. Maxfield, *The Design Warrior's Guide to FPGAs*. Amsterdam: Elsevier, 2004.

[47] A. Z. Elsherbeni and V. Demir, *The Finite-Difference Time-Domain Method for Electromagnetics with MATLAB Simulations*. Chennai: SciTech Publications, 2009.

[48] M. R. Zunoubi, P. Payne, and W. P. Roach, "CUDA implementation of TEz-FDTD solution of Maxwell's equations in dispersive media," *IEEE Antennas Wireless Propag. Lett.*, vol. 9, pp. 756–759, 2010.

[49] L. Savioja, "Real-time 3D finite-difference time-domain simulation of low and mid frequency room acoustics," in *Proceedings of the 13th International Conference on Digital Audio Effects* (DAFx-10), 2010.

[50] S. Chen, S. Dong, W.-X. Liang, "GPU-based accelerated FDTD simulations for double negative (DNG) materials applications," in *International Conference on Microwave and Millimeter Wave Technology*, pp. 839–841, 2010.

[51] R. Shams and P. Sadeghi, "On optimization of finite-difference time-domain (FDTD) computation on heterogeneous and GPU clusters," *J. Parallel Distrib. Comput.*, vol. 71, pp. 584–593, 2011.

[52] S. H. Zainud-Deen, E. Hassan, M. S. Ibrahim, K. H. Awadalla, and A. Z. Botros, "Electromagnetic scattering using GPU based finite difference frequency domain method," *Prog. Electromagn. Res.* B, vol. 16, pp. 351–369, 2009.

5 Passive plasmonic waveguide-based devices

This chapter presents the hybrid plasmonic waveguide (HPW) platform and its components. In particular, the effective mode index, propagation distance, and mode confinement of the planar vertical hybrid plasmonic waveguides (VHPWs) are numerically characterized as functions of dimensions and materials in the near-infrared wavelengths. The chapter demonstrates that the vertical hybrid silver–silica–silicon plasmonic waveguide achieves better propagation characteristics than those of the counterpart silver–silica–silver MIM and silicon-based dielectric-loaded plasmonic waveguide. Moreover, we propose and design various passive waveguide-based components based on the optimized VHPW, including bends, power-splitters, couplers, and ring resonator filters. An important issue is how to implement these HPW-based components and devices in standard complementary metal–oxide–semiconductor (CMOS) electronic–photonic integrated circuits. To meet this requirement, two CMOS-compatible HPW platforms, namely a copper-cap VHPW and a hybrid horizontal copper–silicon dioxide–silicon–silicon dioxide–copper plasmonic waveguide, and devices based on them such as bends and ring resonator filters are subjected to further investigation both in experiments and in theory in the subsequent sections.

5.1 Introduction

In the context of communications and information processing, integration of photonic and nanoelectronic devices on the same chip would lead to a tremendous synergy by combining the ultra-compactness of nanoelectronics with the super-wide bandwidth of photonics. Such integration will benefit considerably from the application of nanotechnology for data-processing, sensing, medical, health-care, and energy purposes. In recent years, it has been demonstrated theoretically and experimentally that plasmonic devices, taking advantage of plasmon-enabled tight modal confinement, promise to overcome the size mismatch between microscale photonics and nanoscale electronics. Thus, plasmonic technology is a potential platform for the next generation of optical interconnects that will enable the deployment of small-footprint and low-energy integrated circuitry, and holds great promise for on-chip optical interconnects of high integration density at the chip-scale [1–5]. Figure 5.1 illustrates a typical plasmonic link for on-chip data

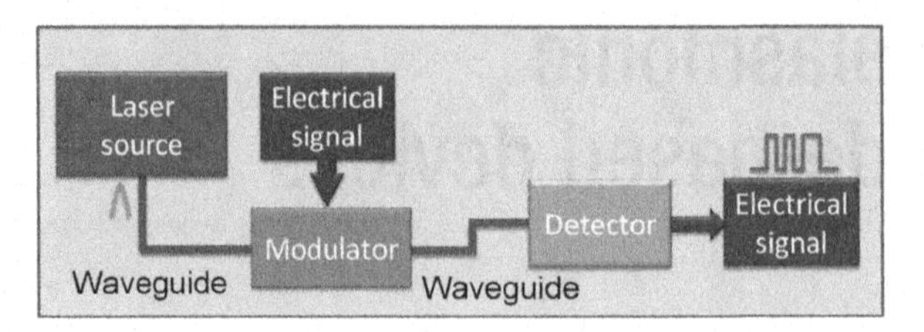

Figure 5.1 A schematic illustration of a typical plasmonic link for on-chip data transfer.

transfer. The plasmonic laser source generates the plasmonic signal, which will be guided through straight or bent waveguides. Then, this plasmonic signal will be modulated by a plasmonic modulator. For practical reasons relating to the purpose of integration, it is preferable to use an electro-optic plasmonic modulator, which allows one to encode the electrical signals into the optical data stream. The modulated signal will pass through the subwavelength plasmonic photodetector to convert the propagating electromagnetic waves into electrical signals.

Current advances in nanofabrication and full-wave electromagnetic simulation techniques have permitted the realization of a wide variety of plasmonic waveguide devices. Several types of plasmonic waveguide platform differing in terms of the topology, material composition, and propagation mechanisms have been developed. Each waveguide platform possesses different waveguiding characteristics, such as structures of coupled nanowires [6–10], metallic stripes [11, 12], metallic grooves [5, 13], metal–insulator–metal (MIM) structures [14–17] and dielectric-loaded plasmonic waveguides [18–21]. A key challenge in designing a plasmonic waveguide platform is the tradeoff between the optical confinement and the propagation length. For example, the MIM configuration has received particular attention, thanks to its subwavelength confinement whereby the modal size is solely determined by the distance between two metal stripes. However, due to the ohmic loss of the metal, the propagation distance of the MIM is typically of the order of several micrometers at frequencies around the telecommunication wavelength of 1.55 μm, and it decreases with the slot width (and hence with the modal size). Waveguide-based long-range surface plasmon polaritons such as those of thin metal stripes, on the other hand, can support plasmonic waves traveling for a propagation length of up to several millimeters, but their mode confinement is significantly large compared with the operating wavelength [22]. Moreover, subwavelength confinement and low propagation loss play an essential role in achieving the nanoscale integration of practical photonic–electronic integrated circuits [5]. The HPW has recently attracted substantial research interest because of the exciting potential for integrating metallic and semiconductor nanostructures to fully exploit the advantages both of metals (light concentration and high electrical conductivity) and of semiconductors (light emission and photocurrent generation). As a result, this waveguide yields the best tradeoff between the propagation loss and power confinement. These waveguides consist of a very thin gap of low-index insulator material separated from a higher-index semiconductor nanowire and a metal layer. Theoretical and experimental analyses show that the gap can

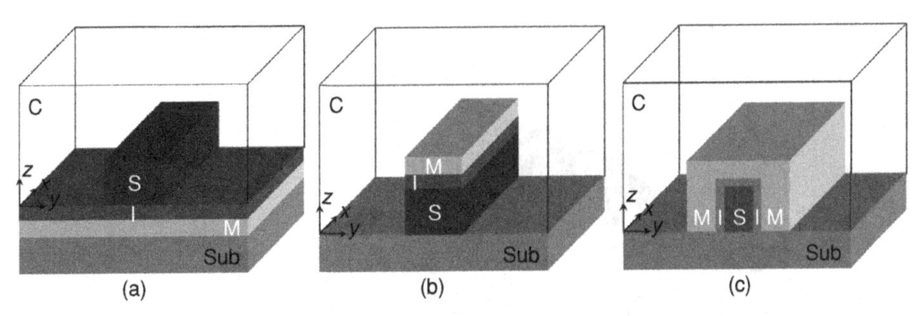

Figure 5.2 Sketches of different hybrid plasmonic waveguides. (a) The vertical hybrid metal–insulator–semiconductor plasmonic waveguide (VHPW). (b) The metal-cap vertical hybrid metal–insulator–semiconductor plasmonic waveguide. (c) The horizontal hybrid metal–insulator–semiconductor–insulator–metal plasmonic waveguide (HHPW). C stands for cladding medium, S stands for a high-index semiconductor nanowire, I stands for a low-index insulator layer, M stands for the plasmonic metal, and Sub denotes the substrate support for the waveguide structure.

support a low-loss compact hybrid mode whose propagation length is a strong function of the high-index semiconductor nanowire dimensions and low-index gap size [23, 24]. The high-index semiconductor nanowire is typically patterned to form a rectangular, square, or cylindrical core, which supports a photonic-like mode. By adjusting the high-index semiconductor nanowire dimensions and the low-index insulating gap size, the composite waveguide mode can be varied from plasmonic to photonic, making a wide range of modal characteristics such as strong confinement and long propagation distance. Such a particular configuration allows the design of high-performance nanoscale semiconductor-based passive plasmonic waveguide components [23–33] as well as lasers at a deep subwavelength scale [34, 35]. Different types of hybrid plasmonic waveguides arise depending on the production process. These types include the vertical hybrid metal–insulator–semiconductor plasmonic waveguide [23, 24, 27, 28], metal-cap vertical hybrid metal–insulator–semiconductor plasmonic waveguide [29, 30] (also known as metal-cap hybrid plasmonic waveguide), and horizontal hybrid metal–insulator–semiconductor–insulator–metal plasmonic waveguide [31–33]. In Fig. 5.2(a), a conventional vertical hybrid metal–insulator–semiconductor plasmonic waveguide is sketched. The insulator layer is sandwiched between a high-index semiconductor nanowire and a plasmonic material layer. The plasmonic material can be one of several metals (gold, silver, copper, or aluminum) or silicides (nickel silicide, platinum silicide, titanium silicide, or tungsten silicide), a highly doped semiconductor, or doped graphene. A schematic illustration of the metal-cap hybrid plasmonic waveguide is shown in Fig. 5.2(b). It can be seen that this structure can also be considered as a reversed orientation of the waveguide in Fig. 5.2(a) due to the different fabrication processes. On the other hand, the prototype of the horizontal hybrid plasmonic waveguide based on the metal–insulator–semiconductor–insulator–metal (MISIM) combination is sketched in Fig. 5.2(c). This hybrid plasmonic waveguide is considered as an extension of the conventional metal–semiconductor–metal plasmonic waveguide. The MISIM waveguide permits one to reduce the propagation loss in comparison with

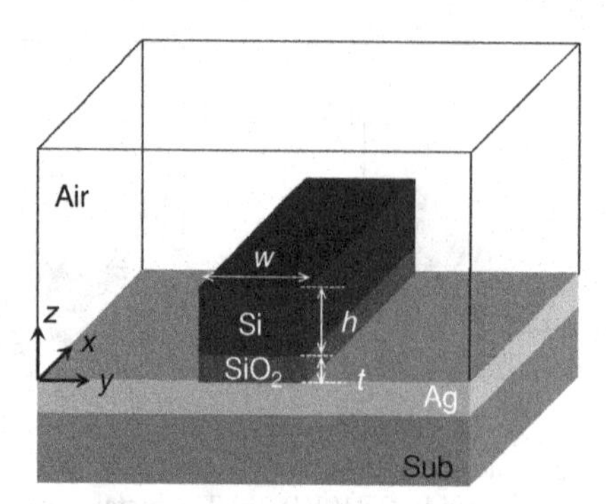

Figure 5.3 A schematic illustration of the vertical hybrid Ag–SiO_2–Si plasmonic waveguide.

that of the MSM plasmonic waveguide counterpart while still retaining the waveguide confinement.

In order to realize a compact integrated nanoscale electronic–photonic circuit to perform various tasks such as guiding, filtering, and switching a light signal at near-infrared wavelengths, it is necessary to develop and design several compact key components and devices offering flexible and reconfigurable functionality. In this chapter, we will focus on the passive hybrid plasmonic waveguide-based components. The three configurations of HPW platform which will be investigated here are the silver-based vertical hybrid metal–insulator–semiconductor plasmonic waveguide (abbreviated to VHPW), the copper-cap VHPW, and the highly confined copper-based horizontal hybrid MISIM plasmonic waveguide (abbreviated to HHPW). Silver is used as the plasmonic material due to its low ohmic loss in the near-infrared frequency. In the meantime, the choice of copper is due to the benefit of the fully complementary metal–oxide–semiconductor (CMOS)-compatible plasmonic material for the standard silicon CMOS technology as a requirement from the current development of electronic–photonic integrated circuits.

5.2 The vertical hybrid Ag–SiO_2–Si plasmonic waveguide and devices based on it

5.2.1 Theoretical background

The schematic structure of the vertical hybrid Ag–SiO_2–Si plasmonic waveguide (VHPW) is shown in Fig. 5.3. This waveguide consists of a low-index SiO_2 stripe ($n_{SO_2} = 1.44$) sandwiched between a rectangular high-index Si stripe ($n_{Si} = 3.48$) and a silver film ($\lambda = 1.55$ μm, $n_{Ag} = 0.145 + 11.36i$ [36]). The SiO_2 stripe has dimensions of w nm width and t nm thickness. The width and height of the high-index silicon nanowire are w and h, respectively. The thickness of the silver film is taken as 100 nm largely to ensure that there are no effects of the silver film thickness on the plasmonic mode inside the SiO_2 stripe because the plasmonic skin depth in silver remains roughly constant at 20 nm for near-infrared wavelengths [16]. Glass is chosen as the substrate and air as the cladding medium. The operating wavelength is the telecommunication wavelength of 1.55 μm. Since the propagation direction is assumed to be the x direction, the dominant electric field is in the z direction (E_z), and this must be used to excite the surface plasmon along the waveguide. The propagation properties of a plasmonic waveguide are generally characterized through the effective index, propagation length, and confinement factor. Since the propagation direction in the waveguide is along the x axis and the electric field can be expressed in the form $E = E_0(y, z)e^{i(\gamma x - \omega t)}$, where γ is a propagation constant, as a complex number. To determine this propagation constant, the eigenmodes of the proposed plasmonic waveguides at a given wavelength λ are calculated by using the full-vectorial finite-difference (FVFD) method [37]. Note that, in order to guide the plasmonic mode inside the SiO_2 stripe or the slot, the E-field component must be p-polarized and perpendicular to the surface of the silver film. By solving the eigenvalue equations we obtain the complex propagation constant γ, from which the propagation length L_p and the effective index n_{eff} of the propagating mode are calculated through the equation

$$\gamma = \alpha + i\beta = L_p^{-1} + i2\pi n_{eff}\lambda_0^{-1}. \tag{5.1}$$

From the value of n_{eff}, the dispersion relation can be represented as $\omega = \omega(\beta) = \omega(n_{eff})$. On the other hand, the propagation length L_p is defined by the $1/e$ power decay length of the plasmonic propagating mode. The confinement factors are numerically characterized by the confined power P_c and confined intensity I_c. The confined power P_c represents the power guided inside the SiO_2 stripe, which is measured as a percentage of the optical power focused inside the SiO_2 stripe with respect to the total waveguide optical power [38], and thus is calculated as

$$P_c = \frac{\iint_{SiO_2} \boldsymbol{P} \cdot d\boldsymbol{S}}{\int_{-\infty}^{\infty}\int_{-\infty}^{\infty} \boldsymbol{P} \cdot d\boldsymbol{S}}. \tag{5.2}$$

On the other hand, the confined intensity is defined as an average optical intensity I_c inside the slot, and calculated as $I_c = P_c/(wt)$ where w is the width of the Si nanowire (Fig. 5.3), t is the thickness of the SiO_2 stripe, and the guided area is the SiO_2 stripe.

5.2.2 The dependence of the propagation characteristics on the thickness of the SiO_2 stripe

Figure 5.4(a) shows the mode effective index n_{eff} and the propagation length L_p as functions of the thickness t of the SiO_2 stripe for three Si-nanowire thicknesses h of 100, 150, and 200 nm, where the width w of the SiO_2 stripe and Si nanowire was chosen as 200 nm to ensure that only single-mode propagation at $\lambda = 1.55$ μm occurred. We observed that increasing the thickness of the SiO_2 stripe or decreasing the thickness of the Si nanowire reduces the mode effective index and increases the propagation length. However, for Si nanowires thinner than 100 nm (regardless of the thickness of the SiO_2 stripe), the fundamental mode turns progressively into a plasmonic wave that propagates along the silver/air interface (n_{eff} =1.06), leading to poor vertical confinement. To quantitatively understand the confinement in the VHPW structure, we studied the confined optical power P_c and confined intensity I_c, which are shown in

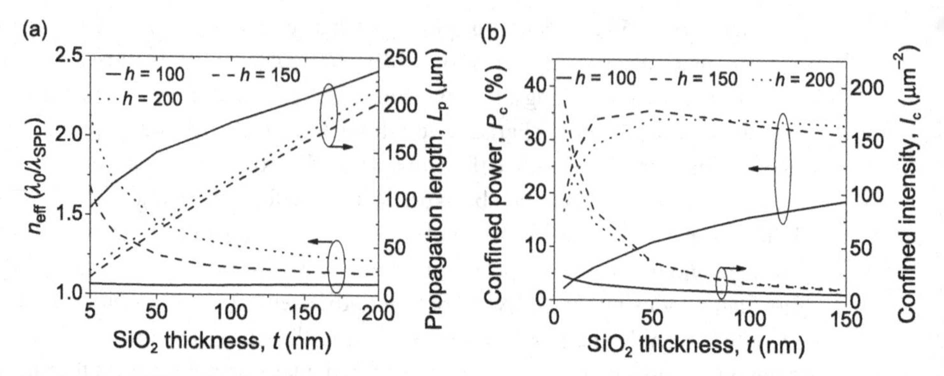

Figure 5.4 (a) The effective index n_{eff} and propagation length L_{p}, and (b) the confined power and confined intensity, as functions of the thickness of the SiO_2 stripe t for different Si-nanowire thicknesses h. Adapted from Ref. [24].

Fig. 5.4(b) as functions of the SiO_2 stripe thickness for three different Si-nanowire thicknesses of 100, 150, and 200 nm. We see that the power confined inside the SiO_2 stripe increases with increasing SiO_2-stripe thickness, and reaches a saturated value when the thickness is larger than 15 nm, the Si-nanowire thickness being 150 nm or larger. We also note that the confined power becomes very small when the thickness h of the Si nanowire is reduced to 100 nm. The corresponding confined intensity is also shown on the left-hand axis of Fig. 5.4(b). This value, on the other hand, decreases exponentially as the SiO_2-stripe thickness increases. For dimensions of $t = 150$ nm, 36% of the total optical power can be confined in the VHPW, and an average optical intensity of 36 μm^{-2} can be achieved.

5.2.3 The dependence of the propagation characteristics on the dimensions of the Si nanowire

The geometric dimensions have a significant impact on the propagation characteristics of the fundamental mode in the VHPW. Thus, the optimization of the Si-nanowire dimensions to achieve the best tradeoff between the confinement factor and the propagation length must be done. Note that the thickness of the SiO_2 stripe is arbitrarily chosen to be 50 nm.

In Figs. 5.5(a) and (b) we plot the mode effective indexes, the propagation lengths, and the confinement factor as functions of the waveguide width w for three Si-nanowire thicknesses h of 120, 150, and 180 nm. As can be seen in Fig. 5.5(a), the effective indexes increase monotonically with the waveguide width w and Si-nanowire thickness h. In particular, for very small waveguide widths the mode effective index n_{eff} approaches that of the plasmonic mode at the silver/air interface. Regarding the propagation length, a small value of w below 150 nm is not a good choice because it introduces very poor confinement, as shown in Fig. 5.5(a). The propagation does not vary much on increasing the width w. Moreover, the confinement factor is expressed as a concave function, where the maximum confinement factor is observed. Therefore, the dimensions of the

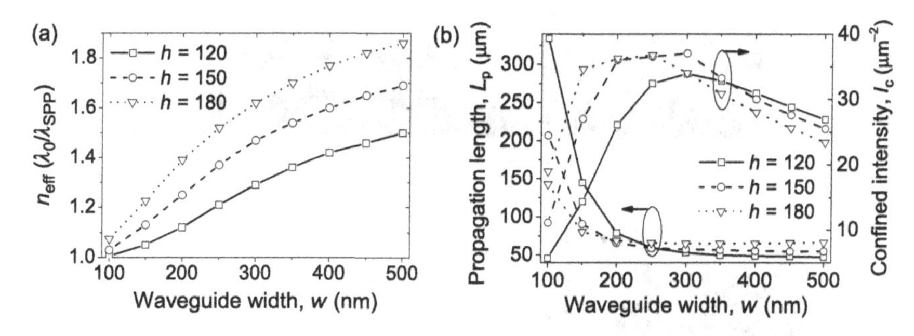

Figure 5.5 (a) The mode effective index n_{eff} and (b) the propagation length L_{p} and confined intensity I_{c} as functions of the waveguide width w for three Si-nanowire thicknesses h of 120, 150, and 180 nm. The thickness of the SiO_2 stripe is taken to be 50 nm.

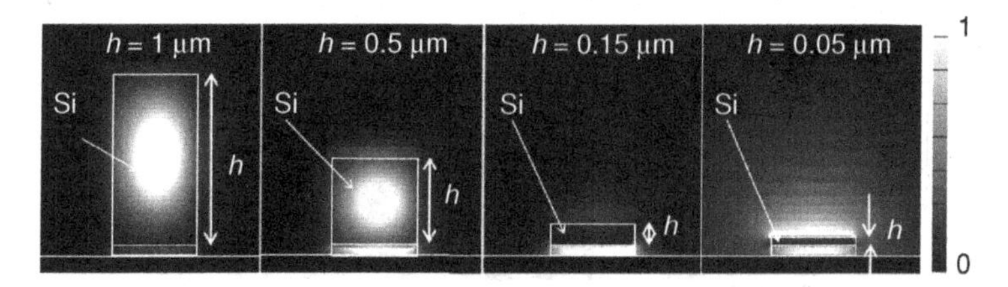

Figure 5.6 The cross-sectional energy distribution for different Si-nanowire thicknesses with a constant width $w = 300$ nm. The best mode confinement is achieved at the Si-nanowire thickness of 150 nm ($h = 0.15$ μm).

Si nanowire can be optimized to provide the best tradeoff between the mode confinement and the propagation length.

As can be seen from Fig. 5.3, the VHPW is built by adding an Si-nanowire-based photonic waveguide into the dielectric-loaded plasmonic waveguide (SiO_2 stripe above the Ag film). As a result, the combination plasmonic–photonic mode is formed. This mode may vary from a low-loss photonic-like mode to a hybrid plasmonic/photonic-like mode or a plasmonic-like mode, depending on the size of the high-refractive-index nanowire. The detailed properties of these modes have been discussed in Ref. [23]. As an example, we show the electromagnetic energy distribution as a function of the Si-nanowire thickness when its width is fixed at 300 nm (Fig. 5.6). A large value of the Si-nanowire thickness makes the photonic mode dominant in the waveguide. A thinner Si nanowire results in weak confinement of the energy inside the SiO_2 stripe. As observed, the VHPW achieves the best energy confinement inside the SiO_2 stripe at 150 nm Si-nanowire thickness.

Figure 5.7(a) shows the normalized electric field of the dominant $|E_z|$ component (in the y–z plane) along the propagation direction for a distance of 50 μm. The result shows that the field decreases exponentially along the propagation distance, which is consistent with the propagation length obtained from the effective-refractive-index calculation. We also present, in Fig. 5.7(b), the normalized electric field of the dominant $|E_z|$ component over the y–z cross-section plane of the VHPW for $w = 300$ nm and an operating wavelength of $\lambda = 1.55$ μm. It reveals that the electric field of the fundamental

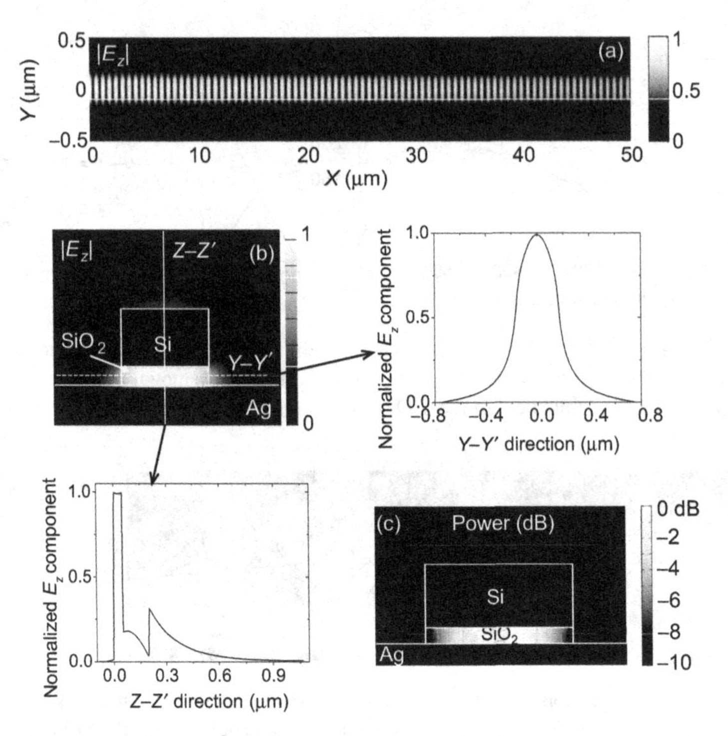

Figure 5.7 (a) The $|E_z|$ component (in the x–y plane) along the propagation direction for a distance of 50 μm, (b) The $|E_z|$ component in the y–z plane together with the $|E_z|$ component in the horizontal direction (Y–Y' line) and in the vertical direction (Z–Z' line). (c) The power distribution in dB observed in the cross-sectional plane.

mode is strongly confined inside the SiO_2 stripe in both waveguides. The electric field $|E_z|$ component along the vertical direction (Z–Z') implies that a high field quantity resides inside the 50-nm gap of the SiO_2 stripe. In the meantime, the $|E_z|$ component along the horizontal direction (Y–Y') is represented as a Gaussian curve. Moreover, the power distribution in dB, shown in Fig. 5.7(c), demonstrates that the energy of the fundamental mode is well confined in the SiO_2 stripe according to a cross-section of 50 nm$\times$ 300 nm. The power-confinement level can be better than -5 dB.

In order to understand the dependence of the propagation characteristics on the operating wavelength, we study the propagation length and the confined intensity of the optical mode as a function of wavelength in the near-infrared spectrum range from 1 μm to 2 μm. For the purpose of simplification, the two dimensions of the Si nanowire are taken to be $w = 200$ nm and $h = 150$ nm, because a reasonable propagation length and high confinement of the intensity of the optical mode can be achieved with these dimensions. As observed in Fig. 5.8, the propagation length non-monotonically varies on increasing the operating wavelength. Surprisingly, there is a peak according to the long propagation length observed at the wavelength of 1.1 μm. In general, the propagation length is represented as a convex function where the shorter propagation length is observed at a wavelength of around 1.4 μm. In contrast, the confined intensity I_c can

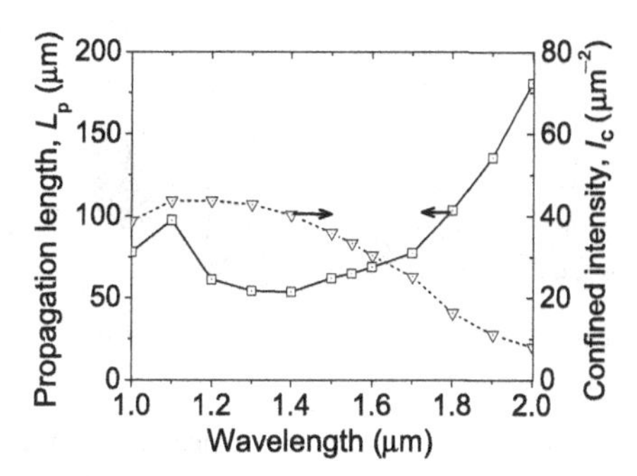

Figure 5.8 The dependence of the propagation length and the confined intensity on wavelengths in the near-infrared spectrum. The dimensions of the Si nanowire are taken to be $w \times h = 200\,\text{nm} \times 150\,\text{nm}$. The optical wavelength to achieve the best tradeoff of low waveguide propagation loss and confinement is roughly 1.1 μm.

be expressed as a concave function, and the maximum confined intensity is achieved at wavelengths in the range from 1.1 μm to 1.2 μm. Therefore, for the Si–SiO_2–Ag hybrid plasmonic waveguide, the wavelength to achieve the best tradeoff of the propagation length and confined intensity is 1.1 μm.

5.2.4 The propagation characteristics of the vertical hybrid, metal–insulator–metal, and dielectric-loaded plasmonic waveguides

In order to highlight the excellence as a waveguiding platform (better tradeoff between the loss and confinement) of the VHPW in comparison with other counterparts, including the vertical Ag-based metal–insulator–metal (VMIM) waveguide and the Si-based dielectric-loaded plasmonic waveguide (DLPW), we compare the mode effective index, propagation length, and confined intensity of the proposed VHPW with those of the VMIM and DLPW. All these kinds of waveguide platform are classified as subwavelength plasmonic waveguides due to their structural similarity, and are thus very efficient components for nanoscale electronic–photonic integrated circuits. For simplicity, the active loaded nanowire is assumed to have a square shape, e.g. Ag nanowire for the VMIM waveguide and Si nanowire for the VHPW and DLPW. It is evident that the different loaded nanowires, shapes, and materials of which these types of waveguide are made result in different waveguide propagation characteristics in terms of the mode effective index, propagation length, and confinement of the optical power. The structures of these three waveguides are shown in Fig. 5.9. The VMIM waveguide (Fig. 5.9(b)) is quite similar to the VHPW, except that the Si nanowire in the VHPW is replaced by an Ag nanowire of identical dimensions in the VMIM structure. For the DLPW structure, as shown in Fig. 5.9(c), the SiO_2 stripe is not present, and the Si nanowire is loaded on top of the Ag film.

We have studied the propagation length, the effective index, and the confinement of the fundamental mode as a function of the dimension w in the range from 100 nm to 500 nm of the square loaded nanowire in these three waveguides at the telecommunication wavelength of $\lambda = 1.55\,\mu\text{m}$. Figure 5.10 shows a plot of the effective index

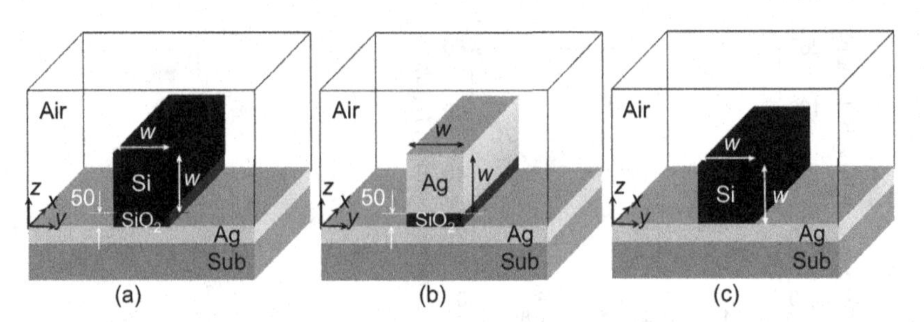

Figure 5.9 Different subwavelength plasmonic waveguides: (a) the vertical hybrid plasmonic waveguide (VHPW), (b) the metal–insulator–metal (MIM) waveguide, and (c) the Si-based dielectric-loaded plasmonic waveguide (DLPW).

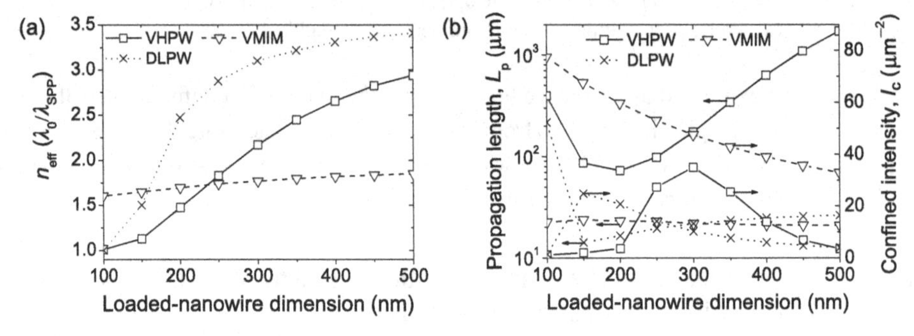

Figure 5.10 A comparison between the propagation characteristics of the VHPW, VMIM waveguide, and Si-based DLPW: (a) the effective refractive index and (b) the propagation length and the confinement factor as functions of the dimension of the square loaded nanowire.

n_{eff} for the mode inside the SiO_2 stripe, $n_{eff} = k_{SPP}/k_0 = \lambda/\lambda_{SPP}$, as a function of w. It appears that n_{eff} of the VMIM waveguide increases only slightly as the waveguide width grows, whereas in the VHPW and DLPW, the increase of n_{eff} is more pronounced due to the high index of the Si nanowire. The different behavior of n_{eff} is due to the difference between the optical constant of Ag and that of hybrid Si/SiO_2 and Si at near-infrared wavelengths. Figure 5.10 shows the propagation length L_p and the confined intensity factor I_c for these three waveguides. We observe that the VHPW has a roughly three times longer propagation length than the VMIM and DLPW waveguides, although the VMIM waveguide demonstrates a slightly better power confinement than that of the VHPW. As a result, the VHPW achieves a lower propagation loss combined with a given confinement than does the VMIM waveguide, so the VHPW is preferable for nanoscale electronic–photonic integrated circuits. This feature of the VHPW is due to the presence of the hybrid semiconductor/oxide interface, which has a much lower absorption loss, and due to the fact that the electric field mainly spreads over the oxide layer, thus having less contact with the metal surface than in the VMIM waveguide. Moreover, we find that the propagation length and the confinement are nonlinear functions of the waveguide width. This nonlinear dependence allows us to obtain an optimal width of $w = 300$ nm for the Si nanowire at which the VHPW supports single-mode operation and achieves the best tradeoff between the propagation length and the confinement.

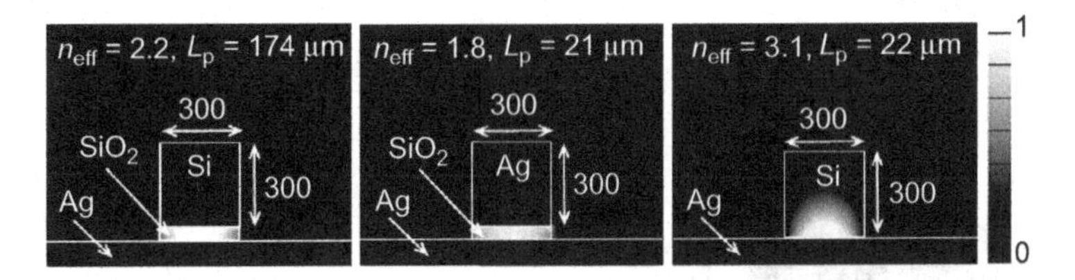

Figure 5.11 The normalized intensity of the electric-field profile versus the effective index and the propagation length of three waveguides studied. The dimension of the square Si nanowire is taken to be 300 nm.

In Fig. 5.11, we calculate the effective index and the propagation length as well as the normalized electric-field intensity of these three waveguides for $w = 300$ nm at an operating wavelength of $\lambda = 1.55$ μm. This reveals that, although the VMIM waveguide provides a better confined intensity than the other two due to the SiO_2 gap formed by the two metal walls, the VHPW achieves the best tradeoff of waveguide propagation loss and confinement. In particular, the VHPW may achieve a propagation length seven times longer than that for the others. The electric field of the fundamental mode is strongly confined inside the SiO_2 stripe for the VHPW and the VMIM waveguide, and in the Si nanowire for the DLPW. The outstanding performance of the VHPW can be explained by two factors. Firstly, the low propagation loss is due to the presence of the hybrid Si/SiO_2 interface, which has a much lower absorption loss, and the electric field mainly spreading over the oxide layer and having relatively less contact with the metal surface. Secondly, the dielectric discontinuity at the semiconductor/oxide interface generates polarization charges that interact with the plasma oscillation at the metal/oxide interface. These advantages thus make the VHPW a promising technology for building silicon-based plasmonic waveguide components with excellent optical performance.

5.2.5 Waveguide couplers

For the development of nanoscale electronic–photonic integrated circuits it is crucial to investigate the possible cross-talk or coupling efficiency between two neighboring waveguides. This is an important factor in designing efficient directional couplers to couple between two or more plasmonic waveguides. Such components can be used as switch or modulator components in which the amount of light coupled from one channel to the other is controlled electro-optically. On the other hand, a detailed understanding of the cross-talk between plasmonic waveguides and interconnectors will also allow one to determine how close the two waveguides and/or waveguide-based devices could be placed together in a circuit, so that they do not interfere with each other. As a result, a low level of cross-talk between plasmonic interconnectors is one of the major factors determining the achievable degree of integration of nano-optical circuits. Therefore, we investigate the performance of directional couplers based on two parallel straight VHPWs with an edge-to-edge separation d as sketched in Fig. 5.12(a). Each waveguide has a dimensions $w = 200$ nm, $h = 150$ nm, and $t = 50$ nm. We chose one waveguide to

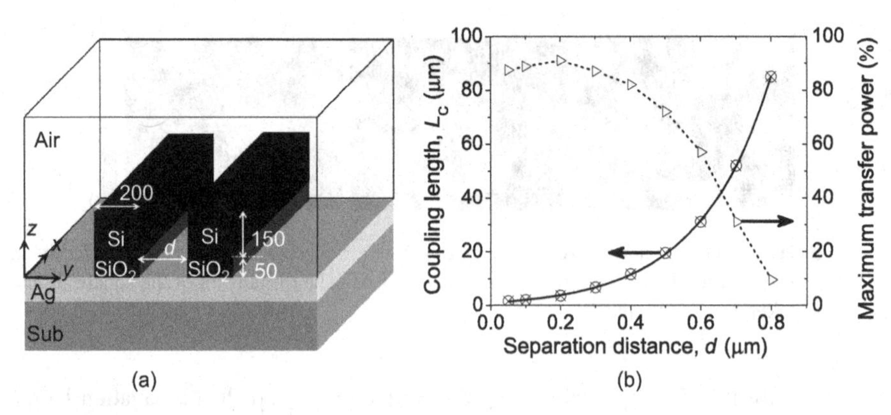

Figure 5.12 (a) The proposed vertical hybrid plasmonic waveguide-based directional coupler. (b) The dependences of the coupling length L_c and maximum transmitted power on the separation distance d between two parallel straight waveguides. Adapted from Ref. [24].

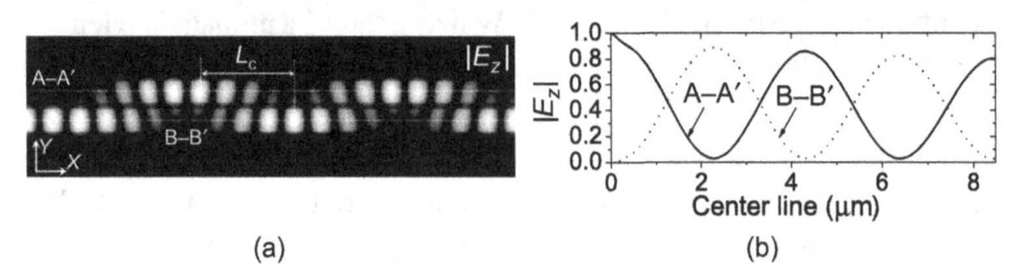

Figure 5.13 (a) The near-field distribution of the E_z component in the x–y plane, at the center of the SiO_2 stripe, is calculated for the case $d = 100$ nm, which provides a coupling length of 2 μm and a maximum power transfer of 89% (left-hand side). (b) The $|E_z|$ component extracted from the lines A–A′ and B–B′.

be 1.55 μm shorter than the other one to ensure the direct excitation of plasmonic waves over one waveguide only. The dependence of the coupling length (the length at which the energy is completely passed from one waveguide to the other) and the maximum power transfer (the percentage of energy passed from one waveguide to the other at the coupling length) on the separation distance is shown in Fig. 5.12(b). The coupling length L_c decreases exponentially as the separation distance d decreases, and their relationship can be expressed by $L_c = 1.72e^{d/0.21} - 0.74$, which is obtained by using an exponential curve-fitting procedure. This result provides a guideline to design the waveguide coupler to achieve a low level of cross-talk between coupled waveguides. As an example of efficient waveguide-coupler design, when a separation distance as small as 100 nm is used, the maximum power transfer of 89% is observed, and the coupling length is 2 μm. The field intensity in the coupler (Fig. 5.13) demonstrates that the signal is efficiently coupled from one waveguide to another over a small separation distance. As observed in Fig. 5.13, after a coupling length L_c the transmitted power is fully switched from one waveguide to the other. This investigation shows that it is possible to realize a compact, high-performance coupler or power splitter with the proposed VHPW structure.

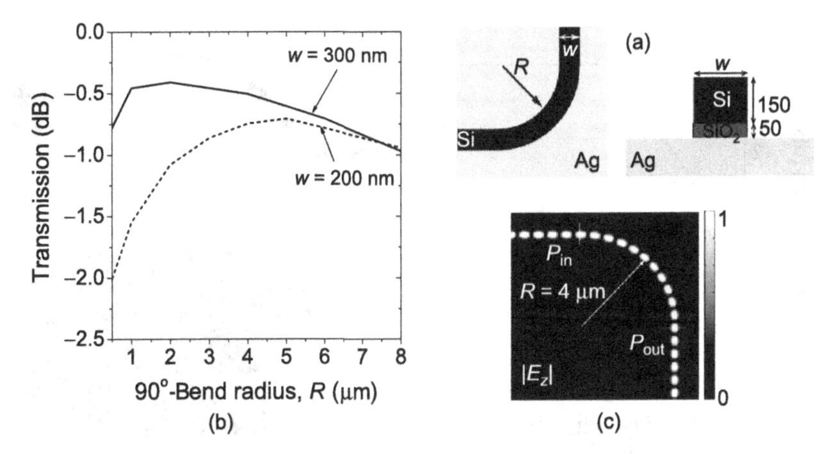

Figure 5.14 (a) Top view and cross-sectional view of the 90° sharp waveguide bend. (b) The power transmitted by the 90° sharp waveguide bend as a function of the bend radius R for two waveguide widths, $w = 200\,\text{nm}$ and 300 nm. (c) The near-field intensity distribution, $|E_z|^2$, for a bend radius of $R = 4\,\mu\text{m}$.

5.2.6 Waveguide bends

The compact size and best tradeoff between the confinement and the propagation loss in the VHPW offer prospects for building various plasmonic waveguide-based components with excellent optical performance. In this section, we investigate the performance of two popular waveguide bend shapes, namely 90°-bend and S-bend VHPW waveguides. Bends are considered necessary building blocks of state-of-the-art nanoscale integrated electronic–photonic circuits, so they have always been of interest in this context. To achieve a denser integrated circuit, the waveguide bends must be small and of low bend loss [39].

90°-Bend waveguides

The power transmitted through a 90° sharp waveguide bend with respect to two sets of waveguide dimensions $[w, h] = [200, 150]$ and $[300, 150]$ nm is studied. Note that the thickness of the SiO_2 stripe is maintained at $t = 50\,\text{nm}$, at which the high-intensity single mode can be efficiently guided in a small area of $50\,\text{nm} \times 200$ nm or $50\,\text{nm} \times 300$ nm of the SiO_2 stripe in these two cases. The 90°-bend structure is shown in Fig. 5.14(a). The transmitted power through the 90° bend, $T = 10\log_{10}(P_{\text{out}}/P_{\text{in}})$, is obtained as a ratio between the power flow integrals in the SiO_2-stripe core of the waveguide at the beginning, P_{in}, and at the end, P_{out}, of the bend section. These transmitted powers as a function of the bend radius, Fig. 5.14(b), are numerically calculated at $\lambda = 1.55\,\mu\text{m}$ by using the finite-difference time-domain method (FDTD) [40].

It is observed that the maximum of the transmitted power for $w = 200\,\text{nm}$ and 300 nm is achieved at bend radii of $R = 5\,\mu\text{m}$ and $2\,\mu\text{m}$, respectively. The transmitted power is better when the bend radius is greater, since a smaller bend radius causes a higher radiation loss, and the loss is greater when the radius is less than $1\,\mu\text{m}$ and $3\,\mu\text{m}$ for $w = 200\,\text{nm}$ and 300 nm, respectively. However, the transmission power starts

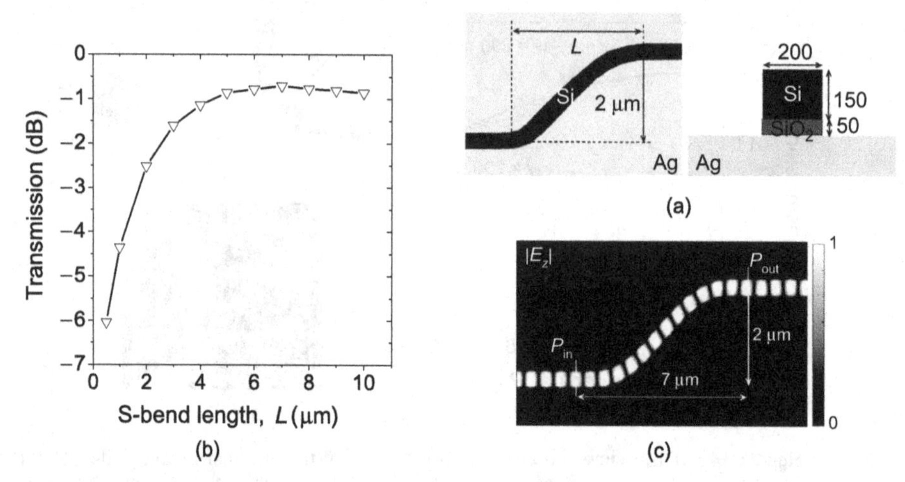

Figure 5.15 (a) Top and cross-sectional views of the simulated structure together with its dimensions. (b) Power transmission through an S-shaped waveguide bend with a lateral distance $D = 2\ \mu\text{m}$ as a function of the length L. (c) The near-field intensity distribution, $|E_z|^2$, for a bend length of $L = 7\ \mu\text{m}$.

decreasing again when the bend radius exceeds 4 μm with $w = 200$ nm and 6 μm with $w = 300$ nm. This occurs because the ohmic loss increases with the length of the bend. As an example, with $w = 300$ nm, a transmission loss of 0.3 dB can be achieved with the optimal radius of 2 μm. We also show, in Fig. 5.14(c), the electric-field intensity $|E_z|^2$ for $R = 4\ \mu\text{m}$ and $w = 200$ nm. This electric field is taken at the mid-plane of the SiO_2 stripe. The result again confirms that the electric field is well confined and efficiently bent in the 90° sharp waveguide bend.

S-bend waveguides

In integrated photonic circuits, it is common to see that each component has a separation distance. Thus the connection between different straight waveguide-based components is highly required to offset them with respect to each other. Apart from the 90° sharp waveguide bend, the S-shaped waveguide bend, abbreviated to S-bend, is a desirable solution for such connections. This kind of S-bend connection may help to reduce the footprint of the integrated circuit while maintaining a low transmission loss when energy is exchanged between two components. The S-bend shape can be represented by different mathematical expressions, for example a general polynomial family of curves (P-curves) has been suggested to link two arbitrary points on the optical circuit. Among them, S-bends based on the cosine and sine functions are the most desirable to implement. They can, respectively, be expressed as $y = [D(1 - \cos(\pi x/L))]/2$ for the cosine function and $y = Dx/L - [D/(2\pi)]\sin(2\pi x/L)$ for the sine function, where D is the separation distance and L the length of the S-bend [41]. According to the work reported in Ref. [20], the optimal dimensions for an S-bend in a planar dielectric waveguide are those given by the cosine function specified previously.

The VHPW S-bend structure is shown schematically in Fig. 5.15(a) with the left-hand diagram showing the top view of the S-bend while the right-hand diagram is a cross-sectional view. For this study, the distance D is simply chosen to be 2 μm to ensure strong isolation between the input and output arms, and the dimensions of Si nanowire are taken to be $h = 150$ nm and $t = 50$ nm. The power transmitted through the S-bend, including the radiation and ohmic loss (material loss of silver), $T = 10\log_{10}(P_{out}/P_{in})$, is obtained as a ratio between the power-flow integrals in the SiO_2-stripe core of the waveguide at the beginning, P_{in}, and at the end, P_{out}, of the bend section. Figure 5.15(b) shows the transmitted power as a function of the bend length. The results show that the total transmission can reach 85% for a length of 7 μm (total insertion loss of 0.7 dB), with an active area of SiO_2 stripe given by $w = 200$ nm and $t = 50$ nm. The electric-field intensity $|E_z|^2$, shown in Fig. 5.15(c), demonstrates that the signal is efficiently guided and bent through the S-bend waveguide. Moreover, the best tradeoff between the ohmic loss and the radiation loss is achieved when the length of the S-shaped bend is 7 μm and the distance $D = 2$ μm.

5.2.7 Power splitters

Power splitters are passive devices used in the field of integrated circuits. They split a defined amount of the electromagnetic power in a waveguide off to a port, enabling the signal to be used in another component, device, or channel. An essential feature of power splitters is that they split only power flowing in one direction. Powers entering the different output ports must be isolated from each other as much as possible.

The results obtained with the Si-based vertical hybrid dielectric-loaded plasmonic waveguide (VHPW) encouraged us to further exploit the potential of a VHPW to realize different functional plasmonic components [42]. As an example, we demonstrate a low-loss Y-splitter that can serve as a power splitter or Mach–Zehnder interferometer. Figure 5.16(a) shows the top and cross-sectional views of the Y-splitter structure consisting of two symmetrical mirrored S-bends. The coordinates of each S-bend were in the S-bend discussion above as $y = [(D/2)(1 - \cos(\pi x/L))]/4$, where D is the separation between the two parallel output arms of the Y-splitter, and L is the distance between the straight sections of the input and output arms. The power transmitted through each output arm of the Y-splitter, T_{upper} (dB) $= 10\log_{10}(2P_{upper}/P_{in})$, is obtained as the ratio of the power-flow integral in the SiO_2-stripe core of the waveguide at the beginning, P_{in}, to the one at the upper output arm, P_{upper}, or the lower output arm, P_{lower}, of the bend section, where the weighting factor of 2 represents the case of two output arms in which the power is divided equally, 50%–50%. These transmitted powers as functions of bend radii are numerically calculated by the FDTD method. For simplicity, we consider a symmetrical Y-splitter power, which means the upper and lower output arms having the same transmitted power, i.e. $T_{upper} = T_{lower}$. The normalized power transmission efficiency of the Y-splitter as a function of the length L at the wavelength of 1.55 μm is plotted in Fig. 5.16(b). The numerical results have been computed for three different distances, $D = 1, 2$, and 4 μm. The results show that the power transmission loss is smaller than 0.4 dB for a small value, $D = 1$ μm, and 0.7 dB for a large one, $D = 4$ μm. The transmitted power depends nonlinearly on the length L, since a smaller length L causes a higher radiation loss, and the loss is greater when the length L is less than

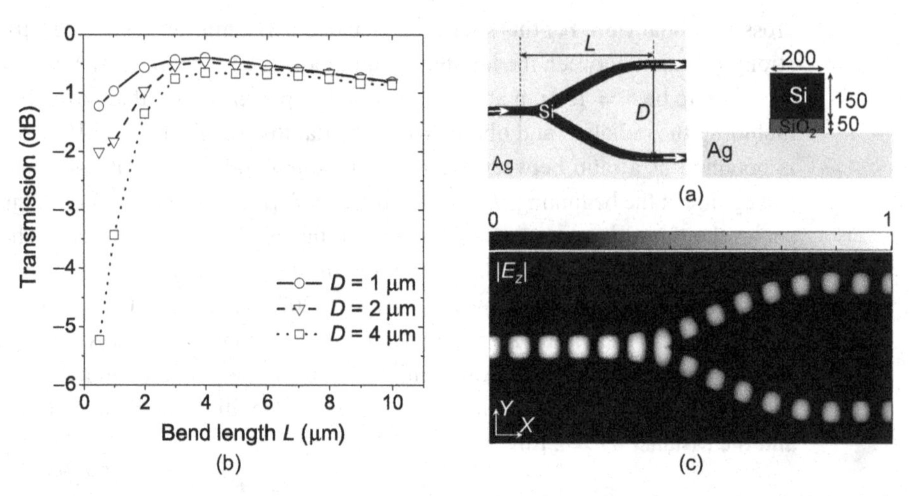

Figure 5.16 (a) Top and cross-sectional views of the vertical hybrid plasmonic waveguide-based Y-shaped power splitter together with its dimensions. (b) The transmitted power efficiency of the output arms as a function of the length L, with respect to different lateral distances $D = 1, 2$, and 4 μm. The dimensions of the Si nanowire are taken to be $w = 200$ nm and $h = 150$ nm, and the SiO_2-stripe thickness is 50 nm. (c) Near-field distributions of the $|E_z|$ component for $D = 4$ μm and $L = 7$ μm. Adapted from Ref. [42].

2 μm. On the other hand, the transmitted power starts decreasing again when the length L exceeds 4 μm. This occurs because the ohmic loss increases with the length of the bend. As a result, the best tradeoff between the ohmic loss and the radiation loss is achieved when the length L of the Y-splitter is approximately 4 μm for different distances D. Moreover, the electric-field intensity $|E_z|$, shown in Fig. 5.16(c) for $D = L = 4$ μm, demonstrates that the light signal at $\lambda = 1.55$ μm is equally divided into the two output arms. The results show that the proposed Y-splitter provides a compact footprint, low loss, and the feasibility to build truly silicon-based plasmonic circuitry.

We are also interested in the optical performance of the asymmetrical Y-splitter. By changing the splitter geometry it is possible to create a splitter having different amounts of transmitted power at the output arm. In this study, we assume that the Y-splitter length is 4 μm and the arm distance of the upper arm is kept constant at 1 μm while the arm distance d of the lower arm is varied from 0 to 2 μm. We show, in Fig. 5.17(a), the transmitted power at the output arm as a function of d. Note that the arm distance is measured from the center of each arm to the middle line interfacing between two arms. The illustration of the asymmetrical Y-splitter power sketched in Fig. 5.17(b) relates to two cases: $d = 0$ μm and $d = 1$ μm. It can be seen that changing the symmetry of the Y-splitter by shifting the output arm vertically from the symmetry axis by a distance of d can provide splitting of the output power. In particular, with d varying from 0 to 2 μm, the output-arm transmitted power in comparison with the input-arm power may nonlinearly change from -0.1 dB to -0.55 dB for the upper arm and from -0.25 dB to -1.2 dB for the lower arm. The result shows that there are two positions providing equal transmitted power at the two output arms, namely $d = 1$ μm for the symmetrical

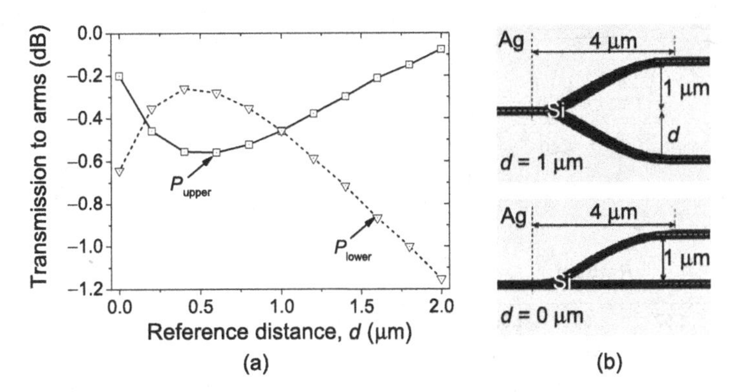

Figure 5.17 (a) The transmitted power as a function of the lower output arm d, for a length $L = 4\ \mu m$ and an upper output-arm distance of 1 μm. The dimensions of the Si nanowire are taken to be $w = 200$ nm and $h = 150$ nm, and the SiO$_2$-stripe thickness is 50 nm. (b) A schematic illustration of two critical cases of an asymmetrical Y-splitter: $d = 0$ and 1 μm.

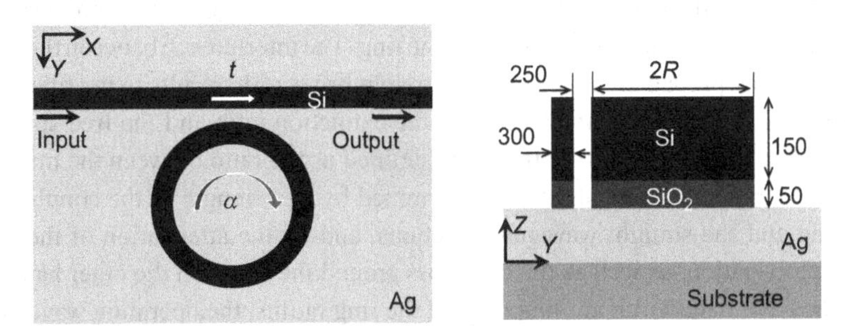

Figure 5.18 The composition and geometry of the VHPW-based ring resonator notch filter (RNF) in top and front views.

case and $d = 0.16\ \mu m$. This investigation provides a useful guideline for the design of different Y-splitters with a fixed value of the transmitted power.

5.2.8 Ring resonator filters

Optical wavelength-selective components based on waveguide-based ring resonator filters are important elements because they are frequently used in nanoscale electronic–photonic integrated circuits. They are required for applications such as optical switching, electro-optical switching, wavelength conversion, and filtering [13, 43–46]. Our main interest here is to demonstrate how the VHPW can be used as a waveguide platform to efficiently realize such high-performance plasmonic ring resonator filters.

Ring resonator notch filters

Given the advantages of the VHPW, we have built a VHPW-based ring resonator notch filter (RNF). It is formed by a ring waveguide placed in the proximity of a straight bus waveguide to allow optical coupling between them. The RNF studied here is sketched schematically in Fig. 5.18. A portion of the optical power propagating in the straight

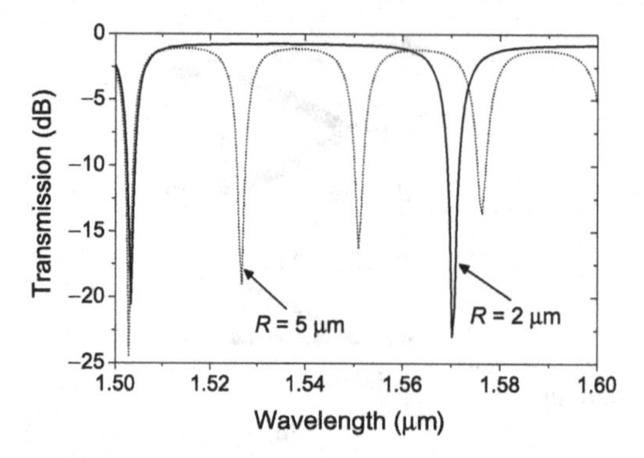

Figure 5.19 The transmission of the ring resonator notch filter as a function of wavelength for ring radii of $R = 2\ \mu m$ and $5\ \mu m$, and a gap width of $g = 250\ nm$.

waveguide (propagating mode) is coupled to the ring and excites a circulating mode of the RNF. The efficiency of this coupling depends on the gap between the ring and the straight waveguide, and on the radius of the ring. The interference between the circulating mode and the propagating mode in the straight waveguide results in the filter properties of the RNF, which are characterized by the extinction ratio and the free-spectral range (FSR). The extinction ratio of the RNF (defined as the ratio between the minimum and maximum transmission outputs) is determined by the strength of the coupling between the ring and the straight waveguide sections, and by the attenuation of the plasmonic wave propagation, as well as the bend loss around the ring. On the other hand, the FSR and the RNF bandwidth are functions of the ring radius, the operating wavelength, and the effective mode index of the ring waveguide. In general, the transmission power P_{out} at the output of the RNF is expected to be periodic with respect to the phase accumulated by the ring mode per circulation and is given by the following expression [47]:

$$P_{out} = \exp\left(-\frac{l}{L_p}\right)\frac{\alpha^2 + t^2 - 2\alpha t\cos\theta}{1 + (\alpha t)^2 - 2\alpha t\cos\theta}, \tag{5.3}$$

where $\theta = (2\pi/\lambda)n_{eff}2\pi R$ is the phase change around the ring. The first (exponential) factor of Eq. (5.3) describes the power loss incurred over the propagation distance between the input and output monitors of $l = 10\ \mu m$ related to the VHPW propagation length $L_p = 57\ \mu m$. Note that the value of L_p is calculated for Si-nanowire dimensions of $w = 300\ nm$ and $h = 150\ nm$, and the SiO_2-stripe thickness of 50 nm. The coefficient α is the mode transmission around the ring, t is the mode transmission through the coupling region in the straight waveguide [47], λ is the free-space wavelength in the near-infrared spectrum, R is the ring radius, and n_{eff} is the waveguide's effective index as a function of different input wavelengths of the excitation signal.

Figure 5.19 shows the transmission coefficient as a function of the wavelength, calculated with the FDTD method for the optimized gap distance of 250 nm and two different ring radii, $R = 2\ \mu m$ and $5\ \mu m$. It appears that the RNF achieves a transmission loss less than 0.8 dB and 1.1 dB together with very high extinction ratios (ERs) of 21 dB and 20 dB for $R = 2\ \mu m$ and $5\ \mu m$, respectively. Moreover, the bandwidth $\Delta\lambda_{FWHM}$ calculated from the full spectral range at $T_{HM} = (T_{max} + T_{min})/2$ is 6 nm at

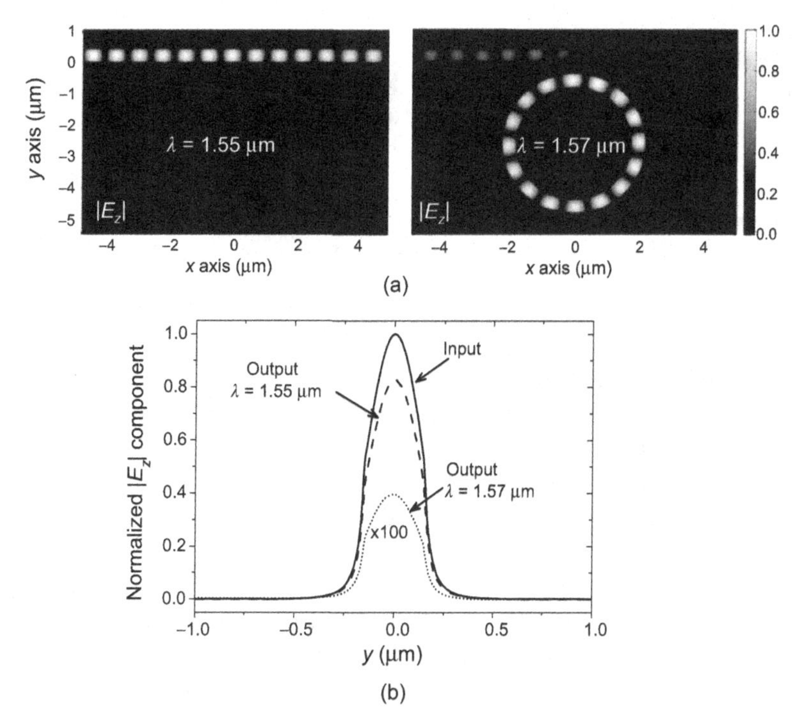

Figure 5.20 (a) The electric-field intensity distribution $|E_z|^2$ in the ring resonator (RR) with respect to the non-resonance wavelength $\lambda = 1.55$ μm and the resonance wavelength of $\lambda = 1.57$ μm. (b) Normalized output intensities $|E_z|^2$ along the y axis in the middle of the SiO_2 stripe with respect to the input and output channel at $\lambda = 1.55$ μm and $\lambda = 1.57$ μm (value multiplied by 100). The RR dimensions are $R = 2$ μm and $g = 250$ nm.

the resonance wavelength of $\lambda = 1.57$ μm with the ring radius of 2 μm. A large FSR of 66 nm is also achieved with the same RNF structure, which may accommodate up to 11 channels.

To show the operating regimes of the proposed RNF at $R = 2$ μm, we plot, in two dimensions, the electric-field intensity distribution $|E_z|^2$, in the x–y plane and in the middle of the SiO_2 stripe (25 nm above the Ag film), for two wavelengths: a non-resonance wavelength of $\lambda = 1.55$ μm and a resonance wavelength 1.57 μm (Fig. 5.20(a)). The left-hand image shows $|E_z|^2$ at 1.55 μm, for which case almost all the light is transmitted to the output channel. The right-hand image is for 1.57 μm, in which case we begin to see light resonate more strongly in the ring of the RNF. We also calculate, in Fig. 5.20(b), the normalized intensities $|E_z|^2$ along the y axis and in the middle of the SiO_2 stripe of the input and output channels for the two previous wavelengths. Note that, because the intensity value at $\lambda = 1.57$ μm is too small, it is multiplied by 100 to show the curve more clearly. The intensity at the output waveguide is normalized with respect to the input intensity. The results obtained again demonstrate a high contrast of 24 dB at the RNF output for the wavelengths corresponding to constructive and destructive interference.

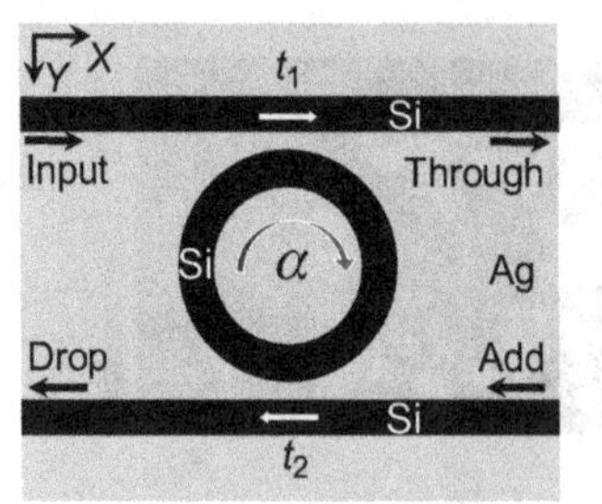

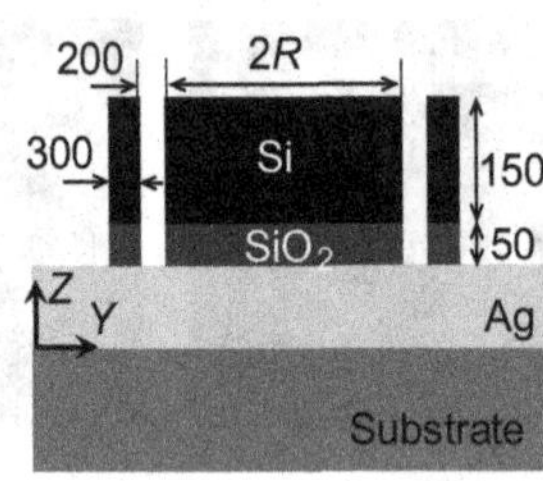

Figure 5.21 The composition and geometry of the VHPF ADF in top and front views.

Ring resonator add–drop filters

The second wavelength-selective component investigated is an add–drop filter (ADF). This component has been used widely in wavelength-division multiplexing systems for multiplexing and routing different channels of optical signals in data-transmission applications. The layout of the proposed VHPW-based ring resonator ADF is shown in Fig. 5.21.

The optical performance of the ADF generally depends on the materials, the composition of the structure, the ring radius, and the gaps between ring and bus waveguides [44, 47]. As a result, the power transmitted from the input to the through port P_T and drop port P_D is expressed by the following equations:

$$P_T = \exp\left(-\frac{l_T}{L_p}\right)\frac{(\alpha t_2)^2 + t_1{}^2 - 2\alpha t_1 t_2 \cos\theta}{1 + (\alpha t_1 t_2)^2 - 2\alpha t_1 t_2 \cos\theta},$$

$$P_D = \exp\left(-\frac{l_D}{L_p}\right)\frac{(1 - t_1{}^2)(1 - t_2{}^2)\alpha}{1 + (\alpha t_1 t_2)^2 - 2\alpha t_1 t_2 \cos\theta}, \tag{5.4}$$

where $\theta = (2\pi/\lambda)n_{\text{eff}}2\pi R$ is the phase change around the ring. The first (exponential) factor in the above equations describes the power loss incurred over the propagation distance from the input to the through-port, l_T, and drop-port, l_D, monitors, respectively, for the VHPW propagation length $L_p = 57\,\mu\text{m}$ at $w = 300\,\text{nm}$ and $h = 150\,\text{nm}$; α is the mode transmission around the ring; t_1 and t_2 are the mode transmissions through the coupling regions in the bus waveguides [44, 47]; λ is the free-space wavelength; R is the ring radius; and n_{eff} is the VHPW effective index as a function of different input wavelengths of the excitation signal. Note that the power transferred from the input to the add port is theoretically equal to zero. In this design, the ring radius is arbitrarily chosen to be $R = 2\,\mu\text{m}$. At this given ring radius, the gap between the straight and the ring waveguides has been optimized to 200 nm to achieve the best possible performance of the ADF. The power transmission of this ADF is shown in Fig. 5.22 as a function of the wavelength at the through, drop, and add ports. The observed transmission loss is 0.9 dB, while the FSR of the ADF is similar to that of the previous RNF for the same ring radius. The ERs of the ADF are higher than 12 dB both for the through channel and for the drop channel. The channel isolation, defined as the level difference of the output power in the through, drop and add ports at the same wavelength, is approximately 10 dB at the resonance wavelength of $\lambda = 1.566\,\mu\text{m}$, and 14 dB at

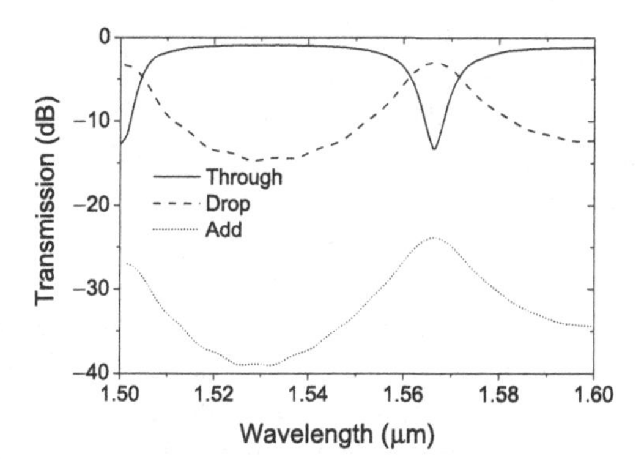

Figure 5.22 Transmission at the through, drop, and add ports of the ADF for a ring.

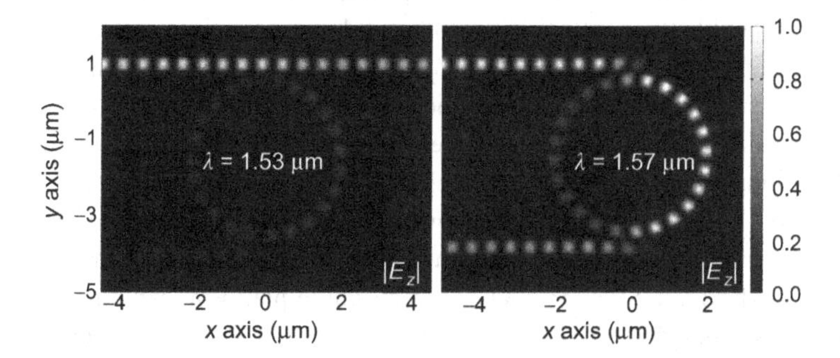

Figure 5.23 The near-field intensity distribution $|E_z|^2$ at 25 nm above the silver film for the ADF with $R = 2$ μm calculated at the non-resonance wavelength of 1.53 μm and at the resonance wavelength of 1.566 μm.

the non-resonance wavelength. Moreover, the bandwidth of −3 dB at the resonance wavelength of $\lambda = 1.566$ μm is achieved at the value of 13 nm, which corresponds to a quality factor of 120. On the other hand, we also observe that the power at the add port approximately reaches zero, which is consistent with the theoretical prediction. The electric-field intensity $|E_z|^2$ in Fig. 5.23 demonstrates that the optical signal is efficiently guided from the input channel to the through channel or the drop channel at different input wavelengths. The left-hand image shows $|E_z|^2$ at 1.53 μm, where almost all the light is transmitted to the through channel. The right-hand image is for 1.566 μm, at which we begin to see light perfectly coupling to the drop channel of the ADF.

5.3 Complementary metal–oxide–semiconductor compatible hybrid plasmonic waveguide devices

Nowadays, complementary metal–oxide–semiconductor (CMOS) technology is standard for constructing electronic–photonic integrated circuits. Silicon photonics has emerged as a viable photonics platform owing to the maturity of the silicon manufacturing procedures and the demonstrated performance of CMOS-compatible photonic components. This technology is envisioned to propel optical interconnect applications that demand high-volume and low-cost production of electronic–photonic integrated circuits (EPICs). As a result, it will impact other application areas, including telecommunications, sensing, spectroscopy, and biology. Therefore, CMOS-compatible Si-based plasmonic components and devices are decidedly essential elements with which to build the EPICs. Although Ag and Au are the most commonly used metals for plasmonic devices due to their low ohmic loss and good resistance to oxidation by the environment, neither of them is compatible with standard silicon CMOS processes. For such metals one usually requires a special process such as electron-beam lithography or focused-ion-beam lithography to fabricate plasmonic devices when realizing an experiment. Therefore, in terms of practical implementation in the existing Si EPICs, it is highly desirable to use CMOS-compatible metals (e.g. Al, Cu, and silicides) and industry-standard lithographic processes. In response to the need for development of CMOS-compatible plasmonic components and devices to fully implement in nanoscale EPICs, starting with a cursory review of CMOS-compatible plasmonic materials, we report the optical constant in terms of the real and imaginary permittivity both of metals and of silicides over the near-infrared spectrum to demonstrate that copper is the CMOS-compatible plasmonic metal of choice. Then, the propagation characteristics of the vertical Cu–SiO_2–Si hybrid plasmonic waveguide platform with a metal cap are investigated both in theory and experimentally. In subsequent sections, relating to the horizontal hybrid Cu–SiO_2–Si–SiO_2–Cu waveguide platform, experimental and theoretical works conducted on passive CMOS-compatible plasmonic elements, including a waveguide bend and submicron ring resonator filter, are reviewed and discussed.

5.3.1 CMOS-compatible plasmonic materials

Ag and Au are the most commonly used metals for plasmonic devices, but neither of them is compatible with CMOS processes. In search of alternatives, we calculate the permittivity values at near-infrared wavelengths in the range from 1 μm to 2 μm for two well-known CMOS metals (Cu and Al) and silicides (NiSi and $CoSi_2$). Their optical constants are taken from Ref. [48] for copper, Ref. [49] for aluminum, and Ref. [50] for silicides. As shown in Fig. 5.24, the real part of the permittivity is an indication of how the electrons are driven by the electromagnetic field. A large negative real part of permittivity is usually desirable for plasmonic devices owing to the small light penetration depth. As the wavelength decreases, the real permittivity decreases in magnitude for all of the CMOS plasmonic materials investigated here. The imaginary part of the permittivity represents the ohmic loss in the metal. Although Al has the largest negative real part of the permittivity, its imaginary part is also very high, which makes its plasmonic waveguide platform less attractive than Cu. The best combination of a large negative real part and a small imaginary part of the permittivity indicates that Cu is a viable metal for CMOS plasmonics, especially at near-infrared wavelengths. Fortunately, current Si-based integrated-circuit technology already uses Cu nanostructures to route and manipulate electronic signals between and within transistors on a chip. This mature

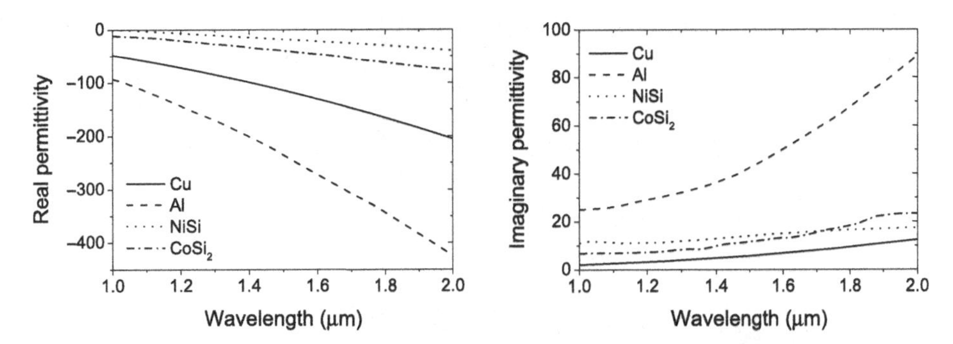

Figure 5.24 Permittivity values of different CMOS-compatible plasmonic metals and silicides: Cu, Al, NiSi, and $CoSi_2$.

processing technology can thus be used to our advantage in integrating plasmonic devices with nanoelectronic circuits.

5.3.2 Vertical hybrid Cu–SiO_2–Si plasmonic waveguide devices

As mentioned above, the VHPW has been touted as a potential candidate to achieve the best tradeoff between strong optical field confinement and a long propagation distance. As a result, various waveguide-based components such as bends, couplers, power splitters, and filters have been studied extensively both in theory and by numerical modeling. Recently, thanks to great advances in the fabrication process, these kinds of structure have been fabricated and characterized [26–28, 34]. In addition, metal-cap VHPWs, also called metal-cap hybrid plasmonic waveguides, have been fabricated using patterned Au as the metal [29, 51]. In general, people have used lithography technology, including electron-beam lithography and focused-ion-beam lithography, to realize different hybrid plasmonic waveguide-based components. As an example, the focused ion beam has been used to fabricate several VHPWs that are prepared by evaporating a high-index semiconductor, ZnS ($n = 2.2$), onto an Ag film separated by a thin 10-nm film of MgF_2 from a quartz substrate [28]. Moreover, the metal-cap hybrid plasmonic waveguide reported in Ref. [29] used patterned Au as the metal. Nanoscale Au patterning requires expensive electron beam lithography (EBL), which is not an industry-standard technology. A self-aligned approach has been proposed in order to avoid the need for EBL [51], but it is viable only for Si waveguides formed by the local oxidation process, which suffer from blunt side walls. Instead, a technology developed for the top-down fabrication of nanowire transistors can realize nanoscale Si beams using industry-standard lithograhy [52]. Using the technology, fully CMOS-compatible vertical Cu–SiO_2–Si hybrid plasmonic waveguides can be realized with the Si core width. In terms of CMOS compatibility, Al or Cu should be used as the metal. At the telecommunications wavelength of 1550 nm, Cu is found to be much better than Al since it provides much lower propagation loss in the same waveguide configuration (as discussed in the previous section and also in Ref. [32]). Considering the above two points, a Cu–SiO_2–Si VHPW has been proposed and experimentally demonstrated on silicon-on-insulator (SOI) substrates using standard Si-CMOS technology [30, 53].

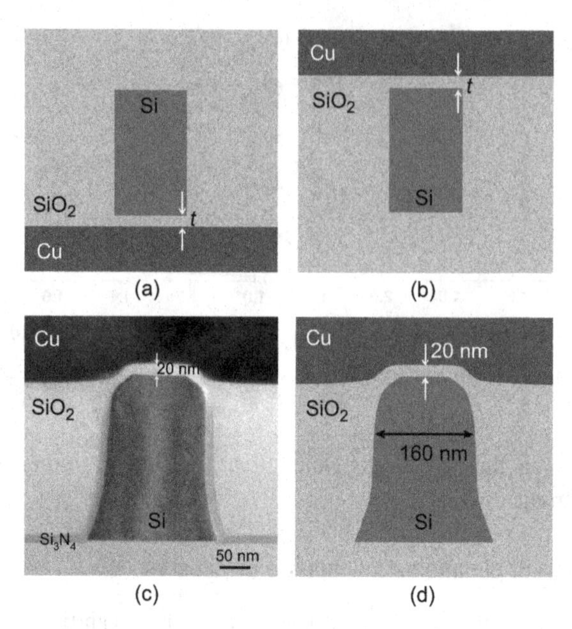

Figure 5.25 (a) A Cu-based vertical hybrid plasmonic waveguide (VHPW) following the previously studied waveguide. (b) A Cu-cap VHPW for the standard Si-CMOS technology. (c) A cross-sectional transmission electron microscope (XTEM) image of a given fabricated Cu-cap VHPW for the SiO_2 gap of 20 nm. (d) The Cu-cap VHPW imported from the XTEM image of the fabricated waveguide for the numerical simulation.

In particular, we discuss the propagation characteristics of the copper-based hybrid plasmonic waveguide both in theory and in experiments.

The VHPW with the CMOS-compatible plasmonic metal of copper at the bottom as the substrate cannot be realized by using standard Si-CMOS technology (Fig. 5.25(a)). Next, as shown in Fig. 5.25(b), we adopt the experimental realization of the conductor gap silicon plasmonic waveguide (i.e. the metal-cap VHPW) for the proposed Cu–SiO_2–Si hybrid plasmonic waveguide platform. Note that the proposed fabrication has used the self-aligned approach with a very wide Cu layer in order to avoid blunt side-wall effects. The waveguide platform is fabricated on SOI wafers with a 340-nm-thick top Si layer and SiO_2 buried at a depth of 2 μm. During patterning the Si core of the Cu-cap VHPW, the tapered couplers, and the input/output Si photonic waveguides must be made simultaneously by using UV lithography. The exposure condition is intentionally varied to result in different critical dimensions of the Si core. A photoresist-trimming process is carried out to further reduce the critical dimension. Using this method, several Cu-cap VHPWs with different Si-core widths are fabricated using the same mask. Then, SiO_2/Si_3N_4 deposition, SiO_2 chemical mechanical polishing (CMP) (using Si_3N_4 as the CMP stopping layer), SiO_2/Si_3N_4 deposition again, SiO_2 window opening (using Si_3N_4 as the SiO_2 dry-etching stopping layer), wet etching of remaining Si_3N_4 in the windows, thermal oxidization to grow a thin SiO_2 layer, Cu deposition, and Cu CMP (to remove Cu outside the windows) are carried out sequentially. The details of the fabrication can be

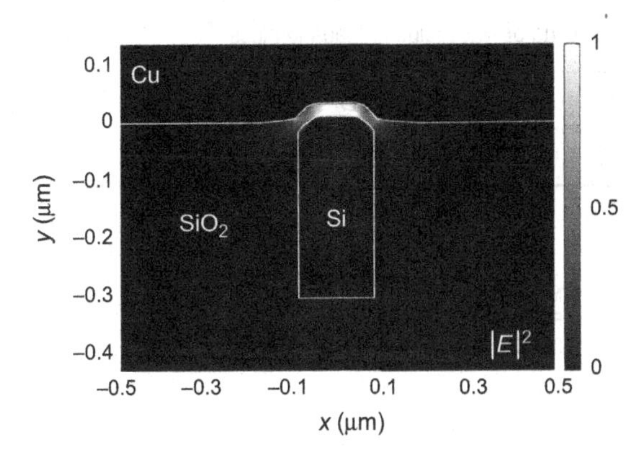

Figure 5.26 The $|E|^2$ intensity distribution of the fundamental quasi-TM mode in the Cu-cap VHPW calculated using the full-vectorial finite-difference (FVTD) method.

found elsewhere [30, 53]. The thickness (t_{SiO_2}) of the thin SiO_2 layer between the Si core and the metal is a critical parameter to determine the property of the hybrid plasmonic waveguide (see Section 5.2.2), thus it needs to be carefully controlled in order to get the precise thickness desired. Figure 5.25(c) shows a cross-sectional transmission electron microscopy (XTEM) image of a given fabricated Cu-cap VHPW. This cross-sectional waveguide consists of a 20-nm-thick SiO_2 layer sandwiched between Si-core nanowire of width 165 nm at middle plane and thickness 304 nm, as well as a wide Cu cap layer on top. It is worthy of note that the exact shape and dimension of the fabricated waveguide need to be used in the modeling and simulation of the propagation characteristics. Owing to the imperfection of the fabrication process the Si core can be rectangle-, triangle-, or polygon-like in shape. Figure 5.25(d) sketches the Cu-cap VHPW structure used in the theoretical calculation, which is imported from the XTEM image of the fabricated waveguide.

Since the Cu-cap VHPW is coupled to the conventional Si photonic waveguide via tapered couplers, the waveguides can be measured as a conventional Si waveguide chip [54, 55]. A beam of 1550-nm transverse electric (TE)-polarized light from a tunable laser is coupled into the input silicon waveguide through a lensed polarization-maintaining fiber. Straight VHPWs of different lengths are fabricated and measured. According to the linear curve fitting of the propagation loss as a function of waveguide length, we obtain a measured value of the propagation loss as 0.122 dB μm^{-1}. In order to validate the experimental result, both types of full-wave numerical simulation, i.e. full-vectorial finite-difference (FVFD) and finite-difference time-domain (FDTD) methods [37, 40], are used to calculate the propagation loss.

In Fig. 5.26, we show the $|E|^2$ intensity distribution of the fundamental quasi-TM mode of the Cu-cap VHPW calculated by using the FVFD method. The result indicates that the electric-field intensity is strongly enhanced in a narrow 20-nm-thick SiO_2 gap and confined along the curved interface of the Cu cap.

The FVFD method is used to calculate the complex value of the mode effective refractive index (Section 5.2.1), where the propagation loss is deduced from the propagation

Table 5.1 Measurement and modeling of waveguide propagation loss

Case	Measurement	FDTD	FVTD
Loss (dB μm^{-1})	0.122	0.13	0.123

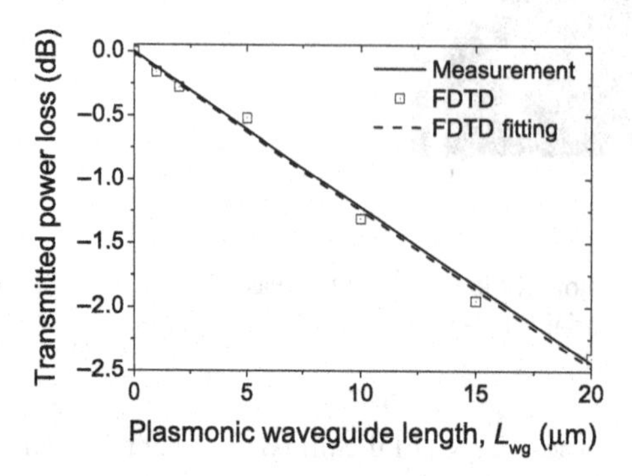

Figure 5.27 A comparison between the modeling and measurement of the transmitted power (propagation loss) of a Cu-cap VHPW as a function of the waveguide length.

length at which the energy of the SPP decays to 1/e of its original value due to ohmic losses. The relation between the propagation loss (in dB m^{-1}) and the imaginary part of the complex value k of the mode effective index can be expressed by the formula

$$\text{Loss (dB m}^{-1}) = \frac{40\pi \log_{10}(e)k}{\lambda_0}, \tag{5.5}$$

where λ_0 (m) is the wavelength used to calculate the propagation mode.

We also compare the propagation losses arising from the measurement and FDTD calculation for different waveguide lengths. The result is plotted in Fig. 5.27. Using linear curve fitting for the FDTD calculation, a slope of 0.13 dB μm^{-1} is achieved. Good agreement between measurement and theoretical predictions (FDTD and FVFD method) is observed, and is summarized in Table 5.1.

As remarked from Figs. 5.25(c) and (d), the Si cores are not of a perfect rectangle shape, but have round shoulders at the upper two corners due to the imperfection of the Si dry-etching process. Because the plasmonic mode is strongly confined near the upper surface of the Si core, the round shoulder makes the effective Si-core width smaller than the width measured for the middle height of the Si core, especially for components with a narrower Si-core width. As an example, shown in Fig. 5.25(c), the width of the flat region of the Si-core top is only 90 nm, whereas at the cutting plane of the middle height the width of the Si core is measured to be 165 nm. In order to further understand this effect on the propagation characteristics, we calculate the complex effective index of the rectangular Si-core shape for two different widths, w = 165 nm and 90 nm. The results are presented in Table 5.2, together with those of the benchmark waveguide imported from the XTEM image. It is evident that the shape and dimension of the Si core exert an effect on the propagation characteristics of the Cu-cap VHPW. The width taken at the

Table 5.2 Waveguide propagation losses for different shapes

Case	XTEM shape	Rectangle ($w_{Si} = 165$ nm)	Rectangle ($w_{Si} = 90$ nm)
Loss (dB μm^{-1})	0.122	0.113	0.096
n_{eff}	2.1	2.15	1.8

middle height of the Si core provides a real effective index and propagation loss closer to the imported XTEM waveguide shape than the width taken at the flat region of the Si-core top. Therefore, for simplicity, in the modeling and design of Cu-cap-VHPW-based components, the width taken at the middle height of the Si-core rectangle shape is a reasonable value to choose.

5.3.3 Horizontal hybrid Cu–SiO_2–Si–SiO_2–Cu plasmonic waveguide devices

We want to finish this chapter by taking a brief look at CMOS horizontal hybrid Cu–SiO_2–Si–SiO_2–Cu plasmonic waveguides. The motivation for this waveguide platform is inspired by the combination of the best of both types of waveguide platforms: the metal–insulator–metal (MIM) and hybrid metal–insulator–semiconductor waveguides. As we know, among the different plasmonic waveguide platforms, the MIM has received special attention thanks to its truly subwavelength confinement where the modal size is determined solely by the distance between two metal stripes. However, due to the ohmic loss of the metal, the propagation distance of the MIM is typically of the order of several micrometers at frequencies around the telecommunications wavelength of 1550 nm, and it decreases with the slot width (and hence the modal size) [14–17, 57, 58]. As discussed in the previous sections, the hybrid plasmonic waveguide (HPW) has attracted substantial research interest because the motivation of this structure is twofold: (i) the exciting potential option for integrating metallic and semiconductor nanostructures to fully exploit the advantages of both metals (light concentration and high electrical conductivity) and semiconductors (light emission and photocurrent generation); and (ii) the possibility of attaining optimal tradeoff between propagation loss and power confinement. Therefore, taking the best properties of both MIM and HPW, we have developed a horizontal hybrid Cu–SiO_2–Si–SiO_2–Cu plasmonic waveguide (HHPW) platform upon which the design of a number of plasmonic components can be based. In particular, the propagation characteristics of this HHPW are numerically characterized, designed, and optimized for different refractive indexes of copper provided by various references, waveguide shapes, and dimensions. These results are also validated by experiment. Moreover, several waveguide-based components, including bends and waveguide ring resonators, are also of interest to further discuss here.

Waveguide structures

The horizontal hybrid plasmonic waveguide (HHPW) studied in this part is considered as a combination of the hybrid metal–insulator–semiconductor and metal–insulator–metal

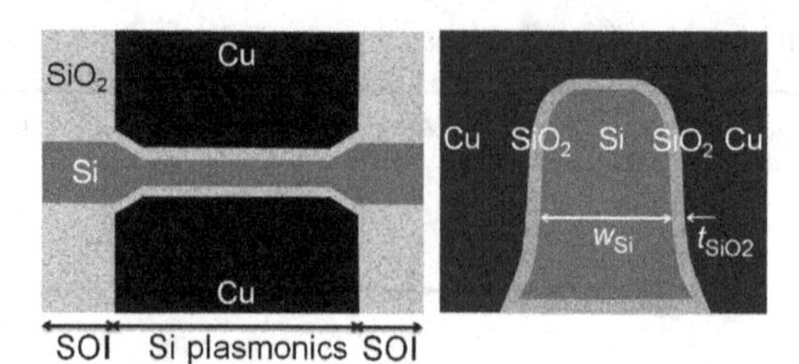

Figure 5.28 Cross-sectional and top views of a fully CMOS-compatible hybrid $Cu–SiO_2–Si–SiO_2–Cu$ plasmonic waveguide. An integration of Si-based plasmonics into silicon-on-insulator (SOI) photonics is proposed as a promising hybrid platform for nanoscale electronic–photonic integrated circuits.

plasmonic waveguides. To implement it with the standard silicon CMOS technology, the HHPW is known as a metal–insulator–semiconductor–insulator–metal waveguide, which is practically realized as a combined hybrid $Cu–SiO_2–Si–SiO_2–Cu$ structure [31, 32]. The cross-sectional view and top view are shown in Fig. 5.28. To facilitate the realization of an on-chip hybrid photonic–plasmonic platform, the integrated circuit should not only take advantage of components that have already been developed for Si photonics, but also feature low-loss interconnects for guiding light between the miniaturized components. The proposed integration scheme between Si photonic and Si plasmonic platforms is shown on the left-hand side of Fig. 5.28. The Si waveguide is designed to support the TE mode, which is coupled to the HHPW via a tapered coupler and then converted to the gap mode in the HHPW. This coupler has been discussed in detail in Ref. [32].

A standard silicon-on-insulator (SOI) wafer is used, with a rectangular Si-core nanowire of thickness 340 nm and width w. It is surrounded by a thin SiO_2 layer of thickness t. The deposition and the chemical polishing of the copper layer are the last steps. In general, the fabrication process is similar to that mentioned in Section 5.3.2. The detail of this fabrication process is also discussed elsewhere in Refs. [31, 32]. In Fig. 5.29, we summarize different fabrication processes for a particular case. Two fabricated waveguides are shown in Fig. 5.30, which presents the real cross-sectional transmission electron microcopy (XTEM) images of two fabricated plasmonic waveguides for different dimensions of Si-core width and SiO_2-layer thickness.

The fundamental TE mode is launched to excite the plasmonic mode at the wavelength of 1.55 μm, where the dominant field component is along the horizontal direction of the waveguide. We use the FVFD method to analyze the propagation characteristics of the waveguide [37]. The refractive indexes of silicon and silicon dioxide taken from [49] are used hereafter as $n_{\mathrm{Si}} = 3.48$ and $n_{\mathrm{SiO_2}} = 1.45$, respectively, while the complex permittivity of Cu is $\approx -122 + 6.2\mathrm{i}$ from Ref. [48]. We depict, in Fig. 5.31, the electric field of the optical mode in a cross-section for the waveguide dimensions $w_{\mathrm{Si}} = 50\,\mathrm{nm}$ and $t_{\mathrm{SiO_2}} = 25\,\mathrm{nm}$, which implies that the field is highly confined in the nanometer-sized gap of the SiO_2 layer. As a result, a highly effective mode volume is achieved with this waveguide thanks to this high-confinement property.

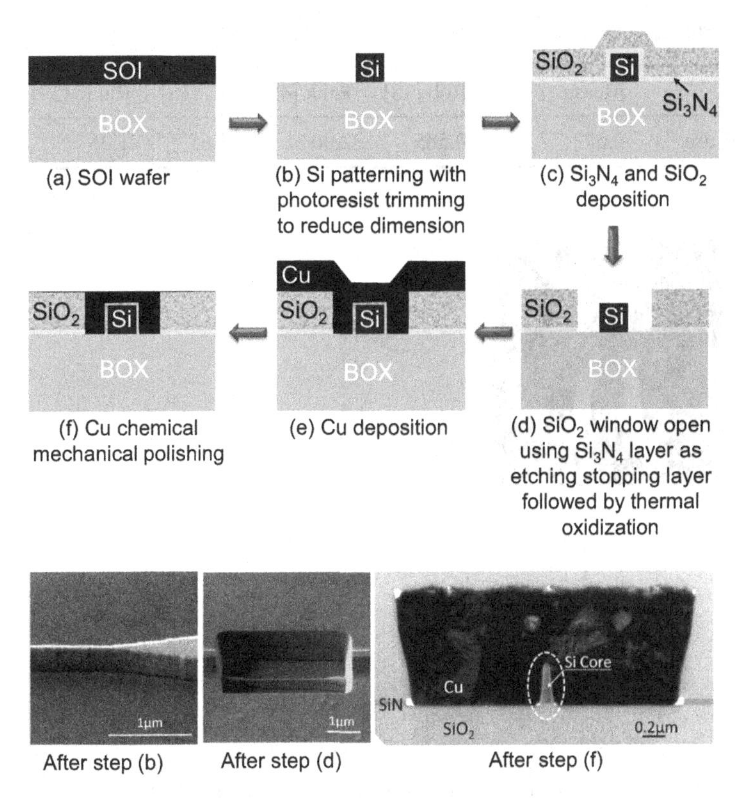

Figure 5.29 A summary of the fabrication flow to realize the hybrid Cu–SiO_2–Si–SiO_2–Cu plasmonic waveguide and its components based on the standard Si-CMOS technology.

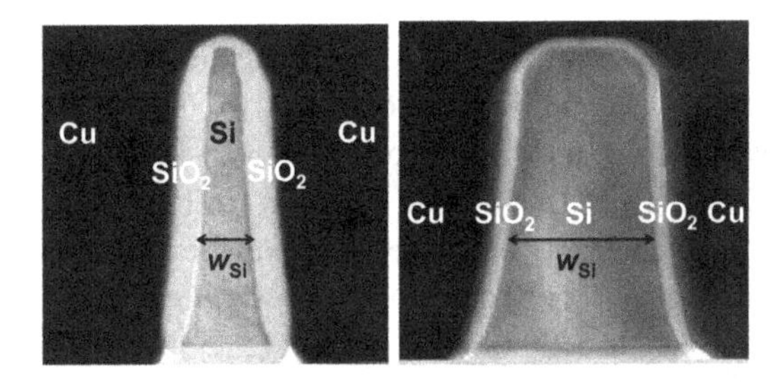

Figure 5.30 Cross-sectional transmission electron microscopy (XTEM) images of two given fabricated waveguides for different dimensions of Si-core width and SiO_2-layer thickness.

Propagation loss for different optical constant references of copper

It is clear that the optical constant, via the complex permittivity of the metal, is an important parameter in calculating the optical properties of plasmonic nanostructures due to the contribution of the attenuation and phase shift when the material is interacting with light. In particular, the imaginary part of the metal's permittivity represents the ohmic loss. Hence data from different references may provide different values for the propagation characteristics of a plasmonic waveguide. The optical constant reported by different sources can be discrepant because of the fabrication conditions of the

Table 5.3 Propagation loss according to different copper data references

Reference	Measurement	Rob [48]	Palik [49]	JC [58]	Rakic [59]
Loss (dB μm^{-1})	0.573	0.585	2.00	1.65	1.06

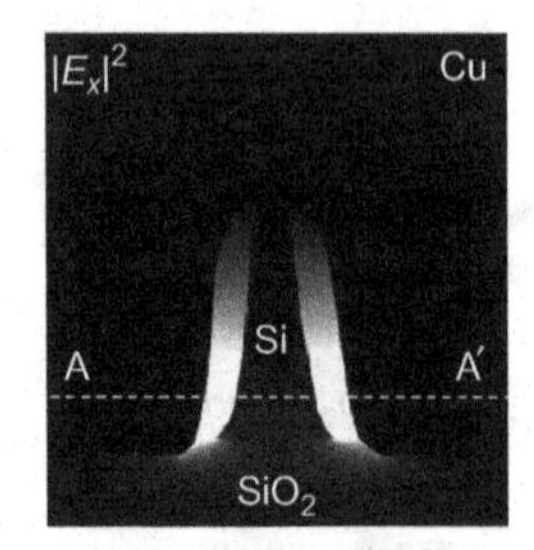

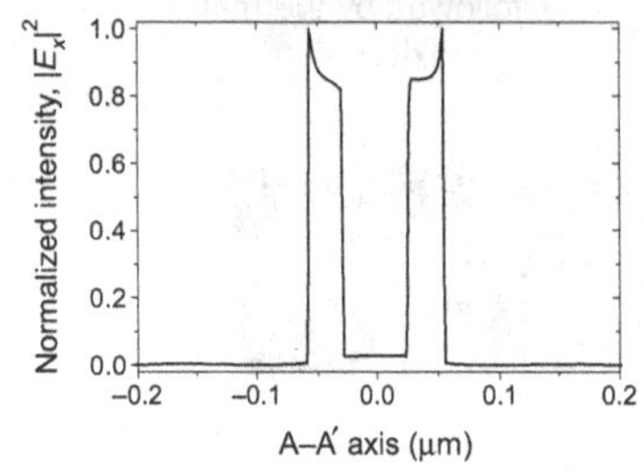

Figure 5.31 The $|E_x|$ field intensity distribution of a given horizontal Cu–SiO_2–Si–SiO_2–Cu plasmonic waveguide in a 2D plot (upper image) and 1D horizontal cross-section (along the line A–A′). The waveguide dimensions are taken to be $w_{Si} = 50$ nm and $t_{SiO_2} = 25$ nm.

copper material produced, such as the purity and surface roughness, the way of making the measurement, and some other conditions. In our work we used the experimental values of the optical constants of copper from Roberts (Rob) [48], Palik's handbook (Palik) [49], Johnson and Christy (JC) [58], and Rakic *et al.* [59]. It is worthwhile to test which reference optical constants of copper will provide the propagation loss which matches the experimental results best. As a result, we model and compare the propagation loss of the proposed Cu–SiO_2–Si–SiO_2–Cu waveguide for different optical constants of copper. In order to be consistent with the dimension of the experimental waveguide, the width of the Si-nanowire core and the thickness of the SiO_2 layer are taken to be 50 nm and 12 nm, respectively. The propagation loss results are compared in Table 5.3, including the measurement data of the propagation loss. Note that the waveguide shape used in the modeling is similar to the one doing the characterization.

It can be observed that the propagation loss significantly changes when different refractive indexes of copper are taken from various references. We also benchmark this modeling result against the measurement on a waveguide, where the cross-sectional transmission electron microscopy (XTEM) image of the measured waveguide is similar to the right-hand-side image of Fig. 5.30. It is evident that the optical constant extracted from Ref. [48] provides a propagation loss that agrees well with measurements. Thus

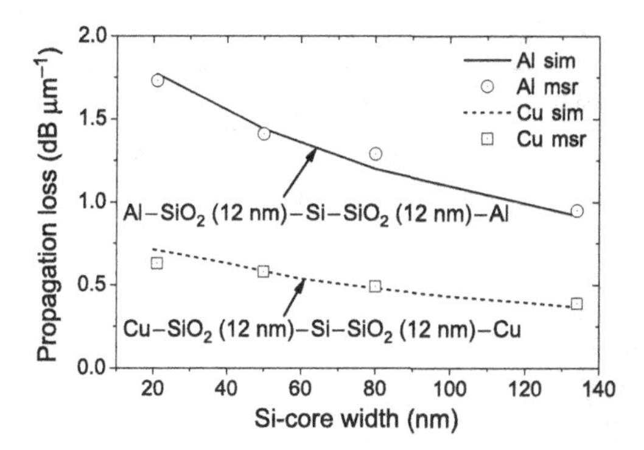

Figure 5.32 Modeling (sim) and measurement (msr) of the propagation loss of the horizontal hybrid Cu–SiO_2–Si–SiO_2–Cu and Al–SiO_2–Si–SiO_2–Al plasmonic waveguides at TE 1550 nm as a function of the Si-core width, w_{Si}. The SiO_2-layer thickness t_{SiO_2} is taken as 12 nm.

the refractive index of copper extracted from this reference will be used in the rest of this section.

We are also interested in whether the Cu-based HHPW platform can perform better than its Al-based counterpart in terms of the propagation loss. We calculate the propagation loss of these two CMOS metal-based HHPW platforms, Cu–SiO_2–Si–SiO_2–Cu and Al–SiO_2–Si–SiO_2–Al. In order to be consistent with the experimental measurement, the SiO_2-layer thickness is taken as 12 nm, and the Si-core width is varied in the range from 20 nm to 140 nm. The theoretical result is calculated by the FVTD method. Figure 5.32 depicts the theoretical and experimental propagation losses of these two waveguides as functions of the Si-core width w_{Si}. Four sets of Cu-based HHPWs from different chips were measured. It can be seen that excellent agreement is obtained between the theoretical prediction and measurement data. For both Cu- and Al-based HHPWs, the propagation loss increases monotonically with decreasing Si-core width, in agreement with the well-known fact that tighter light confinement in HHPWs leads to a shorter propagation distance. As expected, the Cu-based HHPWs exhibit propagation losses more than two times smaller than those of their Al-based counterparts both theoretically and experimentally.

The dependence of the waveguide propagation loss on waveguide shape

In the fabrication process for Si photonics, the realization of a truly rectangular Si nanowire is extremely challenging. The shape of the Si-core nanowire strongly depends on its width; it may vary from a trapezoid to a polygon. In order to understand how the shape influences the propagation characteristics, we benchmark the propagation loss of different cross-sectional waveguide shapes against experimental data.

The XTEM image of the fabricated waveguide (A) with $w_{Si} = 50\,\text{nm}$ and $t_{SiO_2} = 12\,\text{nm}$ and four different shapes (polygon (B), rectangle (C), trapezoid (D), and triangle (E)) are shown in Fig. 5.33. Note that the polygon waveguide has the exact shape and dimensions imported from the XTEM image of the waveguide, and the waveguide width is taken at the middle height of the Si-core nanowire. Two sets of waveguide

Table 5.4 Waveguide propagation loss for different waveguide shapes

	Measurement	Polygon	Rectangle	Trapezoid	Triangle
Loss (dB μm^{-1}) ($w_{Si} = 50$, $t_{SiO_2} = 12$)	0.61	0.585	0.583	0.561	0.573
Loss (dB μm^{-1}) ($w_{Si} = 50$, $t_{SiO_2} = 25$)	0.36	0.366	0.365	0.351	0.38

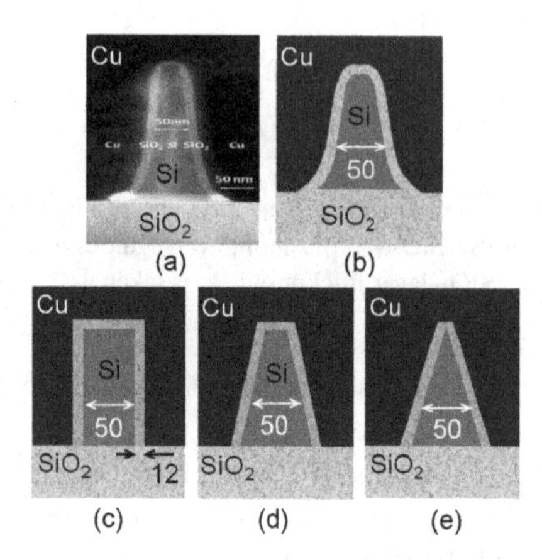

Figure 5.33 Schematic illustrations of different cross-sectional waveguide shapes: an XTEM image of the fabricated waveguide (a) and the theoretical modeling ones: polygon (b), rectangle (c), trapezoid (d), and triangle (e). The horizontal cross-sectional dimensions of the SiO_2–Si–SiO_2 system are $w_{Si} = 50$ nm and $t_{SiO_2} = 12$ nm.

dimensions in nanometers are chosen to investigate, namely [w_{Si}; t_{SiO_2}] = [50; 12] and [50; 25]. The propagation loss is numerically characterized and shown in Table 5.4 for different waveguide shapes. The results are also benchmarked with the experimental measurement. As can be observed, the polygon, rectangle, and trapezoid show small deviations from the measured result. These deviations are clearly observed to be less than 3%. Therefore, for the sake of simplicity, a rectangular-shaped waveguide will be used in the next investigation.

Effects of SiO_2-layer thickness and Si-core width on propagation characteristics

In this part, we study the propagation loss factor and the effective mode index as a function of the SiO_2-layer thickness. The results are plotted and shown in Fig. 5.34. It is observed that the propagation loss and the effective index are strongly dependent on this thickness. The propagation loss and the effective index decrease exponentially when the thickness of the SiO_2 layer increases. This is due to the fact that a thicker oxide layer favors the dominant photonic mode, and, hence, less confinement and a lower propagation loss are achieved. In particular, the propagation loss is 0.4 dB μm^{-1} for a horizontal cross-section for SiO_2–Si–SiO_2 of roughly $\lambda/15$ ($w_{Si} = 50$ nm, $t_{SiO_2} = 25$ nm). Furthermore, this propagation loss can reduce approximately to 0.15 dB μm^{-1} when a cross-section of $\lambda/6$ ($w_{Si} = 50$ nm, $t_{SiO_2} = 100$ nm) is used. The simulated results

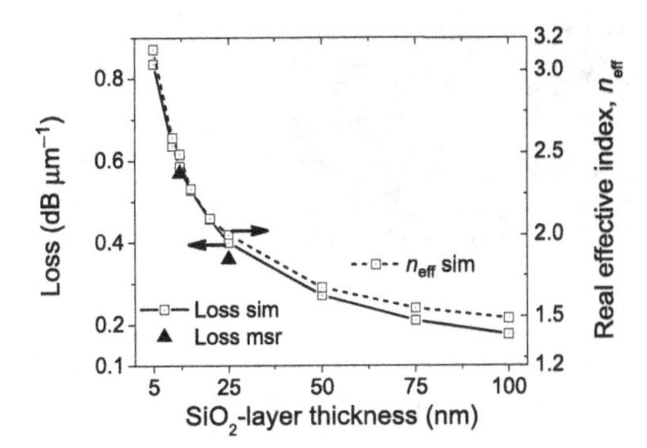

Figure 5.34 The waveguide propagation loss and effective index as a function of the SiO_2-layer thickness; the triangles are the measured values of propagation loss. The Si-core width is chosen as 50 nm. Good agreement between simulation (sim) and measurement (msr) is observed. The propagation loss and the effective refractive index decrease exponentially as the SiO_2-layer thickness increases.

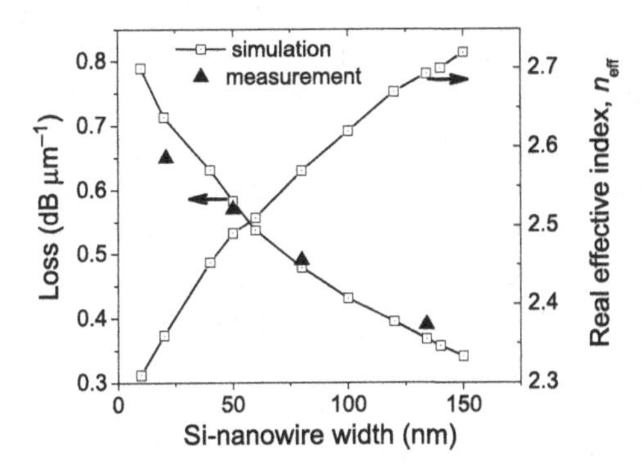

Figure 5.35 The waveguide propagation loss and effective index of the waveguide as a function of the Si-core width; triangles represent the measured values of propagation loss; good agreement between simulation and measurement is observed. Note that the thickness of SiO_2 is chosen as 12 nm to make it compatible with the experimental results.

for the propagation loss, shown in Fig. 5.34, are in good agreement with the measured ones when two sets of SiO_2–Si–SiO_2 dimensions (12–50–12 nm and 25–50–25 nm) are used.

We also investigated, and show in Fig. 5.35, the effect of the Si-core nanowire width on the propagation loss and the effective index of the waveguide. We vary the Si-core width while the SiO_2-layer thickness is kept at 12 nm. A large Si-core nanowire width provides a smaller propagation loss and a larger effective index. However, the confinement is reduced because the photonic mode is dominant in the Si nanowire. In comparison with the results shown in Fig. 5.34, the variation of the Si-nanowire core width has less impact on the propagation characteristics than the SiO_2-layer thickness because a low-index layer of SiO_2 is formed by a metallic layer and a high-index layer of Si. This phenomenon has been discussed in detail in Ref. [23].

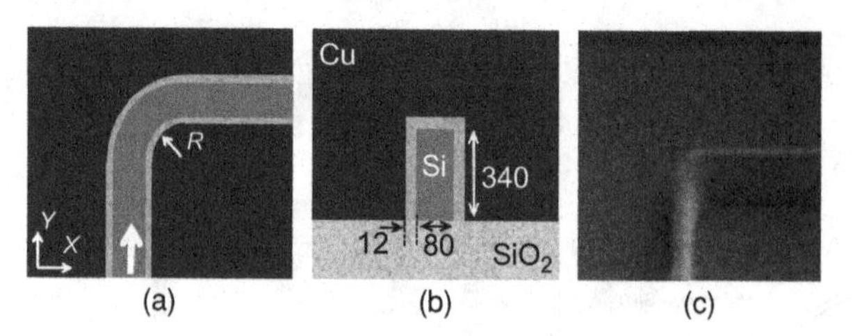

Figure 5.36 (a) Top view and (b) cross-sectional view of the horizontal hybrid plasmonic waveguide sharp 90° bend, and (c) a scanning electron microscopy (SEM) image of the waveguide bend at the fabrication process step before the Cu deposition.

Waveguide sharp-bend performance

Sharp waveguide bends are basic components for high-density electronic–photonic integrated circuits. One of the most exciting aspects of 3D plasmonic waveguides is that they enable light to be bent around a 90° corner at a subwavelength scale; this feature is available neither in conventional dielectric waveguides nor in photonic crystal devices. Therefore, the compact size and highly confined energy with low propagation loss of the HHPW suggest that sharp waveguide bending components with low propagation loss factors could be built. In order to facilitate the validation against experimental data, the width of the Si core and the thickness of the SiO_2 layer are chosen as $w_{Si} = 80\,\text{nm}$ and $t_{SiO_2} = 12\,\text{nm}$, respectively. Hence, the HHPW sharp 90° bend is as sketched in Fig. 5.36. The transmitted power loss through the bend, $T = 10\log_{10}(P_{out}/P_{in})$, is obtained as a ratio between the power-flow integrals in the waveguide cross-section at the beginning, P_{in}, and at the end, P_{out}, of the bend section. The transmitted power loss as a function of the bend radius is numerically computed by the three-dimensional finite-difference time-domain method (3D-FDTD) [40] at the wavelength of 1.55 μm; the results are shown in Fig. 5.37(a). It can be seen that the transmitted power loss is less than 0.3 dB for a bend radius smaller than 50 nm ($< \lambda/30$).

We also note that a minimum of the transmitted power loss of 0.1 dB can be achieved at a bend radius of 10 nm. Moreover, this loss starts increasing when a larger bend radius is used. For example, this value may reach more than 1 dB when the bend radius exceeds 1 μm. This occurs because the ohmic loss increases with increasing length of the bend. Note that the theoretical predictions by the 3D-FDTD are strongly consistent with the measurements reported in Refs. [15, 32]. In addition, we also plot the corresponding electric field for the optimal bend radius of 10 nm in Fig. 5.37(b) and a given large bend radius $R = 500\,\text{nm}$. The electric field is highly squeezed into a tight SiO_2-layer gap of 15 nm, and the electromagnetic fields are efficiently guided and bent in the waveguiding medium.

Next, it is interesting to study the optical performance of HHPW waveguide bend in the broad spectrum of near-infrared wavelengths ranging from 1 μm to 2 μm. The waveguide bend loss factor as a function of wavelength is calculated and plotted in Fig. 5.38 for a bend radius of 10 nm, and $w_{Si} = 80\,\text{nm}$ and $t_{SiO_2} = 12\,\text{nm}$. It is clear that the waveguide bend loss factor decreases exponentially as the operating wavelength

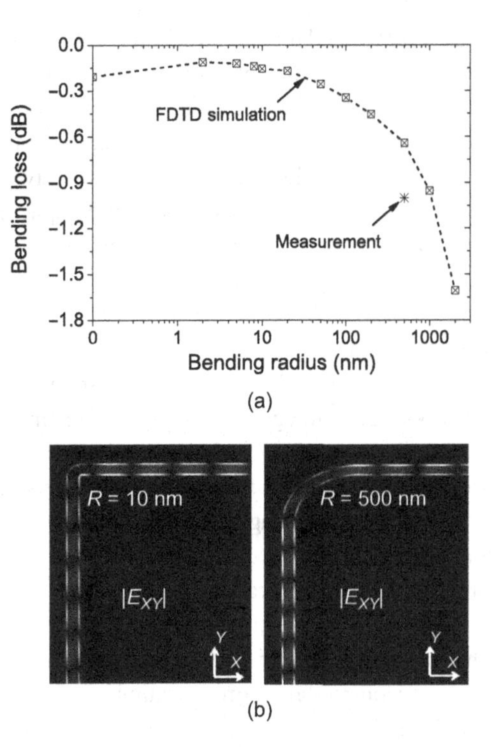

Figure 5.37 (a) The waveguide bend loss as a function of the bending radius with $w_{Si} = 80\,nm$ and $t_{SiO_2} = 12\,nm$. (b) An illustration of the electromagnetic field distribution along the waveguide bends for two different bend radii: small radius $R = 10\,nm$ and large radius $R = 500\,nm$.

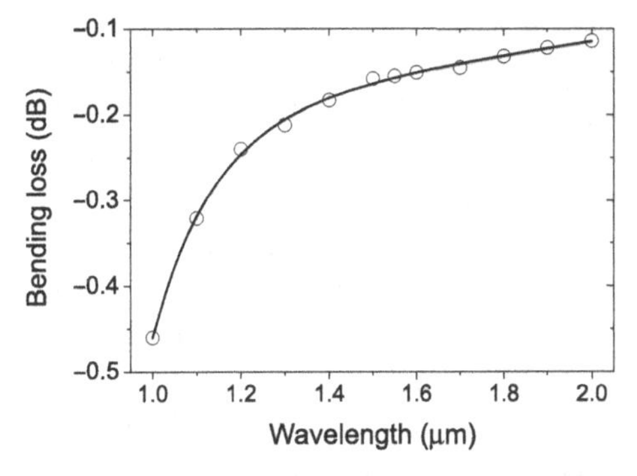

Figure 5.38 The waveguide bend loss as a function of wavelength in the near-infrared spectrum where the bend radius $R = 10\,nm$ is chosen, together with $w_{Si} = 80\,nm$ and $t_{SiO_2} = 12\,nm$.

increases; their relationship can be expressed by an exponential curve-fitting function:

$$\text{Loss (dB)} = -47.2e^{-\lambda/0.17} - 0.62e^{-\lambda/5.71} - 942.5e^{-\lambda/0.11} + 0.32. \quad (5.6)$$

The curve-fitting result is very close to the theoretical prediction. We believe that this exponential variation trend can apply in a similar manner to other dimensions of the Si-nanowire core and the SiO_2 layer.

The result demonstrates that the HHPW bend provides a smaller bend radius and lower loss than the VHPW bend counterpart as reported in previous sections. These ultra-small subwavelength bending radii combined with the low bending loss, high confinement, and CMOS compatibility promise to make these components very attractive for highly dense integrated circuits with a compact footprint. Furthermore, this attractive waveguide bend could also play a key role in the design of various important compact plasmonic-based components, including power splitters, Mach–Zehnder interferometers, and ring resonator filters.

Submicron waveguide ring resonators

The proposed HHPW platform can be used to design various plasmonic components to efficiently perform different tasks in the integrated circuit. For example, a design for a fully CMOS-compatible nanoscale waveguide ring resonator (WRR) has been proposed. The ring resonator (RR) is one of the most versatile components of nanoscale electronic–photonic integrated circuits for realizing many functions, such as those of filters, switches, and modulators, on a single device. Since the footprint of the photonic system increases with the number of devices integrated, there is a strong motivation to minimize the size of each component in order to achieve denser device integration. An added advantage of keeping the ring radius small is that the free-spectral range (FSR) of the resonator becomes larger to accommodate a greater number of communication channels. Our proposed WRR is formed by a ring waveguide placed in the vicinity of a straight bus waveguide to allow optical coupling between them. The efficiency of this coupling depends on the gap between the ring and the straight waveguide, and on the radius of the RR. However, in the current CMOS fabrication process, it is very challenging to realize a gap smaller than 100 nm, even with state-of-the-art electron-beam or ion-beam lithography. To overcome this difficulty, we place an aperture coupler between the RR and a bus plasmonic waveguide instead of a narrow gap. Such an aperture coupler has been theoretically investigated for metal–insulator–metal plasmonic waveguide RRs [60] and experimentally demonstrated for dielectric-loaded plasmonic WRRs [43].

A horizontal cut through the WRR is shown in Fig. 5.39(a). The SEM images obtained halfway through the fabrication process of the Si-nanowire core pattern (shown in the inset) and after copper deposition are shown in Fig. 5.39(b). As a design example, the dimensions of the Si core and the SiO_2 layer are chosen as $w_{Si} = 50\,nm$ and $t_{SiO_2} = 25\,nm$. In addition, the dimensions of the aperture coupler separating the bus and the ring waveguide are $g = 100\,nm$ and $w = 200\,nm$. The choice of these dimensions is to facilitate the comparison with experimental results. It is noted that the ring radius is measured from the center of the ring to the center of the Si layer (Fig. 5.39(a)).

We study the optical performance of two different rings with radii $R = 0.5\,\mu m$ and $0.9\,\mu m$. In Fig. 5.39, the resulting power transmission calculated with the 3D-FDTD method is shown as a function of the wavelength within telecommunication bands. Note that the input and output power are monitored at locations separated by a distance of $10\,\mu m$. It is clear that the excellent power transmission of the WRR is achieved with ring radii as small as $0.5\,\mu m$ and $0.9\,\mu m$. More specifically, these compact WRRs also feature a low insertion loss of less than 2.2 dB and a high extinction ratio of more than

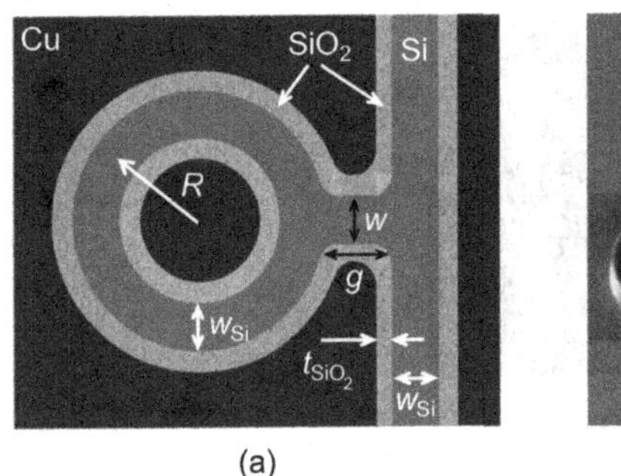

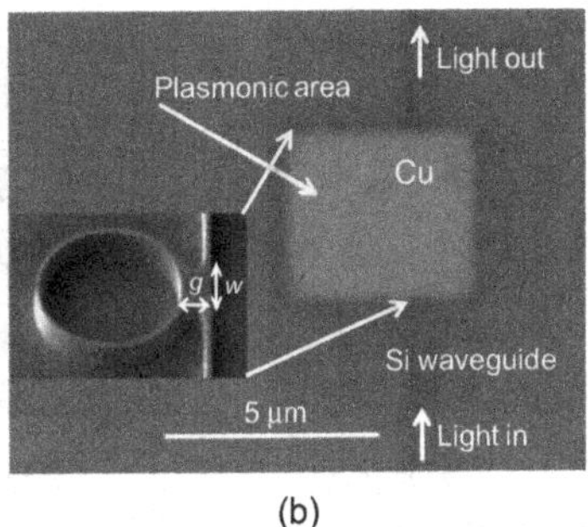

(a) (b)

Figure 5.39 (a) A sketch of a horizontal hybrid Cu–SiO_2–Si–SiO_2–Cu plasmonic waveguide-based ring resonator (WRR). (b) Scanning electron microscope (SEM) images of the fabricated WRR with radius $R = 0.9$ μm; the inset shows the Si-core nanowire WRR pattern during the fabrication process.

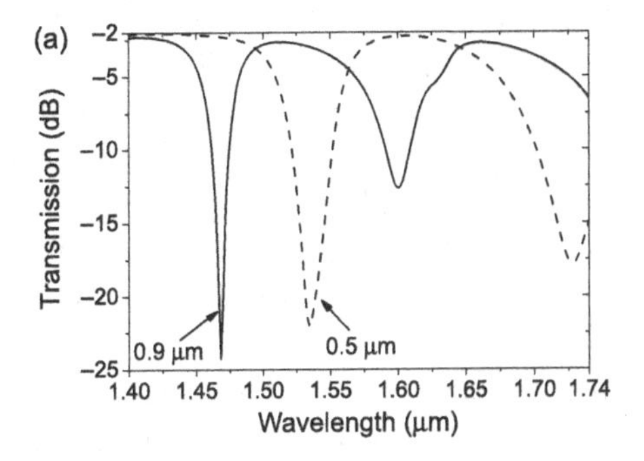

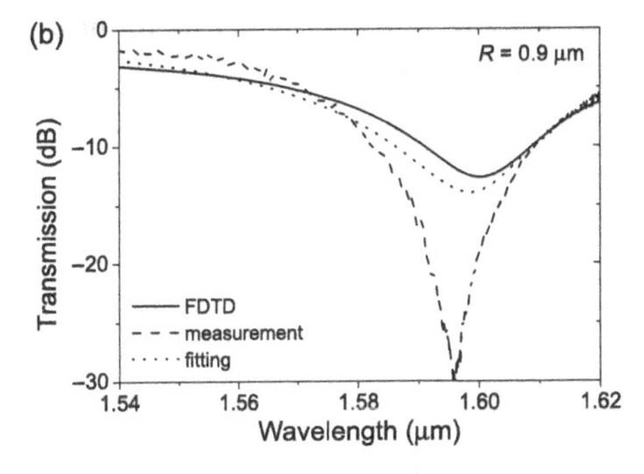

Figure 5.40 (a) The transmitted power of waveguide ring resonators (WRRs) as a function of the wavelength for two different ring radii, $R = 0.5$ μm and 0.9 μm, in the telecommunication wavelengths. (b) A comparison of the modeling, experimental, and analytical results of the transmitted power for a ring radius of around 0.9 μm.

20 dB. Furthermore, the bandwidth, defined by the full width of the resonance curve at half maximum, is 25 nm for $R = 0.9$ μm and 50 nm for $R = 0.5$ μm. Moreover, the FSR, measured as the distance between two transmission resonance peaks, is obtained at 190 nm and 130 nm for ring radii of 0.5 μm and 0.9 μm, respectively. These results show that the 0.9-μm-radius WRR can accommodate up to five high-bit-rate channels within one FSR. In order to validate our design, we have compared the modeling, experimental,

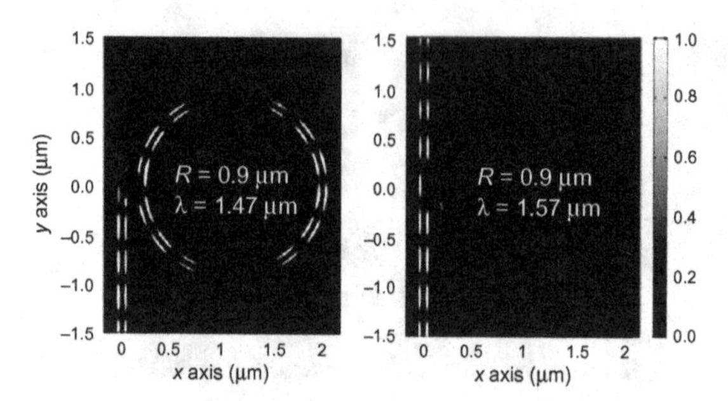

Figure 5.41 The simulated electric field in the 0.9-μm WRR is shown at the turn-on and turn-off wavelengths of 1.47 μm and 1.5 μm, respectively.

and fitting results for the transmitted power for a radius of 0.9 μm (Fig. 5.40(b)). Note that the fitting result is based on the analytical parameters adapted from Eq. (5.3). The modeling, analytical, and experimental results predict very well the resonance peak and the bandwidth at −3 dB, although there is a discrepancy in the value of the extinction ratio. Moreover, the electric-field intensity $|E_x|^2$ shown in Fig. 5.41 demonstrates that the optical signal is efficiently confined and guided in the SiO_2 layer, and that the filtering functionality from the input channel to the output channel is well performed at the resonance wavelength of 1.47 μm and at the non-resonance wavelength of 1.5 μm for a WRR ring radius of 0.9 μm.

The proposed WRR is a best demonstration of an ultra-compact and highly confined submicrometer plasmonic WRR for CMOS-compatible Si-based nanophotonic circuits. Furthermore, this WRR also features a much smaller footprint in comparison with the experimental results for the channel plasmon and dielectric-loaded plasmonic waveguide-based ring resonator counterparts [13, 43, 44].

References

[1] W. L. Barnes, A. Dereux, and T. W. Ebbesen, "Surface plasmon subwavelength optics," *Nature*, vol. 424, pp. 824–830, 2003.

[2] R. Zia, J. A. Schuller, A. Chandran, and M. L. Brongersma, "Plasmonics: The next chip-scale technology," *Material Today*, vol. 9, pp. 20–27, 2006.

[3] E. Ozbay, "Plasmonics: Merging photonics and electronics at nanoscale dimensions," *Science*, vol. 311, pp. 189–193, 2006.

[4] T. W. Ebbesen, C. Genet, and S. I. Bozhevolnyi, "Surface-plasmon circuitry," *Phys. Today*, vol. 61, pp. 44–50, 2008.

[5] D. K. Gramotnev and S. I. Bozhevolnyi, "Plasmonics beyond the diffraction limit," *Nature Photonics*, vol. 4, pp. 83–91, 2010.

[6] S. A. Maier, P. G. Kik, H. A. Atwater *et al.*, "Local detection of electromagnetic energy transport below the difraction limit in metal nanoparticle plasmon waveguides," *Nature Mater.*, vol. 2, pp. 229–232, 2003.

[7] A. Boltasseva, T. Sondergaard, T. Nikolajsen *et al.*, "Propagation of long-range surface plasmon polaritons in photonic crystals," *J. Opt. Soc. Am. B*, vol. 22, pp. 2027–2038, 2005.
[8] H. S. Chu, W. B. Ewe, E. P. Li, and R. Vahldieck, "Analysis of sub-wavelength light propagation through long double-chain nanowires with funnel feeding," *Opt. Express*, vol. 15, 4216–4223, 2007.
[9] H. S. Chu, W. B. Ewe, W. S. Koh, and E. P. Li, "Remarkable influence of the number of nanowires on plasmonic behaviors of the coupled metallic nanowire chain," *Appl. Phys. Lett.*, vol. 92, pp. 103103–103105, 2008.
[10] A. Alu, P. A. Belov, and N. Engheta, "Coupling and guided propagation along parallel chains of plasmonic nanoparticles," *New J. Phys.*, vol. 13, pp. 033026–033048, 2011.
[11] J. Takahara, S. Yamagishi, H. Taki, A. Morimoto, and T. Kobayashi, "Guiding of a one-dimensional optical beam with nanometer diameter," *Opt. Lett.*, vol. 22, pp. 475–477, 1997.
[12] J. C. Weeber, A. Dereux, C. Girad, J. R. Krenn, and J. P. Goudonnet, "Plasmon polaritons of metallic nanowires for controlling submicron propagation of light," *Phys. Rev. B*, vol. 60, pp. 9061–9068, 1999.
[13] S. I. Bozhevolnyi, V. S. Volkov, E. Devaux, J.-Y. Laluet, and T. W Ebbesen, "Channel plasmon subwavelength waveguide components including interferometers and ring resonators," *Nature*, vol. 440, pp. 508–511, 2006.
[14] E. N. Economou, "Surface plasmons in thin films," *Phys. Rev.*, vol. 182, pp. 539–554, 1969.
[15] L. Liu, Z. Han, and S. He, "Novel surface plasmon waveguide for high integration," *Opt. Express*, vol. 13, pp. 6645–6650, 2005.
[16] J. A. Dionne, L. A. Sweatlock, H. A. Atwater, and A. Polman "Planar metal plasmon waveguides: Frequency-dependent dispersion, propagation, localization, and loss beyond the free electron model," *Phys. Rev. B*, vol. 72, pp. 075405–075415, 2005.
[17] N. N. Feng, M. L. Brongersma, and L. D. Negro, "Metal dielectric slot waveguide structures for the propagation of surface plasmon polaritons at 1.55 μm," *IEEE J. Quant. Electron.*, vol. 43, pp. 479–485, 2007.
[18] B. Steinberger, A. Hohenau, H. Ditlbacher *et al.*, "Dielectric strips on gold as surface plasmon waveguides," *Appl. Phys. Lett.*, vol. 88, pp. 094104–094106, 2006.
[19] T. Holmgaard and S. I. Bozhevolnyi, "Theoretical analysis of dielectric-loaded surface plasmon-polariton waveguides," *Phys. Rev. B*, vol. 75, pp. 245405–245416, 2007.
[20] A. V. Krasavin and A. V. Zayats, "Three-dimensional numerical modeling of photonic integration with dielectric-loaded SPP waveguides," *Phys. Rev. B*, vol. 78, pp. 045425–045432, 2008.
[21] H. S. Chu, W. B. Ewe, and E. P. Li, "Tunable propagation of light through a coupled-bent dielectric-loaded plasmonic waveguides," *J. Appl. Phys.*, vol. 106, pp. 106101–106103, 2009.
[22] P. Berini, "Long-range surface plasmon polaritons," *Adv. Opt. Photonics*, vol. 1, pp. 484–588, 2009.
[23] R. F. Oulton, V. J. Sorger, D. A. Genov, D. F. P. Pile, and X. Zhang, "A hybrid plasmonic waveguide for subwavelength confinement and long-range propagation," *Nature Photonics*, vol. 2, pp. 496–500, 2008.
[24] H. S. Chu, E. P. Li, P. Bai, and R. Hegde, "Optical performance of single-mode hybrid dielectric-loaded plasmonic waveguide-based components," *Appl. Phys. Lett.*, vol. 96, pp. 221103–221105, 2010.

[25] M. Z. Alam, J. Meier, J. S. Aitchison, and M. Mojahedi, "Propagation characteristics of hybrid modes supported by metal-low–high index waveguides and bends," *Opt. Express*, vol. 18, pp. 12971–12979, 2010.

[26] J. Tian, Z. Ma, Q. Li *et al.*, "Nanowaveguides and couplers based on hybrid plasmonic modes," *Appl. Phys. Lett.*, vol. 97, pp. 231121–231123, 2010.

[27] F. Lou, Z. Wang, D. Dai, L. Thylen, and L. Wosinski, "Experimental demonstration of ultra-compact directional couplers based on silicon hybrid plasmonic waveguides," *Appl. Phys. Lett.*, vol. 100, pp. 241105–241108, 2012.

[28] V. J. Sorger, Z. Ye, R. F. Oulton *et al.*, "Experimental demonstration of low-loss optical waveguiding at deep sub-wavelength scales," *Nature Comm.*, vol. 2, pp. 1–5, 2011.

[29] M. Wu, Z. Han, and V. Van, "Conductor-gap-silicon plasmonic waveguides and passive components at subwavelength scale," *Opt. Express*, vol. 18, pp. 11728–11736, 2010.

[30] S. Zhu, G. Q. Lo, and D. L. Kwong, "Experimental demonstration of vertical Cu–SiO_2–Si hybrid plasmonic waveguide components on an SOI platform," *IEEE Photonics Technol. Lett.*, vol. 24, pp. 1224–1226, 2012.

[31] S. Zhu, T. Y. Liow, G. Q. Lo, and D. L. Kwong, "Fully complementary metal–oxide–semiconductor compatible nanoplasmonic slot waveguides for silicon electronic photonic integrated circuits," *Appl. Phys. Lett.*, vol. 98, pp. 021107–021109, 2011.

[32] S. Y. Zhu, T. Y. Liow, G. Q. Lo, and D. L. Kwong, "Silicon-based horizontal nanoplasmonic slot waveguides for on-chip integration," *Opt. Express*, vol. 19, pp. 8888–8902, 2011.

[33] J. T. Kim, "CMOS-compatible hybrid plasmonic slot waveguide for on-chip photonic circuits," *IEEE Photonics Technol. Lett.*, vol. 23, pp. 1481–1483, 2011.

[34] R. F. Oulton, V. J. Sorger, T. Zentgraf *et al.*, "Plasmon lasers at deep subwavelength scale," *Nature*, vol. 461, pp. 629–632, 2009.

[35] R. M. Ma, R. F. Oulton, V. J. Sorger, G. Bartal, and X. Zhang, "Room-temperature sub-diffraction-limited plasmon laser by total internal reflection," *Nature Mater.*, vol. 10, pp. 110–113, 2011.

[36] P. B. Johnson and R. W. Christy, "Optical constants of the noble metals," *Phys. Rev. B*, vol. 6, pp. 4370–4379, 1972.

[37] Z. Zhu and T. G. Brown, "Full-vectorial finite-difference analysis of microstructured optical fibers," *Opt. Express*, vol. 10, pp. 853–864, 2002.

[38] V. R. Almeida, Q. Xu, C. A. Barrios, and M. Lipson, "Guiding and confining light in void nanostructure," *Opt. Lett.*, vol. 29, pp. 1209–1211, 2004.

[39] C. Manolatou, S. G. Johnson, S. Fan *et al.*, "High-density integrated optics," *J. Lightwave Technol.*, vol. 17, pp. 1682–1692, 1999.

[40] A. Taflove and S. C. Hagness, *Computational Electrodynamics: The Finite-Difference Time-Domain Method*, 3rd ed. Boston, MA: Artech House, 2005.

[41] A. Kumar and S. Aditya, "Performance of S bends in integrated optic waveguides," *Microwave Opt. Technol. Lett.*, vol. 19, pp. 289–292, 1998.

[42] H. S. Chu, P. Bai, E. P. Li, and W. R. J. Hoefer, "Hybrid dielectric-loaded plasmonic waveguide-based power splitter and ring resonator: Compact size and high optical performance for nanophotonic circuits," *Plasmonics*, vol. 6, pp. 591–597, 2011.

[43] T. Holmgaard, Z. Chen, S. I. Bozhevolnyi, L. Markey, and A. Dereux, "Dielectric-loaded plasmonic waveguide-ring resonators," *Opt. Express*, vol. 17, pp. 2968–2975, 2009.

[44] V. S. Volkov, S. I. Bozhevolnyi, E. Devaux, J.-Y. Laluet, and T. W Ebbesen, "Wavelength selective nanophotonic components utilizing channel plasmon polaritons," *Nano Lett.*, vol. 7, pp. 880–884, 2007.

[45] Q. Xu, B. Schmidt, S. Pradhan, and M. Lipson, "Micrometre-scale silicon electro-optic modulator," *Nature*, vol. 435, pp. 325–327, 2005.

[46] W. Bogaerts, P. De Heyn, T. Van Vaerenbergh *et al.*, "Silicon microring resonators," *Laser Photon. Rev.*, vol. 6, pp. 47–73, 2012.

[47] A. Yariv, "Universal relations for coupling of optical power between microresonators and dielectric waveguides," *Electron. Lett.*, vol. 36, pp. 321–322, 2000.

[48] S. Roberts, "Optical properties of copper," *Phys. Rev.*, vol. 118, pp. 1509–1518, 1960.

[49] D. W. Lynch and W. R. Hunter, "An introduction to the data for several metals, " in *Handbook of Optical Constants of Solids* II, E. D. Palik, Ed. San Diego, CA: Academic, 1991, pp. 341–419.

[50] http://www.sopra-sa.com.

[51] I. Goykhman, B. Desiatov, and B. Levy, "Experimental demonstration of locally oxidized hybrid silicon-plasmonic waveguide," *Appl. Phys. Lett.*, vol. 97, pp. 141106–141108, 2010.

[52] J. W. Peng, S. J. Lee, G. C. A. Liang *et al.*, "Improved carrier injection in gate-all-around Schottky barrier silicon nanowire field-effect transistors," *Appl. Phys. Lett.*, vol. 93, pp. 073503–073505, 2008.

[53] S. Y. Zhu, G. Q. Lo, and D. L. Kwong, "Performance of ultracompact copper-capped silicon hybrid plasmonic waveguide-ring resonators at telecom wavelengths," *Opt. Express*, vol. 20, pp. 15232–15246, 2012.

[54] S. Y. Zhu, Q. Fang, M. B. Yu, G. Q. Lo, and D. L. Kwong, "Propagation losses in undoped and n-doped polycrystalline silicon wire waveguides," *Opt. Express*, vol. 17, pp. 20891–20899, 2009.

[55] S. Y. Zhu, G. Q. Lo, and D. L. Kwong, "Low-loss amorphous silicon wire waveguide for integrated photonics: Effect of fabrication process and the thermal stability," *Opt. Express*, vol. 18, pp. 25283–25291, 2010.

[56] L. Chen, J. Shakya, and M. Lipson, "Subwavelength confinement in integrated metal slot waveguide on silicon," *Opt. Lett.*, vol. 31, pp. 2133–2135, 2006.

[57] Z. Han, A. Y. Elezzabi, and V. Van, "Experimental realization of subwavelength plasmonic slot waveguides on a silicon platform," *Opt. Lett.*, vol. 35, pp. 502–504, 2010.

[58] P. B. Johnson and R. W. Christy, "Optical constants of the noble metals," *Phys. Rev. B*, vol. 6, pp. 4370–4379, 1972.

[59] A. D. Rakic, A. B. Djuriic, J. M. Elazar, and M. L. Majewski, "Optical properties of metallic films for vertical-cavity optoelectronic devices," *Appl. Opt.*, vol. 37, pp. 5271–5283, 1998.

[60] Z. Han, V. Van, W. N. Herman, and P.-T. Ho, "Aperture-coupled MIM plasmonic ring resonators with sub-diffraction modal volumes," *Opt. Express*, vol. 17, pp. 12678–12684, 2009.

6 Silicon-based active plasmonic devices for on-chip integration

Photonic devices integrated in Si optoelectronic circuits offer less power dissipation and larger bandwidth than those of electronic components, but suffer from a larger footprint due to the fundamental diffraction limit of light in dielectric waveguides and the weak optical response of Si. These two limitations may be overcome by utilizing plasmonics owing to the tight optical mode confinement in plasmonic waveguides. Besides the capability of miniaturization of photonic devices on the nanometer scale, plasmonics also provides the potential to design novel photonic devices due to the incorporation of metal and dielectrics. In this chapter, we present Si-based active plasmonic devices developed in our laboratory, including modulators and detectors. These active plasmonic devices can be seamlessly integrated into existing Si optoelectronic circuits using standard CMOS technology.

6.1 Introduction

Silicon photonics, in which photonic devices are fabricated on silicon-on-insulator (SOI) platforms using mature CMOS technology, has been well developed recently for high-performance Si electronic–photonic integration circuits (EPICs) [1]. In particular, integrated Si modulators and Ge-on-Si detectors with performances competitive with those of their counterparts based on III–V semiconductors have been demonstrated [2, 3]. However, due to the fundamental diffractive limit of light in dielectric waveguides as well as the weak optical response of Si, the Si photonic devices suffer from large footprints. For example, Mach–Zehnder-based Si modulators, which are mostly implemented in Si EPICs, require a millimeter-scale length to reach π phase shift. Ring-resonator-based Si modulators have a smaller footprint than the Mach–Zehnder-based ones at the price of a higher temperature sensitivity and lower optical bandwidth [4, 5]. However, they are still quite large compared with nanoscale electronic devices because the minimum bending radius is $\sim$1.5 μm for Si-strip waveguides, and is usually larger than $\sim$5 μm for the Si-rib waveguides in which the electro-optic modulators are implemented, which corresponds to a minimum footprint of $\sim$80 μm^2. The integrated Ge detectors in Si EPICs also require a relatively long length for sufficient light absorption due to the relatively small absorption coefficient of Ge at 1550 nm ($\sim$0.046 μm^{-1}) [3].

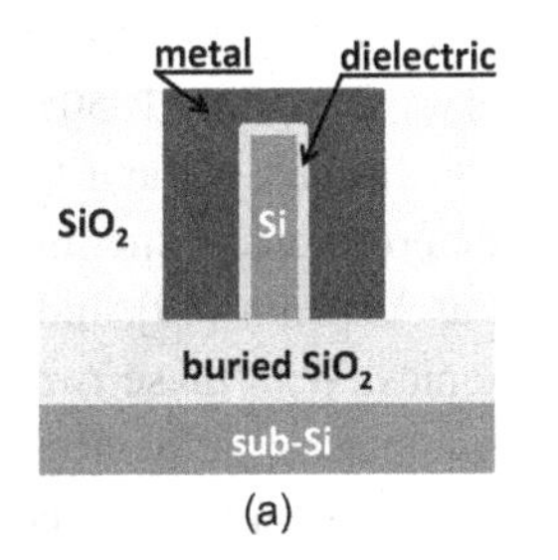

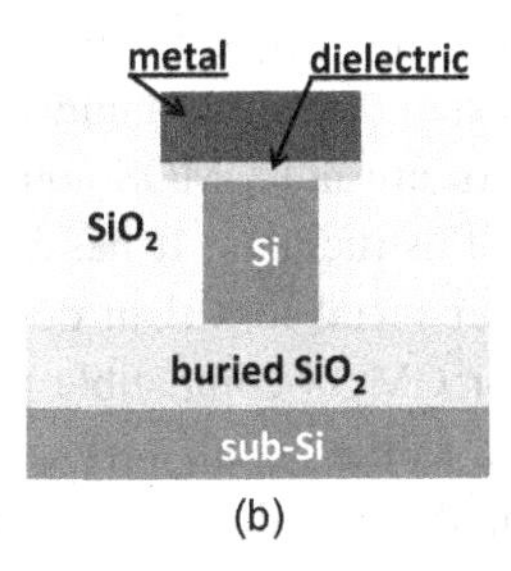

Figure 6.1 Two CMOS-compatible plasmonic waveguide platforms for various passive and active plasmonic devices: (a) a horizontal metal–insulator–Si–insulator–metal (MISIM) plasmonic waveguide and (b) a vertical metal–insulator–Si (MIS) hybrid plasmonic waveguide.

Moreover, although the technology of Ge heteroepitaxy on Si has been well developed, heteroepitaxy of Ge on Si is still an expensive process and will make the whole fabrication flow of Si EPICs complex. For dense Si EPICs in which tens of thousands of modulators and detectors may be integrated in a single chip, it is highly desired to scale down the dimensions of these active photonic devices to be comparable to those of electronic devices.

A technology emerging recently to address this challenge is plasmonics, which deals with surface plasmon polaritons (SPPs) propagating along the metal/dielectric interfaces and provides the capability of tight optical mode confinement far beyond the diffraction limit [6–9]. For integration with optoelectronics, many types of plasmonic waveguides as well as passive photonic devices based on plasmonic waveguides have been proposed and/or demonstrated recently, such as gap-SPP [10], hybrid [11], and dielectric-loaded SPP waveguides [12], etc., just to mention a few examples. Similarly to optical signals in photonics, the SPP signals in plasmonics can also be generated, modulated, and detected by so-called "active plasmonics." This term was coined in a 2004 paper [13]. Since then, many active plasmonic devices have been proposed and/or demonstrated using a range of material systems, including thermo- and electro-optic media, quantum dots, and photochromic molecules. Active plasmonics may roughly be categorized as plasmonic sources (including amplifiers), which generate or amplify SPP signals through electrical or optical excitation [14]; plasmonic modulators, which modulate SPP signals by an external control excitation (optical, thermal, electronic, etc.) [15]; and plasmonic detectors, which convert SPP signals to electrical signals [16]. The active plasmonic devices reported in the literature mostly rely on active materials other than Si or require nonstandard CMOS manufacturing processes, so their implementation in existing Si EPICs is difficult.

In this chapter, we address active plasmonic devices that can be seamlessly integrated into existing Si EPICs using standard CMOS technology. Particular attention is paid to two CMOS-compatible plasmonic waveguide platforms, as shown schematically in Fig. 6.1. One is the horizontal metal–insulator–Si–insulator–metal (MISIM) nanoplasmonic waveguide [17], and the other is the vertical metal–insulator–Si (MIS) hybrid plasmonic waveguide [18]. The metal in plasmonics is usually Ag or Au due to the relatively low metal loss at the telecommunications wavelength of 1550 nm [10–12]. However, both Ag and Au are not CMOS-compatible. Instead, Al and Cu are commonly used in CMOS infrastructures. It has been demonstrated experimentally that Cu provides a much lower metal loss than does Al at 1550 nm [19], indicating that Cu is a suitable metal for CMOS-compatible plasmonics. With these two plasmonic waveguides and using Cu as the metal, various passive plasmonic devices have been demonstrated experimentally, including tapered couplers, which link the conventional Si-channel waveguide and the plasmonic waveguide; power splitters with ultra-compact footprints; waveguide ring resonators (WRRs) with submicron radius; Mach–Zehnder interferometers; and polarization controllers [19–25]. Moreover, both types of plasmonic waveguide allow active functions such as modulation and detection to be implemented by using Si or CMOS-compatible materials such as metal silicides as an active material.

This chapter is organized as follows. Section 6.2 describes ultra-compact electro-absorption (EA) and phase modulators based on the MISIM plasmonic waveguide. Section 6.3 describes athermal ring modulators based on the MIS hybrid plasmonic waveguide. Section 6.4 describes plasmonic detectors using an ultra-thin silicide layer inserted into the MISIM plasmonic waveguide as the absorber. Section 6.5 describes detectors using metallic nanoparticles embedded in an Si p–n junction as the absorber. Finally, some conclusions are drawn in Section 6.6.

6.2 Plasmonic modulators based on horizontal MISIM plasmonic waveguides

6.2.1 The operating principle

A plasmonic modulator based on the MISIM plasmonic waveguide is schematically illustrated in Fig. 6.2. The rectangular Si core of width W_P and length L_P is phosphorus doped with a doping level (N_D) of 5×10^{18} cm^{-3}, and is electrically connected with the n$^+$-doped Si pad through cross waveguides for Si contact. In this proof-of-concept device, thermal oxide is used as the gate dielectric. The modulator has a typical MOS capacitor structure where a voltage can be applied between the Cu cover and the n$^+$-Si pad to induce a highly accumulated electron layer at the SiO_2/Si interface.

Figure 6.3 depicts electron distributions in a Cu/5-nm-SiO_2/n-Si capacitor under different voltages, obtained from one-dimensional MEDICI simulations. The electron concentration is maximized at ~3–5 Å from the SiO_2/Si interface due to the quantum-mechanical effect and then decreases to N_D quickly with the distance from the interface increasing, in agreement with the experimental result [26]. As a first approximation, the distribution can be approximated as a step function and an accumulation layer (AcL) of width t_{AcL} and average concentration N_{AcL} can be defined:

$$N_{AcL} = \frac{\epsilon_0 \epsilon_d}{e t_{ox} t_{AcL}}(V - V_{FB}) = \frac{\epsilon_0 \epsilon_d}{e t_{AcL}} E_d, \tag{6.1}$$

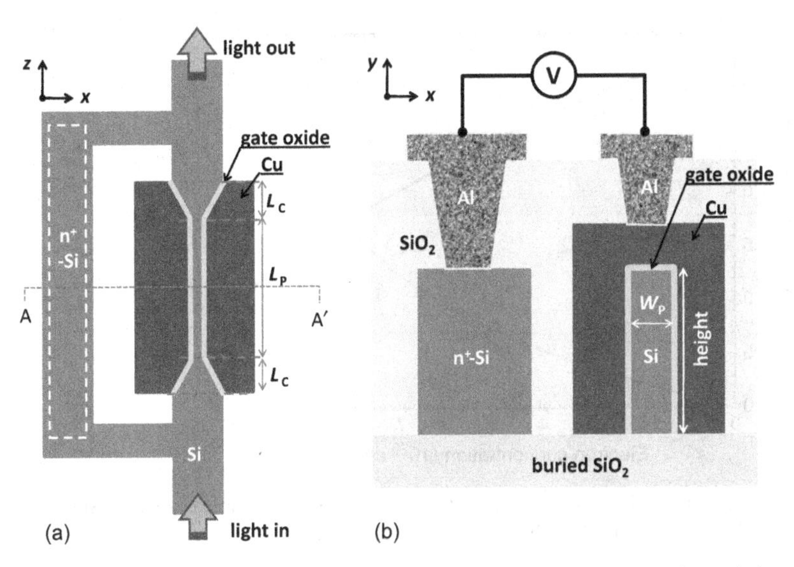

Figure 6.2 A schematic representation of a proof-of-concept Si plasmonic modulator based on the horizontal MISIM plasmonic waveguide: (a) top view and (b) cross-sectional view along A–A′.

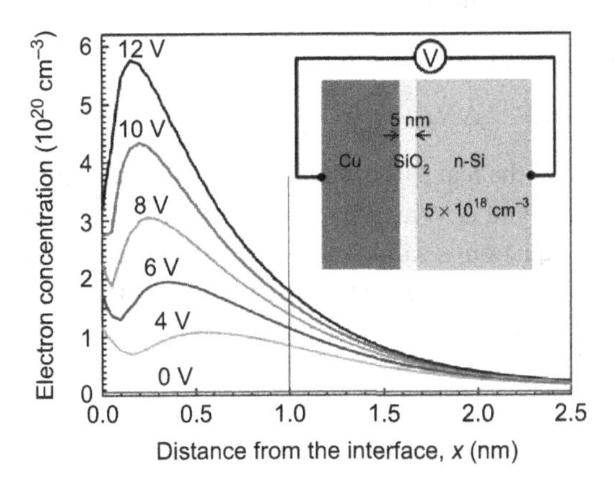

Figure 6.3 The spatial electron distribution in a Cu/5-nm-SiO_2/n-Si capacitor, as shown schematically in the inset, under different voltages, obtained from a one-dimensional MEDICI simulation.

where ϵ_0 is the vacuum permittivity, ϵ_d is the relative permittivity of the gate dielectric, e is the electronic charge, t_{ox} is the gate oxide thickness; t_{AcL} is the AcL thickness, assumed to be 1 nm here, V is the applied voltage, V_{FB} is the flat-band voltage, and E_d is the electric field in the gate oxide. The achievable concentration depends on ϵ_d and the electric breakdown field (E_{BD}) of the gate dielectric. Thermal SiO_2 has an ϵ_d of $\sim$3.9 and an E_{BD} of $\sim$14 MV cm^{-1}, thus the maximum N_{AcL} is $\sim 3 \times 10^{20}$ cm^{-3}. A larger N_{AcL} can be reached when a high-κ dielectric is used. For example, S_3N_4 has an ϵ_d of $\sim$7.6 and an E_{BD} of $\sim$10 MV cm^{-1}, leading to the maximum N_{AcL} of $\sim 4.2 \times 10^{20}$ cm^{-3}; and HfO_2 has an ϵ_d of $\sim$20 and an E_{BD} of $\sim$6 MV cm^{-1}, leading to the maximum N_{AcL} of $\sim 6.6 \times 10^{20}$ cm^{-3}.

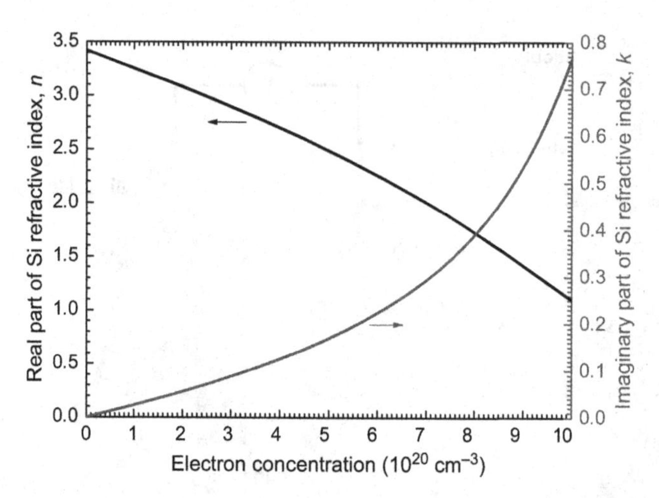

Figure 6.4 The real part (n) and imaginary part (k) of the complex index of Si as a function of the free-electron concentration based on the Drude model.

The refractive-index variation induced by the free-carrier dispersion effect in Si at 1550 nm is commonly expressed by the following equation:

$$\Delta n = -\left[8.8 \times 10^{-22}\,\Delta N_{\mathrm{e}} + 8.5 \times 10^{-18}(\Delta N_{\mathrm{h}})^{0.8}\right],$$
$$\Delta\alpha = 8.5 \times 10^{-18}\,\Delta N_{\mathrm{e}} + 6 \times 10^{-18}\,\Delta N_{\mathrm{h}}, \tag{6.2}$$

where Δn is the modification of the real part of the index, $\Delta\alpha$ is the modification of the absorption coefficient, and ΔN_{e} and ΔN_{h} are the variations in concentration of electrons and holes, respectively. This equation is valid when ΔN_{e} or ΔN_{h} is not very large, e.g. less than $\sim 10^{20}$ cm^{-3}. Δn and $\Delta\alpha$ will deviate significantly from the above prediction when the free-carrier concentration becomes very large. The well-known Drude model could be directly used to estimate the complex permittivity (ϵ) (and hence the complex index of Si) in the region of free-carrier concentration larger than $\sim 10^{20}$ cm^{-3} [27]:

$$\epsilon(\omega) = (n + k\mathrm{i})^2 = \epsilon_\infty - \frac{\Delta N_{\mathrm{e}} e^2/(\epsilon_0 m^*)}{\omega^2[1 + \mathrm{i}/(\omega\tau)]}. \tag{6.3}$$

where $\epsilon_\infty = 11.7$ is the static permittivity of Si, $m^* = 0.272 m_0$ is the effective mass of the electron, and $\tau \sim 7.7 \times 10^{-15}$ s is the electron relaxation time, which can be estimated from the empirical drift mobility of Si. The calculated n and k at 1550 nm ($\omega = 1.2 \times 10^{15}$ s^{-1}) are plotted in Fig. 6.4 as functions of ΔN_{e}.

The modal index of the horizontal MISIM plasmonic waveguides at 1550 nm, i.e. the real effective index (n_{eff}) and the propagation loss (α) in dB μm^{-1}, are calculated using the eigenmode-expansion (EME) method. The structural parameters are set as follows: Si-core height $H = 340$ nm, Si-core width $W_{\mathrm{P}} = 10$–100 nm, SiO_2 gate oxide thickness $t_{\mathrm{ox}} = 1$–10 nm, $t_{\mathrm{AcL}} = 1$ nm, and $N_{\mathrm{D}} = 5 \times 10^{18}$ cm^{-3}. The complex refractive indexes of Cu, SiO_2, and Si at 1550 nm are set as follows: Cu, 0.282 + i11.048 [28], SiO_2, 1.445; Si in depletion, 3.455; and Si with 5×10^{18} cm^{-3} n-type doping, 3.4506 + i0.001 232. Figure 6.5 plots the calculated Δn_{eff} and $\Delta\alpha$ (compared with those in the depletion

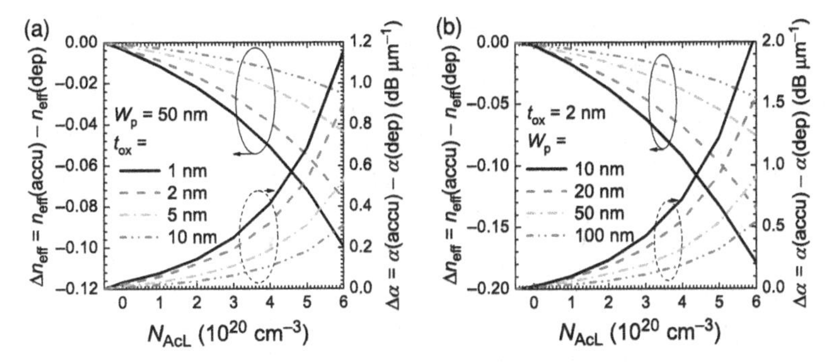

Figure 6.5 The calculated modification of the real part of the modal index (left-hand axis) and the propagation loss (right-hand axis) in the horizontal MISIM plasmonic waveguides as a function of N_{AcL} of the 1-nm-thick AcL: (a) the SiO_2 thickness ranges from 1 to 10 nm while the Si-core width is held at 50 nm; and (b) the Si-core width ranges from 10 to 100 nm while the SiO_2 thickness is held at 2 nm.

condition) as functions of N_{AcL} for plasmonic waveguides with different dimensions. We can see that both Δn_{eff} (negative) and $\Delta\alpha$ increase almost exponentially with increasing N_{AcL}. The plasmonic waveguides with thinner t_{ox} and narrower W_{P} provide larger Δn_{eff} and $\Delta\alpha$ at a certain N_{AcL}, which indicates a larger modulation efficiency at a certain voltage, but they also have larger initial propagation loss (i.e. α in the depletion condition) as well as a larger coupling loss to the input/output Si waveguides, which will lead to a larger insertion loss (IL).

In terms of $\Delta\alpha$, a plasmonic EA modulator can be realized simply by inserting the plasmonic modulator element shown in Fig. 6.2 into an Si-channel waveguide, which offers a broad optical bandwidth [29]. For example, for a plasmonic waveguide with $t_{\mathrm{ox}} = 2$ nm and $W_{\mathrm{P}} = 20$ nm, the propagation loss in the depletion condition is 1.747 dB μm^{-1} and that in the accumulation condition with $N_{\mathrm{AcL}} = 6 \times 10^{20}$ cm^{-3} is 3.302 dB μm^{-1}. Thus an EA modulator based on this plasmonic waveguide of length $L_{\mathrm{P}} = 3$ μm is expected to have an IL of 5.24 dB and an ER of 4.67 dB.

Moreover, in terms of Δn_{eff}, a phase modulator could be designed by inserting the plasmonic modulator element shown in Fig. 6.2 into one arm of a conventional asymmetric Si Mach–Zehnder interferometer (MZI) to convert phase variation into intensity variation. The output power (normalized by the input power) through the MZI can be expressed as [30]

$$P_{\mathrm{out}} = \left[\tau_1^2 + \tau_2^2 + 2\tau_1^2\tau_2^2\cos(\Phi)\right]/4, \tag{6.4}$$

$$\Phi = \frac{2\pi n_{\mathrm{eff}}(Si)}{\lambda}\Delta L - \frac{2\pi n_{\mathrm{eff}}}{\lambda}L_{\mathrm{P}}, \tag{6.5}$$

where $n_{\mathrm{eff}}(\mathrm{Si})$ is the effective index of the Si waveguide, ΔL is the length difference between the two arms which determines the free-spectral range (FSR), and τ_i^2 is the power transmission in arm i ($i = 1, 2$). If the Si-waveguide loss is neglected, we have $\tau_1^2 = 1$ and $\tau_2^2 = 10^{-\alpha L_{\mathrm{P}}/10}$. Figure 6.6 plots the calculated power transmission spectra for an MZI modulator with $\Delta L = 1550$ μm and one arm inserted by a plasmonic

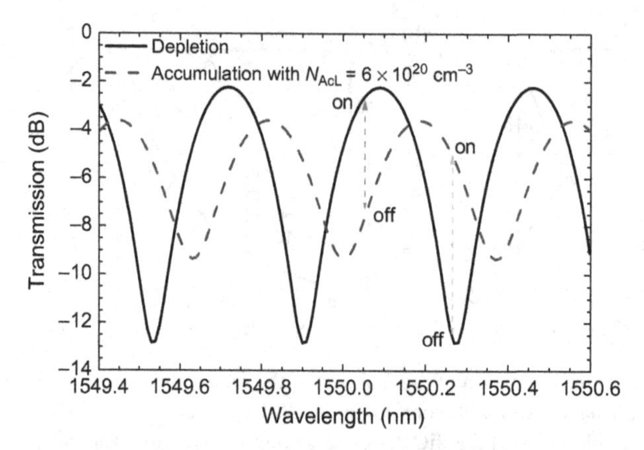

Figure 6.6 Calculated transmission spectra for an asymmetric Si MZI with $\Delta L = 1550$ μm and one arm inserted by a plasmonic phase modulator element with $t_{ox} = 2$ nm, $W_P = 20$ nm, and $L_P = 3$ μm under conditions of depletion and accumulation with $N_{AcL} = 6 \times 10^{20}$ cm^{-3}.

modulator element with $t_{ox} = 2$ nm, $W_P = 20$ nm, and $L_P = 3$ μm under conditions of depletion and accumulation with $N_{AcL} = 6 \times 10^{20}$ cm^{-3}.

At 1550.06 nm, it provides an IL of 2.481 dB and an ER of 4.77 dB. At 1550.26 nm, it provides an IL of 4.863 dB and an ER of 7.383 dB. We can see that at these specific wavelengths both the IL and the ER are significantly improved as compared with the corresponding EA modulator under the same N_{AcL}. Theoretically the ER of an MZI modulator can be infinitely large at some specific wavelengths if the MZI is designed to have $\tau_1 = \tau_2$ initially. The MZI modulator has a narrower optical bandwidth than that of the EA modulator, but the bandwidth is still much broader than that of high-Q ring-resonator-based modulators.

6.2.2 Experimental demonstration

The proposed plasmonic EA and phase modulators are fabricated on SOI wafers using standard CMOS technology. The element shown in Fig. 6.2 is inserted into a single-Si-channel waveguide to behave as an EA modulator as shown in Fig. 6.7(a) or inserted into one arm of an asymmetric Si MZI with ΔL of ~1550 μm to behave as an MZI modulator as shown in Fig. 6.7(b). Figure 6.8(a) shows the cross-sectional transmission electron microscopy (XTEM) image of the final device. The Si core of the plasmonic waveguide is surrounded by a ~6.2-nm-thick SiO_2 layer formed by rapid thermal oxidization and covered by a thick Cu layer formed by sputtering and plating, as shown in Fig. 6.8(b). The current–voltage curves measured for the devices exhibit typical MOS capacitor characteristics. The electric-field ($|E_x|$) profile in this plasmonic waveguide at 1550 nm is shown in Fig. 6.8(c). The calculated n_{eff} and α are 2.626 and 0.881 dB μm^{-1}, respectively.

The diced chips are measured using the conventional fiber–waveguide–fiber method. The input light is TE-polarized light from a broad-band laser (~1520–1620 nm). Using the conventional cutback method, the propagation loss of the MISIM plasmonic

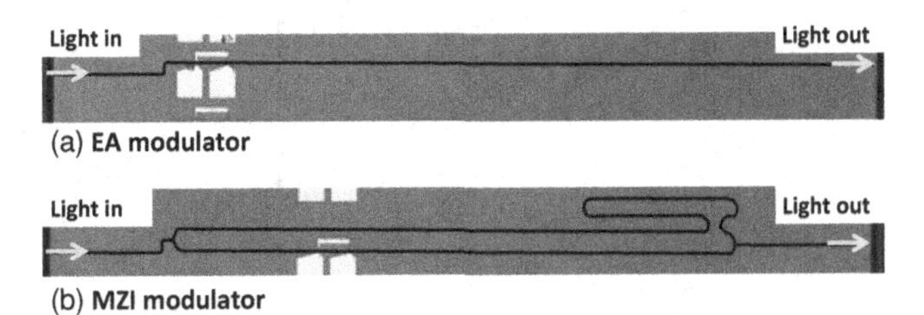

Figure 6.7 A picture of the fabricated plasmonic modulators. (a) A single Si waveguide with the plasmonic modulator inserted to behave as an EA modulator. (b) An asymmetric Si MZI with the plasmonic modulator inserted in one arm to behave as an MZI modulator.

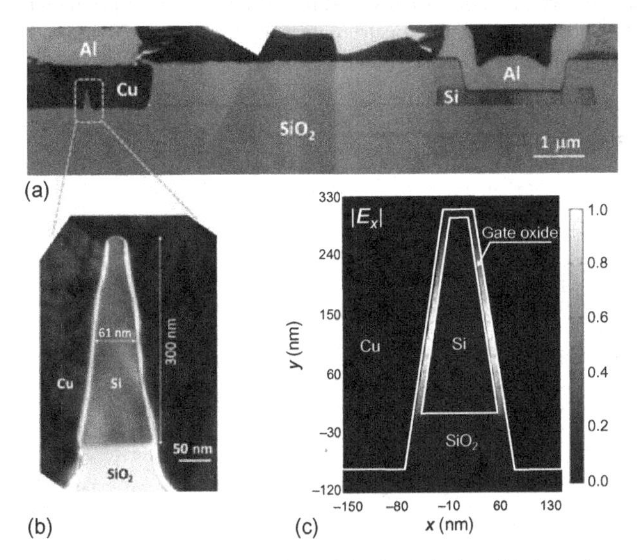

Figure 6.8 (a) An XTEM image of the final device. (b) An enlarged XTEM image around the Si core of the plasmonic waveguide. (c) The corresponding electric field ($|E_x|$) profile at 1550 nm obtained using the EME method; the calculated n_{eff} is 2.626 and α is 0.881 dB μm^{-1}. The measured α is $\sim$0.58 dB μm^{-1}.

waveguide and the coupling loss to the input/output Si channel waveguide are extracted to be $\sim$0.58 dB μm^{-1} and $\sim$0.93 dB per facet, respectively, which are close to the theoretical predictions.

Figure 6.9 plots the output spectra of an EA modulator with $L_{\text{P}} = 3$ μm under voltages ranging from 0 to 8 V, normalized by that of a reference Si waveguide without the plasmonic structure. We can see that the modulation is almost wavelength-independent, reflecting the broad-band nature of the EA modulator. Figure 6.10 depicts the modulation of the output power at 1550 nm versus the voltage for devices with different L_{P}s. At low voltages ($\lesssim$4 V), the modulation is very small, whereas it increases significantly when the voltage becomes larger. As expected, the modulator with longer L_{P} provides a larger modulation at the same voltage, as well as a larger insertion loss.

Figure 6.11 plots output spectra measured on an MZI modulator with L_{P} 1 μm under voltages ranging from 0 to 6 V. They exhibit typical resonance properties with an FSR of

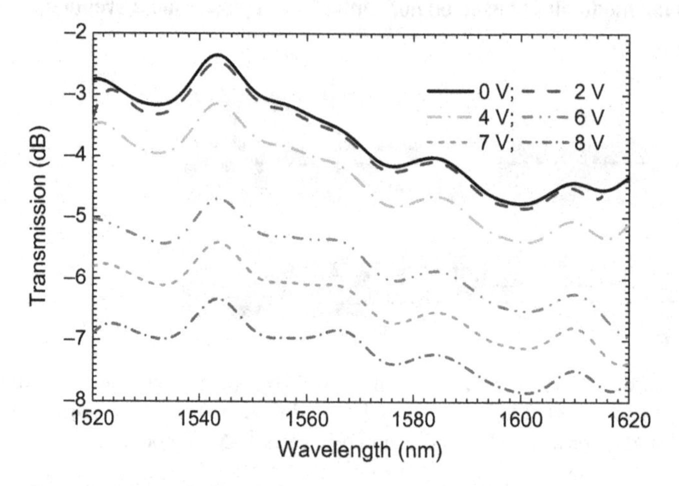

Figure 6.9 Transmission spectra measured on an EA modulator with L_P 3 μm under voltages ranging from 0 to 8 V, normalized by the spectrum measured for a reference Si waveguide without the plasmonic structure.

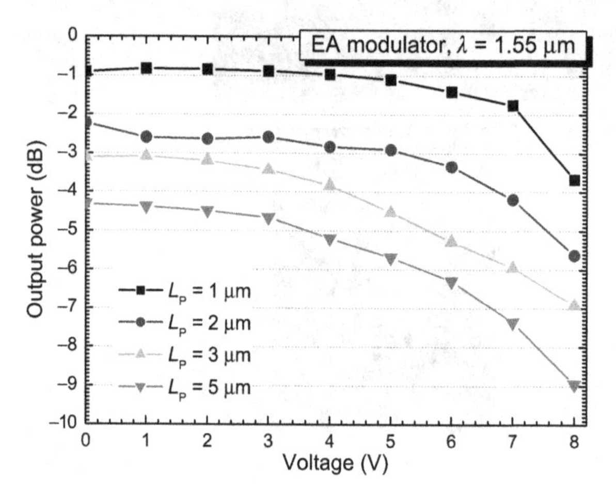

Figure 6.10 The modification of the output power as a function of the applied voltage for plasmonic EA modulators with L_P of 1, 2, 3, and 5 μm.

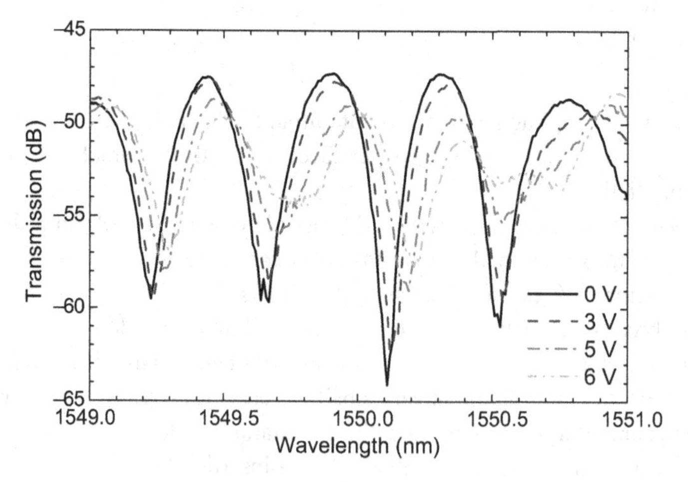

Figure 6.11 Transmission spectra measured on an MZI modulator with L_P 1 μm under voltages of 0, 3, 5 and 6 V.

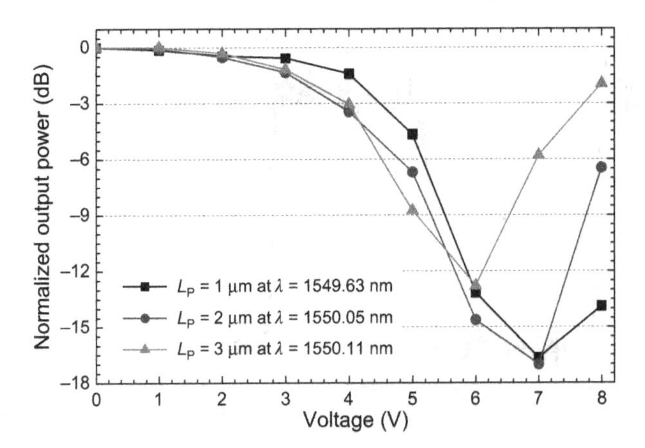

Figure 6.12 Normalized output power versus voltage for MZI modulators with different L_Ps at particular wavelengths. They provide a much larger modulation efficiency than that of the corresponding EA modulator at the price of a narrower optical bandwidth.

$\sim$0.45 nm, in agreement with the theoretical prediction. When a voltage is applied, the phase is shifted and the amplitude is reduced. We can see that the shift is small at low voltage, whereas it becomes large when V is further increased. The π shift is reached at $\sim$6 V bias.

Figure 6.12 depicts normalized output powers versus voltage for MZI modulators with $L_P = 1, 2$, and 3 μm at some specific wavelengths. We can see that a large modulation depth of 12–18 dB is obtained at a voltage of 6–7 V, which is much larger than for the corresponding EA modulators. As mentioned above, the modulation depth can be further improved simply by using a proper MZI design to make the two arms have similar optical losses initially.

Figure 6.13 shows the AC response of an MZI modulator with L_P 1 μm under a 6-V square voltage swing of frequency 10 kHz. It exhibits ER of 9 dB and a relatively large rise and fall time of $\sim$5 μs. This large rise and fall time can be attributed to the long distance between the Si core and the n^+ contact for the present device. It is $\sim$8 μm long and is perpendicular to the Si core, as shown in Fig. 6.2(a), which results in a long transport time, of the order of tens of nanoseconds, and a large series resistance, of the order of tens of kΩ, thus resulting in an apparently slow speed of less than MHz.

6.2.3 Issues and possible solutions

The two apparent main limitations of the present EA and MZI modulators are the large drive voltage requirement (which leads to a large gate leakage current) and their low speed. The first issue arises from the relatively small achievable N_{AcL} because SiO_2 is used as the gate dielectric. Numerical simulation suggests that this problem can be solved, or at least alleviated, by using a high-κ dielectric. Firstly, a high-κ dielectric can provide a large achievable N_{AcL} at a certain voltage. Secondly, a high-κ dielectric usually has a large reflective index, which can lead to a large Δn_{eff} and $\Delta\alpha$ at a certain N_{AcL}. For instance, for a Cu–TiO_2 (2 nm)–Si (20 nm)–TiO_2 (2 nm)–Cu plasmonic waveguide,

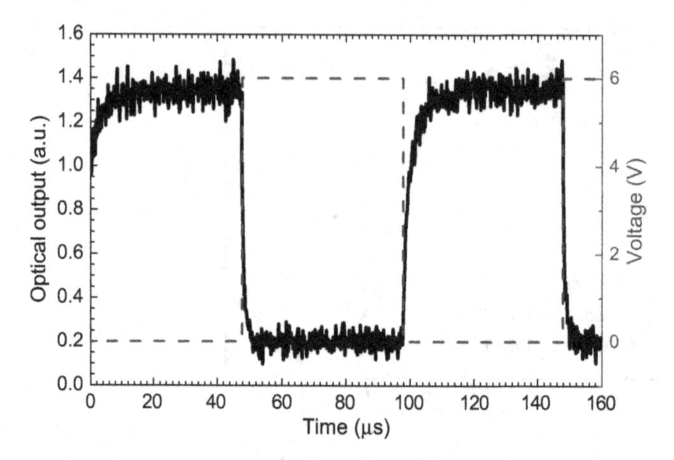

Figure 6.13 The AC response of an MZI modulator with L_P 1 μm under a 6-V voltage swing with a frequency of 10 kHz. The relatively low speed of the present modulator is due to the long distance between the Si core and the n$^+$ contact, and can be significantly improved by structural optimization.

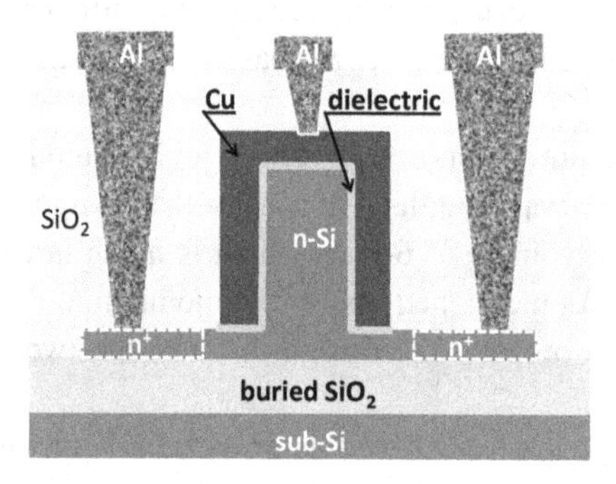

Figure 6.14 A schematic representation of the structure of a plasmonic modulator based on a rib Si waveguide. The short distance between the Si core and the contact can provide a high speed.

numerical simulation predicts that the N_{AcL} of 6×10^{20} cm^{-3} gives a Δn_{eff} of 0.237 and a $\Delta\alpha$ of 2.629, compared with a Δn_{eff} of 0.135 and a $\Delta\alpha$ of 1.555 for the SiO_2 counterpart. Moreover, the required voltage ($V - V_{FB}$) for the N_{AcL} of 6×10^{20} cm^{-3} is 0.54 V for the 2-nm-thick TiO_2 gate dielectric and is 5.57 V for the 2-nm-thick SiO_2 gate dielectric.

For the second issue, the solution is to shorten the distance between the n$^+$ contact and the Si core. A suitable configuration is to use a rib Si waveguide and place the n$^+$ contact on the Si slab parallel to the Si core, as shown schematically in Fig. 6.14. The carrier transport time will be of the order of 10 ps, and the series resistance will be of the order of 100 Ω if the distance is reduced to less than 1 μm. The modulator speed is limited by the RC constant rather than the transition time, which is estimated to be ~80 GHz since the capacitance of the ultra-compact modulator is calculated to be ~20 fF. Moreover, the RF power is predicted to be ~5 fJ per bit when the voltage is reduced to ~1 V and the gate leakage current is below the μA range.

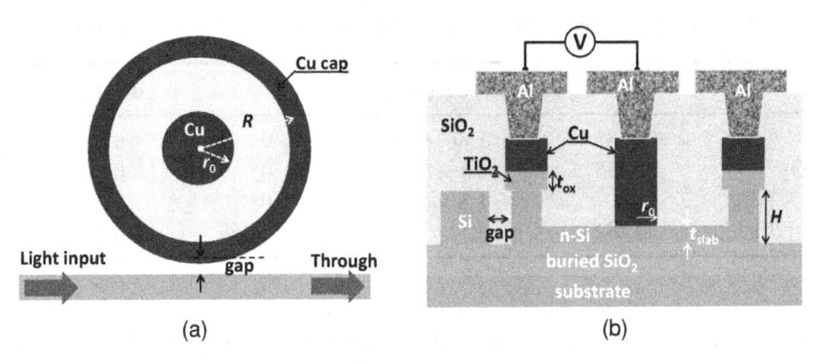

Figure 6.15 (a) Top view and (b) cross-sectional view of the proposed Si plasmonic resonator modulator. The bus waveguide is a conventional single-mode Si-strip waveguide and the resonator is a Cu–TiO_2–Si hybrid plasmonic donut with two electrodes, located at the Cu cap and at the center of the donut, respectively. The structural parameters are indicated.

6.3 Athermal ring modulators based on vertical metal–insulator–Si hybrid plasmonic waveguides

6.3.1 Device structure

The MISIM plasmonic waveguide provides tight optical mode confinement but has a relatively large propagation loss. Waveguide ring resonators (WRRs) based on this plasmonic waveguide have a low Q value of 20–50 [25], thus it is not suitable for ring modulators. On the other hand, the MIS hybrid plasmonic waveguide provides a low propagation loss at the price of relatively loose optical mode confinement. WRRs with Q values of $\sim$200–300 and radii of $\sim$1–2 μm based on this plasmonic waveguide have been demonstrated [21], indicating that ring modulators can be designed on the basis of the MIS hybrid plasmonic waveguide. More importantly, if TiO_2, which has a negative thermo-optic coefficient of about $-1.8 \times 10^{-4}\,K^{-1}$ and is transparent at near-infrared wavelengths [31], is used as the insulator between the Cu cap and the Si core, the WRRs could be athermal. TiO_2 is also used as a gate dielectric in MOS electronics, whose dielectric constant is 4–86, depending on the details of the fabrication process [32]. The Cu–TiO_2–Si structure of the hybrid plasmonic waveguide enables a voltage to be applied between the Cu cap and the Si core to modulate its propagation property.

The structure of the proposed ring modulator is shown schematically in Fig. 6.15. It consists of a Cu–TiO_2–Si hybrid plasmonic donut resonator and a bus waveguide. For easy integration in a dense Si-strip-waveguide-based interconnection system with low insertion loss, the bus waveguide is a conventional Si-strip waveguide positioned near the resonator separated by a gap. For reduction of the overall footprint, a donut rather than a ring is used so that the electrode for Si contact can be positioned at the center of the donut to form a standard MOS capacitor with the other electrode of the Cu cap.

In fabrication, the Si patterns of the bus and donut waveguides are defined by partially dry etching the Si down to t_{slab} using a thin SiO_2 layer as the etching mask, following by dry etching of the remaining Si down to the buried SiO_2 using both SiO_2 and a photoresist as the etching mask. Using this etching method, there is no misalignment issue between

Table 6.1 Optical parameters of Si, SiO_2, TiO_2, and Cu at 1550 nm and room temperature

	n	dn/dT (K^{-1})	k	dk/dT (K^{-1})
Si	3.455	1.86×10^{-4}	0	0
SiO_2	1.455	0.1×10^{-4}	0	0
TiO_2 [18]	2.2	-1.8×10^{-4}	0	0
Cu [20]	0.282	4.1×10^{-4}	11.05	-4.3×10^{-4}

the inner and outer rings of the donut. After Si patterning, a thick layer of SiO_2 is deposited and a ring-shaped window is opened to expose the surface of the Si core. Misalignment between the SiO_2 window and the Si core beneath may be unavoidable. To minimize the misalignment effect, the SiO_2 window (and hence the width of the TiO_2/Cu cap) is intentionally designed to be larger than the Si core beneath by ΔW_P on each side. TiO_2 is then deposited on the Si core through the windows, followed by deposition of Cu and Cu chemical mechanical polishing to remove TiO_2 and Cu outside the windows. The structural parameters indicated in Fig. 6.15 are initially chosen as follows on the basis of our experience: radius $R = 1.5$ μm, Si height $H = 220$ nm, Si-slab height $t_{slab} = 50$ nm, Si core width $W_P = 200$ nm, TiO_2 thickness $t_{ox} = 50$ nm, and TiO_2/Cu cap width wider than the Si core on each side by $\Delta W_P = 50$ nm. The refractive indexes of Si, SiO_2, TiO_2, and Cu depend on wavelength and temperature. For simplification, the indexes and thermo-optic coefficients at wavelength 1550 nm and room temperature are used here, as listed in Table 6.1. The propagation loss of plasmonic waveguides calculated using these optical parameters agrees well with the experimental results measured at room temperature.

6.3.2 Device properties

Thermo-optic properties

The effective modal indices of the curved Cu–TiO_2–Si HPWs are calculated using the EME method. The Si core has an asymmetric rib structure to mimic the donut resonator shown in Figure 6.15(b). The electrical intensity distribution of the fundamental TM mode in the curved plasmonic waveguide is depicted in Fig. 6.16(a), showing that the electric field is enhanced in the TiO_2 layer and in the Si core just beneath the TiO_2 layer. The lateral confinement is well provided by the Si core, and the extended TiO_2 region (laterally over the Si-core region) contains a weak electric field. The effective modal index is calculated to be $2.275 + i0.005\,77$, corresponding to a propagation loss of 0.203 dB μm^{-1}, which is close to the experimental result. The ratios of optical intensity in the TiO_2 layer, the Si rib, the Si slab, and the surrounding SiO_2 cladding layer are 41.3%, 42.0%, 0.04%, and 16.7%, respectively. As expected, the thin Si slab has a negligible effect on the optical mode.

The real (n_{eff}) and imaginary (k_{eff}) parts of effective modal indexes are plotted in Fig. 6.16(b) as functions of temperature in the range of ±20 °C from around room temperature for two curved plasmonic waveguides with TiO_2 thicknesses of 10 and 50 nm. They both show good linearity, allowing thermo-optic coefficients of dn_{eff}/dT

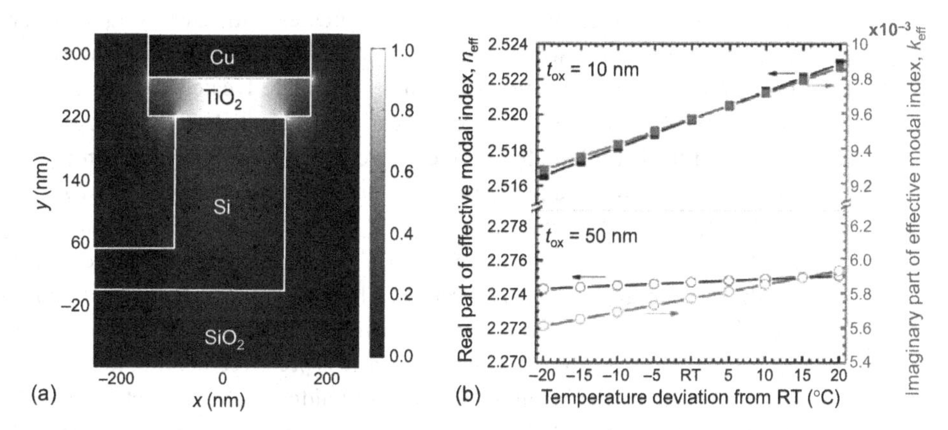

Figure 6.16 (a) The electric intensity distribution of the fundamental TM mode at 1550 nm in the curved Cu–TiO_2–Si plasmonic waveguide with structural parameters as listed in Table 6.1. (b) The real part (n_{eff}) and imaginary part (k_{eff}) of the effective modal index of the curved plasmonic waveguides with TiO_2 thicknesses of 10 and 20 nm as functions of temperature.

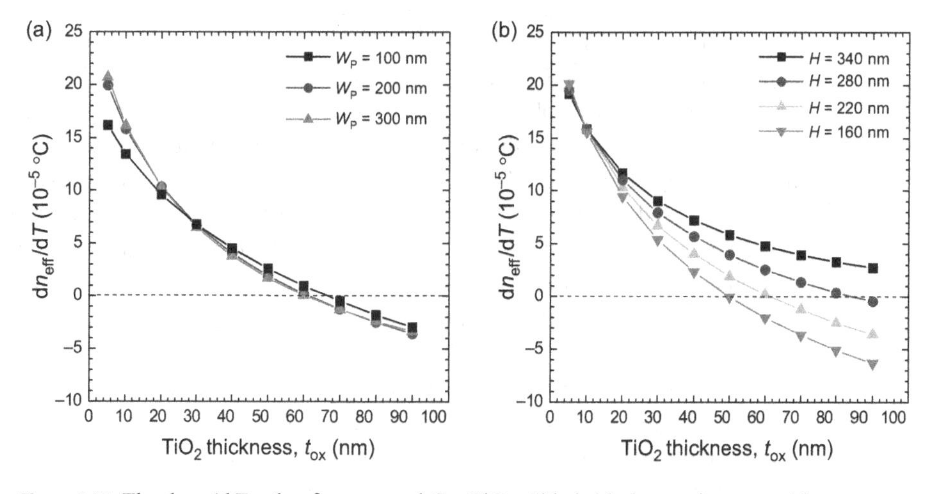

Figure 6.17 The dn_{eff}/dT value for a curved Cu–TiO_2–Si hybrid plasmonic waveguide as a function of the TiO_2 thickness: (a) the waveguide has an Si-core width of 100, 200, and 300 nm with $H = 220$ nm; and (b) the waveguide has an Si height of 340, 280, 220, and 160 nm, with $W_P = 200$ nm. The other structural parameters are as listed in Table 6.1. The athermal point is defined as when $dn_{eff}/dT = 0$.

and dk_{eff}/dT to be defined to describe the thermo-optic effect of the Cu–TiO_2–Si plasmonic waveguides. Since the resonance of a WRR is solely determined by n_{eff}, we may claim that a WRR is athermal if dn_{eff}/dT is zero even though dk_{eff}/dT is not zero. Nevertheless, dk_{eff}/dT is relatively small, for example, $\sim 1.5 \times 10^{-5}\,K^{-1}$ for the 10-nm-TiO_2 waveguide and $8.0 \times 10^{-6}\,K^{-1}$ for the 50-nm-TiO_2 waveguide, as can be read off from Fig. 6.16(b).

Figure 6.17 plots dn_{eff}/dT versus the TiO_2 thickness for bent Cu–TiO_2–Si plasmonic waveguides. As expected, dn_{eff}/dT decreases with increasing t_{ox} simply because the ratio of optical power in the TiO_2 layer increases with increasing t_{ox}. However, the slope becomes smaller with increasing t_{ox}. This is because the hybrid mode shown in Fig. 6.16(a) is a superposition of a pure SPP mode located at the Cu/TiO_2 interface (without the Si core) and a pure optical mode located at the Si core (without the metal), and the hybrid mode becomes more optical-like with increasing t_{ox}. In extreme cases, it will behave as a pure optical mode as in a conventional Si waveguide if t_{ox} is sufficiently large ($\gtrsim$200 nm), and the TiO_2 layer will behave as a cladding layer with an inferior effect on the optical mode. The dn_{eff}/dT versus t_{ox} curve depends on W_P weakly, while it depends on H strongly. There is a critical t_{ox} at which $dn_{eff}/dT = 0$. The critical t_{ox} becomes large when the height of Si waveguides increases, as shown in Fig. 6.17(b). Electrically, a thin gate dielectric is preferred in order to reduce the driving voltage. Optically, the height of the Si waveguide should be thick enough for vertical optical confinement. To balance electric and optical requirements, our modulator is set to $H = 160$ nm and $t_{ox} = 50$ nm. It is athermal, as indicated in Fig. 6.17(b).

Electrical properties

For electrical simulation, the dielectric constant of TiO_2 is set to 80, which is reachable for a high-quality TiO_2 film [32]. The Si core of the resonator is n-type doped with concentrations of 5×10^{18} cm^{-3} in the rib and slab areas and 2×10^{20} cm^{-3} in the contact area for good ohmic contact. The 2D free-carrier distribution in the MOS capacitor under 5 V bias is shown in Fig. 6.18(a). The accumulated electrons are located near the TiO_2/Si interface. To see the distribution clearly, the area near the interface is shown enlarged, as in Fig. 6.18(b). We can see that the electron concentration contours are almost in parallel with the TiO_2/Si interface. Therefore, the 2D distribution may be simplified to a 1D distribution, as plotted in Fig. 6.18(c). As for the MISIM waveguide, a thin accumulation layer of thickness ~1 nm and average N_{AcL} calculated by Eq. (6.1) can be defined.

For transient-state simulation, the gate voltage V is increased from 0 to 8 V with a ramp time of the gate voltage of 10 fs. The free-electron concentration in the 1-nm-thick AcL is plotted in Fig. 6.19 as a function of time. The rise time (t_r) of the electron concentration is defined as the time for the electron concentration to increase from 10% to 90%, and the fall time (t_f) is defined as the time for the electron concentration to drop from 90% to 10%. In the case of the Si n$^+$ contact just below the electrode, as shown in Fig. 6.15(a), the sum of these times ($\tau_s = t_r + t_f$) is ~29 ps. The modulation speed estimated from the inverse of τ_s is ~34 GHz. The speed can be improved by shortening the distance between the AcL and the n$^+$ contact. In the case that the n$^+$ contact in the Si slab is extended to the Si rib, both t_r and t_f decrease, as shown by the dashed curve in Fig. 6.19. τ_s is read off to be ~9.3 ps, corresponding to a modulation speed of ~107 GHz. However, the propagation loss will increase from 0.27 dB μm^{-1} to 0.31 dB μm^{-1} when the doping level in the Si slab is increased from 5×10^{18} cm^{-3} to 2×10^{20} cm^{-3} (while keeping the doping level in the Si rib 5×10^{18} cm^{-3}). To balance the loss and the speed, the n$^+$ contact may be extended to a certain location between the Si rib and the electrode.

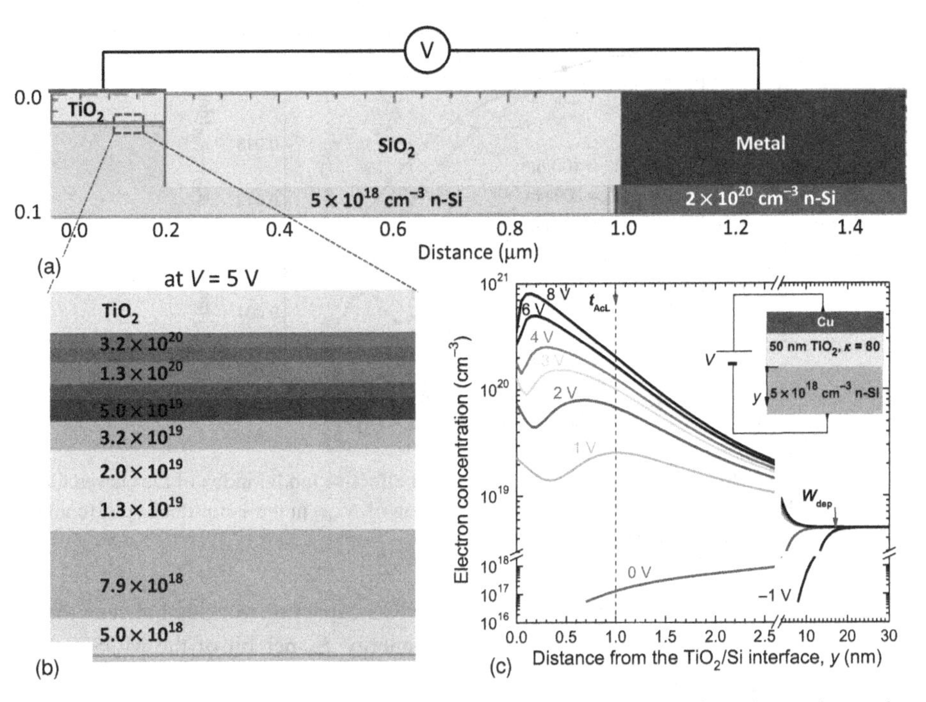

Figure 6.18 (a) The two-dimensional electron distribution of a Cu–TiO_2–Si MOS capacitor under 5 V bias. Free electrons are accumulated near the TiO_2/Si interface. (b) The electron concentration contour near the TiO_2/Si interface; different shadings represent different concentrations. (c) The one-dimensional electron distribution along the y axis of a Cu–TiO_2–Si capacitor as shown schematically in the inset under different biases. The depletion width W_{dep} and the accumulation layer thickness t_{AcL} are also indicated.

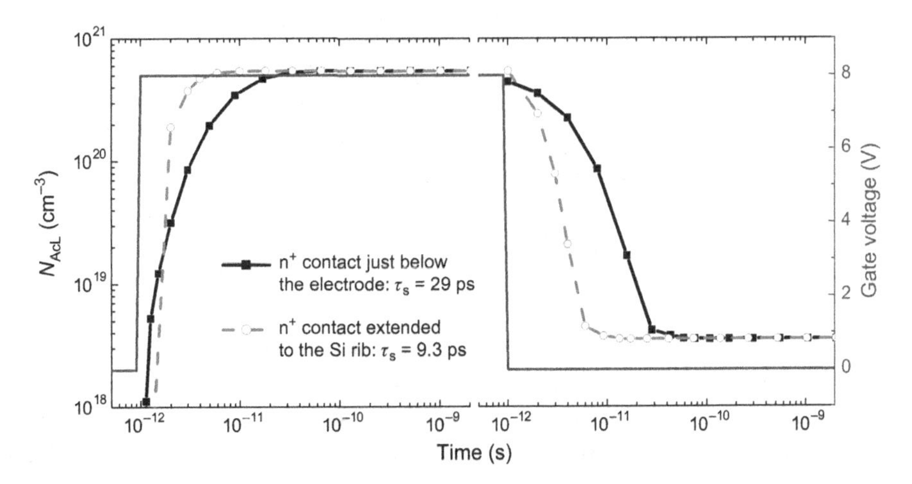

Figure 6.19 (a) The transient response of the electron concentration in the 1-nm-thick AcL of the Cu–TiO_2–Si MOS capacitor for gate voltage varied between 0 and 8 V. The rise and fall times are defined as 10% to 90% time periods. The solid curve is for a capacitor with n^+ contact just below the electrode as shown in Fig. 6.15(a), and the dashed curve is for a capacitor with the n^+ contact extended to the Si rib.

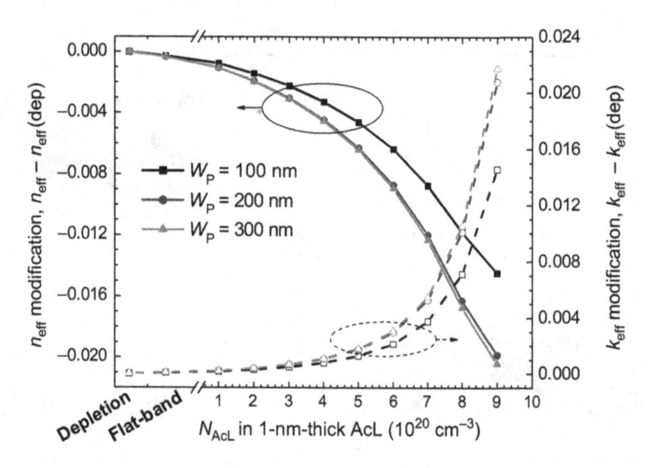

Figure 6.20 Modification of the calculated effective modal index of the curved Cu–TiO_2–Si hybrid plasmonic waveguides as a function of N_{AcL} in the 1-nm-thick AcL (compared with that in the depletion state).

The proposed EO modulator is a MOS capacitor working between the depletion and accumulation states. The switching energy E_s per bit of the MOS modulator can be roughly estimated as

$$E_s = \frac{1}{2} C_{dep} V_{dep}^2 + \frac{1}{2} C_{accu} V_{accu}^2, \tag{6.6}$$

where C_{dep} and C_{accu} are capacitances under the depletion and accumulation states, respectively. Because C_{dep} is smaller than C_{accu} and V_{dep} is smaller than V_{accu}, E_s is mainly determined by the second term of Eq. (6.6), namely the accumulation state. C_{accu} can be approximated to the gate oxide capacitance as

$$C_{accu} \approx A \frac{t_{ox}}{\epsilon_0 \epsilon_d},$$

where A is the active area. For the modulator with A of 1.76 μm^2 and C_{accu} of ~25 fF, E_s is estimated to be ~0.45 pJ in the case of a driving voltage of 6 V.

Electro-optic properties

The effective modal index is calculated as a function of N_{AcL} in the 1-nm-thick AcL for curved Cu–TiO_2–Si waveguides with W_P of 100, 200, and 300 nm (the Si height H is 160 nm), as shown in Fig. 6.20. We can see that the real part of the effective modal index decreases and the imaginary part increases almost exponentially with increasing N_{AcL}, which can be attributed to the increase of modal confinement in the 1-nm-thick AcL, as shown in Fig. 6.21(a), which plots the 2D electric-field ($|E_y|$) distribution of the fundamental TM mode in a curved Cu–TiO_2–Si waveguide with $N_{AcL} = 1 \times 10^{20}$ cm^{-3}. To see the field distribution in the 1-nm-thick AcL clearly, normalized 1D $|E_y|$ distributions along the y axis at $x = 0$ (i.e. the dashed line in Fig. 6.21(a)) are plotted in Fig. 6.21(b) for waveguides with $N_{AcL} = 1 \times 10^{20}$ cm^{-3} and 8×10^{20} cm^{-3}, respectively. In the case of $N_{AcL} = 8 \times 10^{20}$ cm^{-3}, the electric field in the 1-nm-thick AcL is

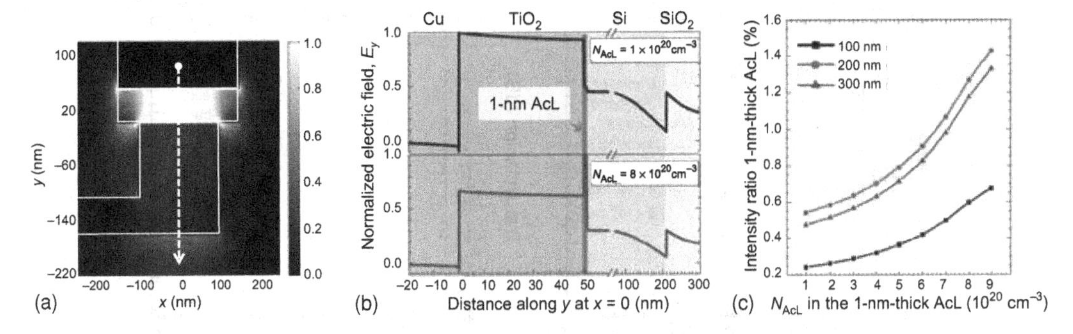

Figure 6.21 (a) The electric-field ($|E_y|$) distribution of the fundamental TM mode in the curved Cu–TiO_2–Si hybrid waveguide; (b) the $|E_y|$ distribution along the y axis at $x = 0$ (the dashed line in (a)) in the cases of $N_{AcL} = 1 \times 10^{20}\,cm^{-3}$ and $8 \times 10^{20}\,cm^{-3}$; and (c) the optical intensity ratio in the 1-nm-thick AcL as a function of N_{AcL} for waveguides with W_P of 100, 200, and 300 nm.

dramatically enhanced. Since the continuity of the electric displacement normal to the interfaces, E_y, is roughly inversely proportional to the permittivity, as a result, the optical field intensity ratio in the 1-nm-thick AcL increases dramatically with increasing N_{AcL}, as shown in Fig. 6.21(c). The modulation efficiency depends on W_P. The waveguide with W_P 100 nm provides a smaller modulation efficiency than do the waveguides with larger W_P, while the waveguide with W_P 200 nm provides modulation efficiency similar to that of the waveguide with W_P 300 nm. This is because the lateral confinement is weak when W_P is small (e.g. 100 nm), whereas it becomes saturated when W_P becomes large ($\gtrsim$200 nm). On the other hand, the modulator with small W_P has a smaller active area, which indicates high speed. To balance the optical and electric performance, W_P is set to 200 nm for our modulator.

It has been demonstrated experimentally that the transmission spectrum of a hybrid plasmonic waveguide based WRR can in general be expressed as [18, 21]

$$T(\lambda) = \frac{\alpha^2 + t^2 - 2\alpha t \cos\theta}{1 + (\alpha t)^2 - 2\alpha t \cos\theta}, \tag{6.7}$$

where $\theta = (2\pi/\lambda)n_g 2\pi R$ is the phase change around the ring, α^2 is the power-loss factor per round trip around the ring, $t = |t| \exp(i\theta)$ is the self-coupling coefficient, and n_g is the group index. α^2 for different states can be calculated from k_{eff} read off from Fig. 6.20. n_g in the depletion state is calculated to be ~3.782, which is similar to the experimental result [21]. Because it is difficult to calculate the n_g value accurately, we simply assume that the modification of n_g induced by the free-carrier effect is the same as the n_{eff} modification, namely $\Delta n_g = \Delta n_{eff}$, thus n_g for different states and different N_{AcL}s can also be read from Fig. 6.20. Since n_g is larger than n_{eff}, this assumption may underestimate the modification efficiency of our modulator. The self-coupling coefficient t is related to the cross-coupling coefficient k as $t^2 + k^2 = 1$. The coupling between the bus waveguide and the resonator depends on the gap between them and on the effective difference in modal index between the bus waveguide and

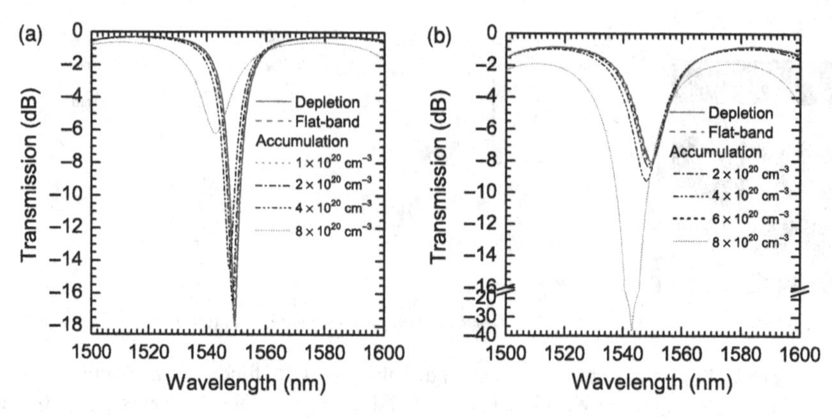

Figure 6.22 Transmission spectra of the plasmonic donut modulator for depletion, flat-band, and accumulation states with N_{AcL} ranging from $1 \times 10^{20}\ \mathrm{cm}^{-3}$ to $8 \times 10^{20}\ \mathrm{cm}^{-3}$, calculated using Eq. (6.5): (a) at under-coupling with $|t| \sim 0.8$ and (b) at over-coupling with $|t| \sim 0.5$.

the curved plasmonic waveguide [33]. The $|t|$ value can be varied in a large range from over-coupling ($|t| < \alpha$) to under-coupling ($|t| > \alpha$), depending on the detailed structural parameters of WRRs [21]. The relationship between the coupling strength and structural parameters has been well studied [33], and will not be discussed here. Instead, we simply set $|t|$ to a certain value and $\theta = 0$ (i.e. we assume that there is no phase shift due to coupling). Figure 6.22(a) plots the calculated spectra of our modulator in the case of $|t| = 0.8$. Because α is $\sim$0.75 in the depletion state, the resonator is subject to under-coupling. The extinction ratio is $\sim$18 dB. With N_{AcL} increasing, n_{g} decreases and α increases simultaneously, which leads to a blue-shift of the resonance wavelength and a decrease of the extinction ratio, respectively. In contrast, for conventional Si WRR modulators only the shift of the resonance wavelength occurs, because their α value remains almost unchanged. From Fig. 6.22(a) we can see that the EO induced modulation is enhanced in the wavelength region larger than the resonance due to the α modification, while in the wavelength region smaller than the resonance, the EO induced modulation is weakened. Figure 6.22(b) plots the calculated spectra in the case of over-coupling, assuming $|t| = 0.5$. The extinction ratio is $\sim$8 dB in the depletion state. With increasing N_{AcL}, the resonance wavelength is blue-shifted and the extinction ratio increases. The extinction ratio becomes very large when α approaches $|t|$. The EO induced modulation is enhanced in the wavelength region smaller than the resonance and is weakened in the wavelength region larger than the resonance due to the α modification. The results indicate that the performance of our modulator depends on $|t|$ strongly. Because of the large propagation loss of the plasmonic waveguide (and hence a small α value) as compared with the case of conventional Si-waveguide-based WRRs, our modulator requires strong coupling (and hence a small $|t|$ value) between the bus waveguide and the resonator.

The spectral property shown in Fig. 6.22 is verified by FDTD simulation. To enhance the coupling between the bus waveguide and the resonator, a race-track-shaped resonator is used. The gap between the bus waveguide and the resonator is set to 10 nm and the

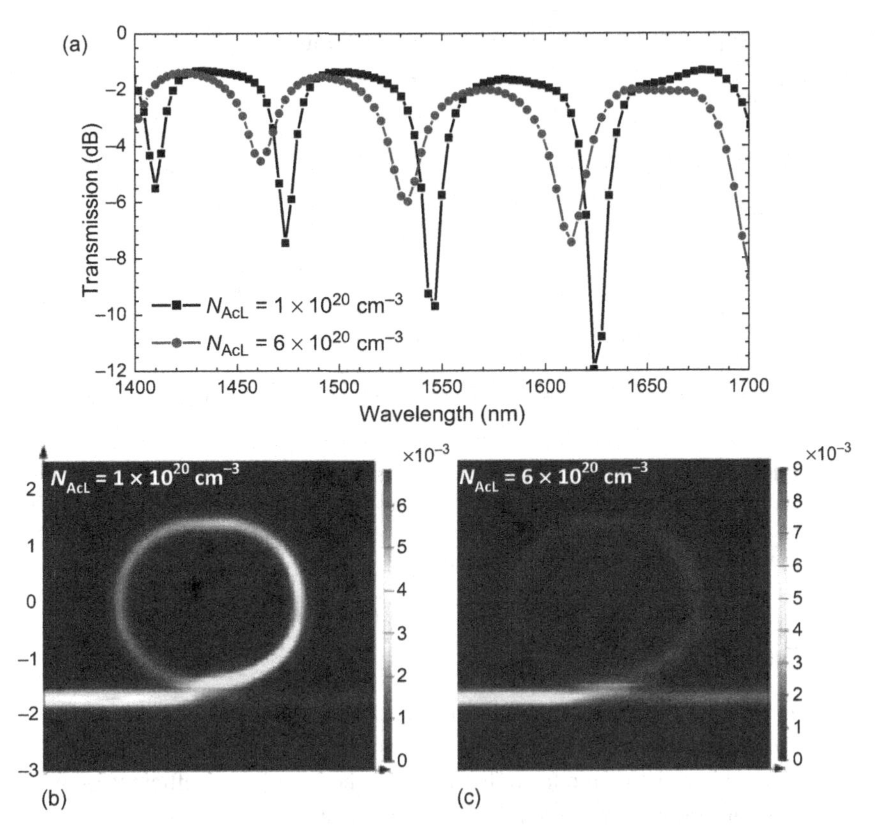

Figure 6.23 (a) Transmission spectra of the plasmonic modulator under accumulation with N_{AcL} of $1 \times 10^{20}\,\mathrm{cm}^{-3}$ and $6 \times 10^{20}\,\mathrm{cm}^{-3}$, obtained from FDTD simulation. (b) The optical power density in the modulator with $N_{\mathrm{AcL}} = 1 \times 10^{20}\,\mathrm{cm}^{-3}$ at $\lambda = 1546\,\mathrm{nm}$. (c) The optical power distribution in the modulator with $N_{\mathrm{AcL}} = 6 \times 10^{20}\,\mathrm{cm}^{-3}$ at $\lambda = 1546\,\mathrm{nm}$. The output power is modulated from about -10 dB to about -2 dB at this wavelength.

directional coupling length is set to 500 nm. The total footprint of the modulator is $\sim 8.6\ \mu\mathrm{m}^2$, and the active area is $A \sim 1.96\ \mu\mathrm{m}^2$. Figure 6.23(a) plots the transmission spectra for the modulator under accumulation with $N_{\mathrm{AcL}} = 1 \times 10^{20}\,\mathrm{cm}^{-3}$ and $6 \times 10^{20}\,\mathrm{cm}^{-3}$, which corresponds to a voltage swing from 1.1 V to 6.8 V. In agreement with that which is predicted in Fig. 6.22(a), the resonance wavelength is blue-shifted and the extinction ratio is reduced for increasing N_{AcL}. At 1546 nm, the output power is modified from about -10 dB to about -2 dB by increasing N_{AcL} from $1 \times 10^{20}\,\mathrm{cm}^{-3}$ and $6 \times 10^{20}\,\mathrm{cm}^{-3}$. The corresponding optical power distributions in the modulator are shown in Figs. 6.23(b) and 6.23(c), respectively. One can see that there is optical enhancement in the resonator due to the resonance in the case of $N_{\mathrm{AcL}} = 1 \times 10^{20}\,\mathrm{cm}^{-3}$, whereas the modulator is in an off-resonance state and light propagates through the bus waveguide in the case of $N_{\mathrm{AcL}} = 6 \times 10^{20}\,\mathrm{cm}^{-3}$. As mentioned above, the modulation depth can be further improved by adjusting the coupling between the bus waveguide and the resonator.

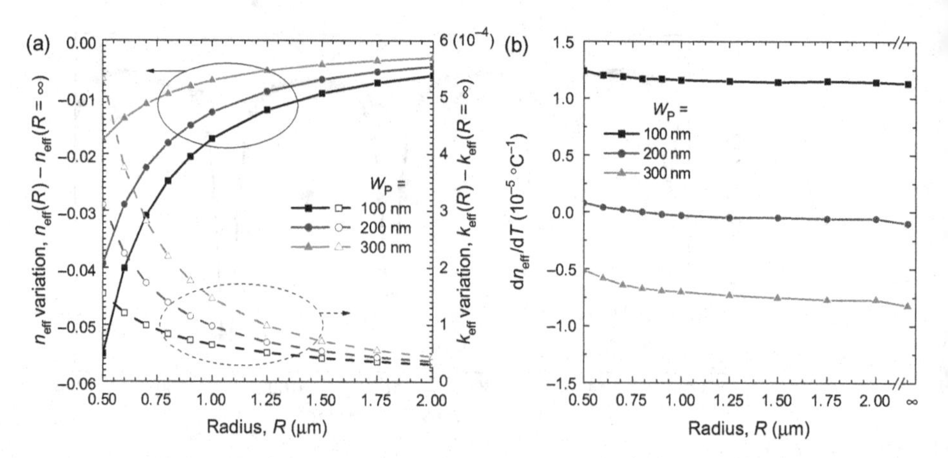

Figure 6.24 (a) The effective modal index versus radius of curved Cu–TiO_2–Si plasmonic waveguides with different W_Ps (compared with that of the corresponding straight plasmonic waveguide, namely $R = \infty$). (b) The thermo-optic coefficient of the real part of the effective modal index, dn_{eff}/dT, versus the radius.

6.3.3 Tolerance

In the above analysis, the radius of the plasmonic resonator is set to $R = 1.5$ μm. The radius (and hence the footprint of modulator) can be further reduced. Figure 6.24 plots the effective modal index and the thermo-optic coefficient as functions of the bending radius for curved Cu–TiO_2–Si plasmonic waveguides with different values of W_P. The real part of the effective modal index decreases and the imaginary part increases with decreasing R due to an outward shift of the optical mode [34], especially for the plasmonic waveguide with larger W_P. A noticeable variation of the effective modal index occurs when R is less than ~1.0 μm for the waveguide with $W_P \lesssim 200$ nm. This indicates that the bending radius of our modulator can be reduced to ~1.0 μm without noticeable performance degradation. Moreover, the thermo-optic property remains almost unchanged for a bending radius as small as 0.5 μm, as shown in Fig. 6.24(b).

In fabrication, the central line of the TiO_2/Cu cap may deviate from the central line of the Si core beneath due to misalignment, as shown schematically in the inset of Fig. 6.25(a). In the case that the TiO_2/Cu cap is intentionally designed to be wider than the Si core beneath by 50 nm on each side, the excess width on one side will be ($50\,\text{nm} - \Delta W'_P$) and that on the other side will be ($50\,\text{nm} + \Delta W'_P$) if the deviation is $\Delta W'_P$. Figure 6.25 plots the real and imaginary parts of the effective modal index (compared with that without the deviation, namely $\Delta W'_P = 0$) as a function of $\Delta W'_P$ for curved Cu–TiO_2–Si hybrid waveguides with W_P of 100, 200, and 300 nm. Both the real and the imaginary part of the modal index decrease with increasing $|\Delta W'_P|$. In the $\Delta W'_P$ range from −10 nm to 20 nm, the variation of the effective modal index with $\Delta W'_P$ is very small, especially for waveguides with larger W_P. This indicates that our EA modulator has a relatively large misalignment tolerance of ~15 nm in the fabrication of the TiO_2/Cu cap.

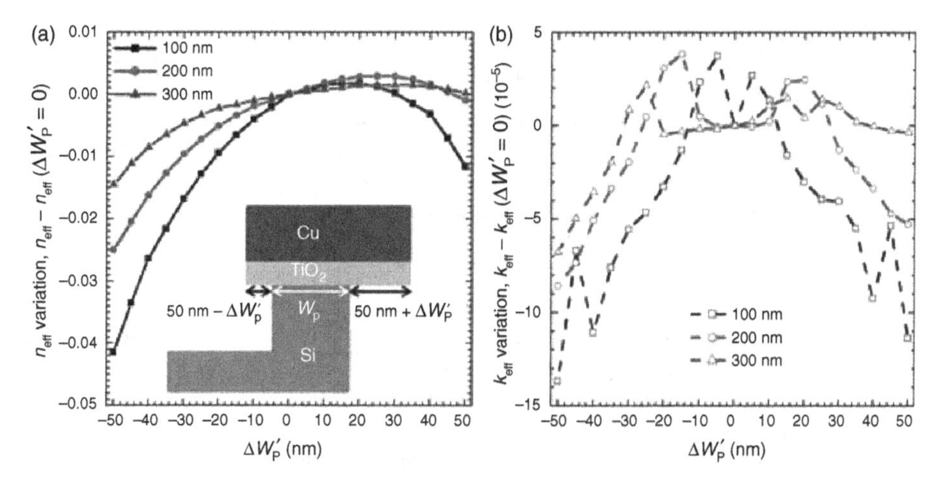

Figure 6.25 (a) The real part of the effective modal index for curved Cu–TiO_2–Si plasmonic waveguides versus the deviation of the center line of the TiO_2/Cu cap from the Si core beneath, $\Delta W'_{\mathrm{P}}$, as shown schematically in the inset, compared with that without the deviation, namely $\Delta W'_{\mathrm{P}} = 0$. (b) The imaginary part of the effective modal index versus $\Delta W'_{\mathrm{P}}$.

6.4 Schottky-barrier plasmonic detectors

6.4.1 Device structure

To shrink the Ge detector, which is commonly used in Si photonics, on the nanometer scale, a precise cavity or antenna structure should be developed to confine the light to a small volume [35, 36], which will make the fabrication processes complex and difficult. On the other hand, a silicide Schottky-barrier detector (SBD) has been demonstrated as an attractive alternative to detect a near-infrared signal [37], in which light absorbed by the silicide layer excites hot carriers in the silicide layer and these hot carriers may emit over the Schottky barrier (Φ_B) to silicon and be collected as photocurrent. Because silicides have a large absorption coefficient, typically around 20–50 μm^{-1} at 1550 nm, the silicide SBD can inherently be much smaller than the Ge counterpart for sufficient absorption, thus enabling scaling down to the nanometer scale. More importantly, the fabrication of a silicide SBD is fully CMOS-compatible, thus enabling seamless integration into existing Si EPICs. However, the conventional silicide SBD suffers from a major shortcoming of low responsivity due to the low internal quantum efficiency, namely the low probability of emission of photoexcited carriers from silicide to Si [38]. To improve the internal quantum efficiency, one needs to reduce the Schottky-barrier height and to decrease the silicide thickness to much less than the hot-carrier attenuation length in silicide, which is typically $\lesssim$3 nm. However, the reduction of Φ_B leads to an increase of the dark current unless the detector is operated at cryogenic temperature, and decreasing the silicide thickness will degrade the light absorption efficiency even with the absorbing silicide layer on the Si waveguide [39]. For example, an Si-waveguide-based $NiSi_2$/p-Si Schottky detector requires a long $NiSi_2$ layer on the Si waveguide for a certain light absorption ratio when the $NiSi_2$ thickness is decreased

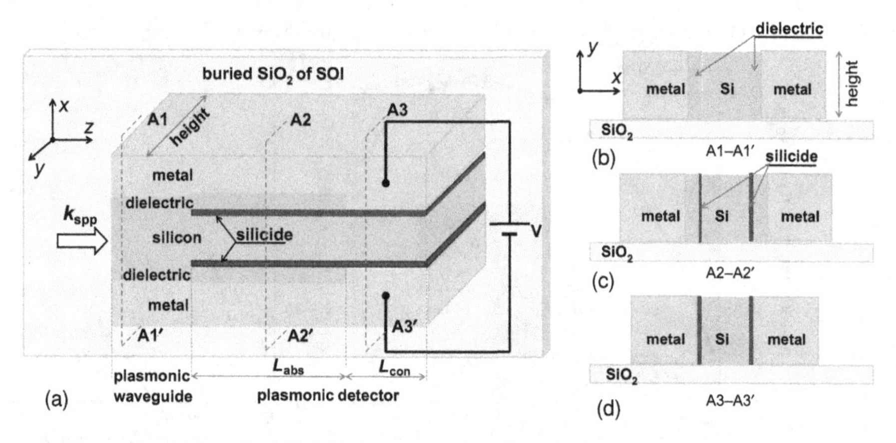

Figure 6.26 The design of a silicide Schottky-barrier detector inserted into a horizontal MISIM plasmonic waveguide. (a) A 3D view of the whole device. (b) A cross-sectional view of the MISIM plasmonic waveguide. (c) A cross-sectional view of the detector's absorber part, which is a metal–insulator–silicide–Si–silicide–insulator–metal waveguide. (d) A cross-sectional view of the detector's contact part, which is a metal–silicide–Si–silicide–metal waveguide.

from 16 nm to 5 nm [40]. This indicates that one of the main advantages of a silicide SBD (i.e. the ultra-compact footprint) will be partially lost when the absorbing silicide layer becomes very thin.

In order to overcome or at least alleviate the above two problems, a new design of silicide detector is proposed, as shown schematically in Fig. 6.26(a), which combines the Si-waveguide-based silicide SBD and the MISIM plasmonic waveguide. The detector contains three parts, whose cross-sections are shown in Figs. 6.26(b)–(d). The first part is the MISIM waveguide. The second part has a horizontal metal–insulator–silicide–Si–silicide–insulator–metal configuration for absorption of the SPP signal, which is defined as the absorber of the detector. The third part has a horizontal metal–silicide–Si–silicide–metal configuration to connect the external electrodes, which is defined as the contact of the detector. The silicide thickness in the third part is much thicker than that in the second part to prevent diffusion of metal into the Si core.

6.4.2 SPP power absorption

The absorber of the proposed detector has a Cu–insulator–silicide–Si–silicide–insulator–Cu configuration, which can also be regarded as a plasmonic waveguide in general. SiO_2 or Si_3N_4 is proposed as the insulator. Figure 6.27 maps the propagation loss (in dB μm^{-1}) of such a waveguide versus n and k of the silicide. The parameters of the waveguide are set as follows: $W_{silicide} = 5$ nm, $W_{Si} = 50$ nm (thus, the actual width of the Si core is 45 nm, assuming that the formation of W-thick silicide will consume $W/2$-think Si), and $W_{SiO_2} = 10$ nm. The n and k of the silicide vary in the ranges 1–6 and 0–5.5, respectively, which cover most known silicides. For comparison, the nanoplasmonic slot waveguide without the silicide layer (i.e. the standard Cu–SiO_2–Si–SiO_2–Cu

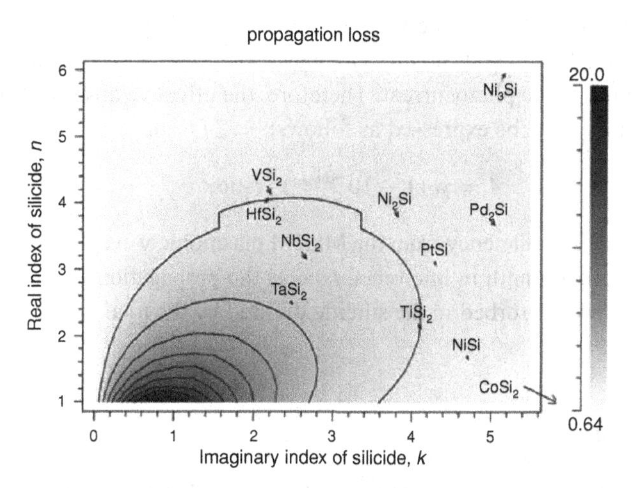

Figure 6.27 The calculated propagation loss of a Cu–SiO_2 (10 nm)–silicide (5 nm)–Si (45 nm)–silicide (5 nm)–SiO_2 (10 nm)–Cu waveguide as a function of the real (n) and imaginary (k) parts of complex index of silicide.

waveguide) has a calculated propagation loss of ~0.85 dB μm^{-1}. We can see that the propagation loss of the waveguide with silicide depends strongly not only on the k value of the silicide, but also on the n value of the silicide. This behavior can be explained by the continuity of electric displacement normal to the interfaces.

Unlike in the case of a conventional plasmonic waveguide for SPP propagation, for which the smaller the propagation loss the better, the requirement for the absorber is that the larger the propagation loss the better. From Fig. 6.27 we can see that $TaSi_2$ is the best silicide among the known compounds. Fortunately, $TaSi_2$ has been well studied and is fully compatible with standard CMOS technology [41, 42]. Moreover, we can see from Fig. 6.27 that a silicide with small permittivity can provide a very large propagation loss. For example, a silicide with an index of 1 + 1.5i can provide a propagation loss ~3.5 times larger than that of $TaSi_2$. Here, we define such a silicide as an ideal silicide. Since the complex refractive index of a silicide is determined mainly by the silicide's material properties, it is expected that the complex index of a silicide may be tailored by alloying two or even three silicides, i.e. the ternary or quaternary silicide may have a different index from those of each of the individual silicides. Although there is a lack of information about the optical properties of ternary or quaternary silicides, such an ideal silicide (i.e. one with a small permittivity at 1550 nm) may be reachable owing to the wide range of possible ternary or quaternary silicides. Moreover, owing to the same mechanism, the larger the permittivity of the insulator the better for the SPP absorption. A propagation-loss map of a Cu–Si_3N_4–silicide–Si–silicide–Si_3N_4–Cu waveguide versus n and k of the silicide has also been generated, which looks very similar to Fig. 6.27 (not shown here) but with a larger propagation loss. For instance, the propagation loss for a silicide with an index of 1 + 1.0i increases from 20 to 35 dB μm^{-1} and that for $TaSi_2$ increases from 4.07 to 7.87 dB μm^{-1} on replacing SiO_2 by Si_3N_4 as the insulator. Thus, Si_3N_4 is chosen as the insulator and $TaSi_2$ as the silicide for the proposed detector.

In the above plasmonic waveguide with a silicide, the SPP power can be absorbed in the Cu, insulator, silicide, and Si core. Obviously, only that absorbed in the silicide region can contribute to the photocurrent. Therefore, the effective absorption (A) of an absorber of length L_{abs} can be expressed as follows:

$$A = \gamma_C(1 - 10^{-\alpha L_{abs}}) \cdot \text{ratio}, \tag{6.8}$$

where γ_C is the coupling efficiency from the MISIM plasmonic waveguide to the detector; L_{abs} is the absorber length in micrometers; α is the propagation loss in dB μm^{-1}, and "ratio" is the power absorbed in the silicide divided by the total power absorbed in the absorber.

6.4.3 Quantum efficiency

The proposed detector has a horizontal silicide–semiconductor–silicide structure, whose band diagram is shown schematically in the inset of Fig. 6.28. Under a positive bias (V), the left silicide/Si Schottky contact is reverse biased and the right silicide/Si Schottky contact is forward biased for electrons, and vice versa for holes. We assume that the left and right silicide layers absorb SPP power equally owing to the symmetrical structure of the proposed detector. The hot electrons excited in the left silicide layer are emitted through Φ_n to the Si core and then collected by the right silicide layer, whereas the hot holes excited in the right silicide layer are emitted through Φ_p to the silicon and then collected by the left silicide layer. Owing to the narrow Si-core width, the Si core is fully depleted under the bias and the constant electric field across the Si core is given by V/W_{Si}. The Schottky effect predicts a decrease of both Φ_n and Φ_p by [43]

$$\Delta\Phi = \sqrt{\frac{eV}{4\pi\epsilon_0\epsilon_{Si}W_{Si}}}. \tag{6.9}$$

In the case of $W_{Si} = 45$ nm and $V = 1$V, $\Delta\Phi$ is ~0.05 eV. For the $TaSi_2$ SBD, Φ_n is ~0.6 eV and Φ_p is ~0.5 eV. Therefore, the effective Schottky-barrier heights under 1 V bias become ~0.55 eV for electrons and ~0.45 eV for holes, respectively. Lower effective Φ_n and Φ_p can be obtained at a higher voltage and/or for a narrower Si-core width. Figure 6.28 depicts the internal quantum efficiency (η_i) of a thin-film silicide/Si Schottky-barrier detector for various Φ_B, calculated using Scales's model [38]. We can see that η_i increases dramatically with Φ_B lowering and silicide thickness reducing. The maximum η_i reaches $\eta'_{i,\infty} \cong (h\nu - \Phi_B)/(h\nu)$ in the case of $t/L \to 0$.

The external quantum efficiency (η_e) is given by $\eta_e = A\eta_i$. Figure 6.29 depicts the absorption and the external quantum efficiency as functions of the $TaSi_2$ thickness. The other parameters of the detector are set as follows: $W_{Si} = (50\,\text{nm} - W_{TaSi_2})$, $W_{SiN} = 10$ nm, and $L_{abs} = 1$ or 2 μm, and the hot-carrier attenuation length (L) is assumed to be 100 nm. In the case of $L_{abs} = 1$ μm, the increase in A and the decrease in η_i with increasing W_{TaSi_2} lead to a maximum η_e at $W_{TaSi_2} = 2$–3 nm, whereas in the case of $L_{abs} = 2$ μm, η_e is mainly limited by η_i, thus it increases monotonically with decreasing W_{TaSi_2}. For a $TaSi_2$ detector with $W_{TaSi_2} = 2$ nm and $L_{abs} = 2$ μm at the 1-V bias (thus,

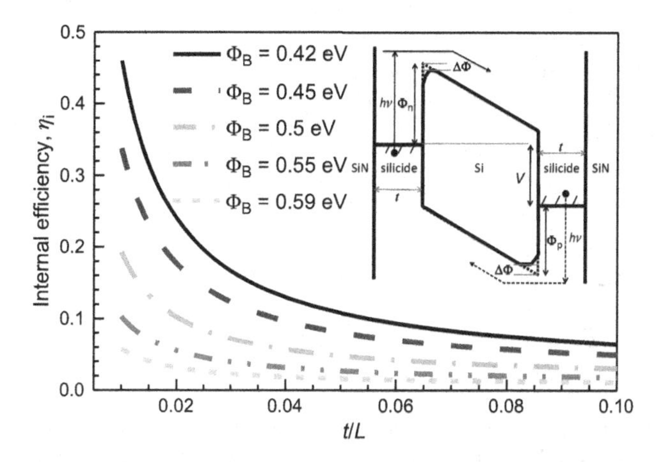

Figure 6.28 The calculated internal quantum efficiency (η_i) as a function of t/L for various barrier heights (Φ_B) at $h\nu = 0.8$ eV, where t is the silicide thickness and L is the hot-carrier attenuation length. The inset shows a schematic sketch of the band diagram of the silicide/Si/silicide structure under a positive bias. The applied voltage can substantially reduce both Φ_n and Φ_p. Both hot electrons and hot holes can contribute to the photocurrent.

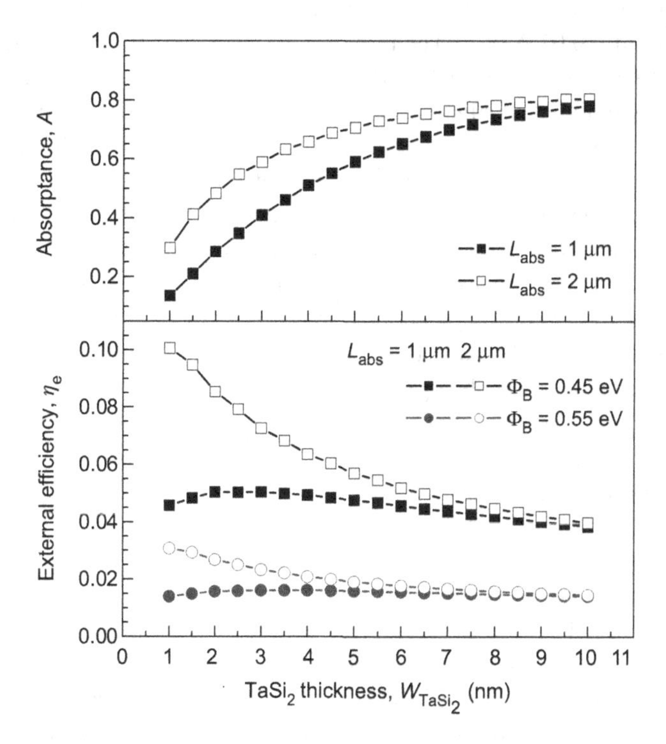

Figure 6.29 The calculated absorption (top) and external quantum efficiency (bottom) as a function of $TaSi_2$ thickness. The other parameters of the detector are set as $W_{Si} = (50\,\text{nm} - W_{TaSi_2})$, $W_{SiN} = 10\,\text{nm}$, $L_{abs} = 1$ or $2\,\mu\text{m}$, $L = 100\,\text{nm}$, $\Phi_B = 0.45$ or 0.55 eV, and $h\nu = 0.8$ eV.

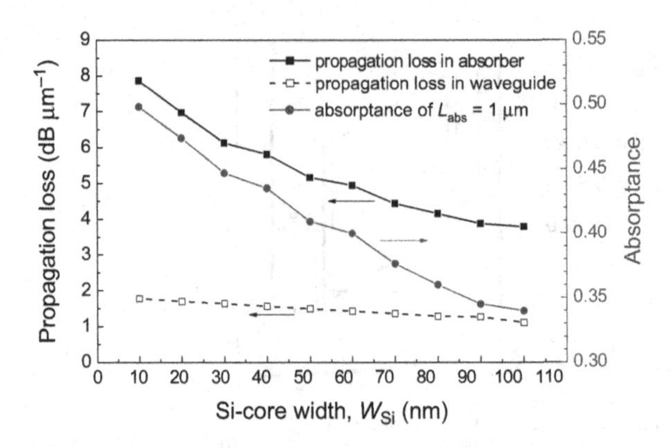

Figure 6.30 The calculated propagation loss (left) and absorption in the silicide region (right) as functions of the Si-core width (W_{Si}). The other parameters are set as $W_{TaSi_2} = 3$ nm, $W_{SiN} = 10$ nm, and $L_{abs} = 1$ μm. The propagation loss for a corresponding nanoplasmonic slot waveguide without silicide is also shown (as the lower dashed line) for comparison.

the effective Φ_B is 0.55 eV for electrons and 0.45 eV for holes), the overall η_e (given by $\eta_e = (\eta_e(\text{electron}) + \eta_e(\text{hole}))/2$) is calculated to be ~0.056, which corresponds to a responsivity of ~0.07 A W^{-1}. The responsivity increases with increasing applied voltage due to the lowering of Φ_B, but the dark current also increases. On the other hand, for the detector with the ideal silicide (assuming that it has the same barrier height as $TaSi_2$), A is ~0.69 in the case of $L_{abs} = 1$ μm and $W_{silicide} = 2$ nm, so the overall η_e is ~0.08 and the responsivity is ~0.1 A W^{-1} at a bias of 1 V.

Figure 6.30 depicts the propagation loss (the left axis) and the SPP power absorption in the silicide region (the right axis) as a function of the Si-core width (W_{Si}) for absorbers with fixed W_{SiN} of 10 nm and fixed W_{TaSi_2} of 3 nm. Here, the real Si-core width in the absorber is $W_{Si} - W_{TaSi_2}$, as mentioned above. For comparison, the propagation loss of a Cu–Si_3N_4–Si–Si_3N_4–Cu plasmonic slot waveguide is also shown. Both the propagation loss and the power absorption in the silicide region increase monotonically with decreasing W_{Si}. Therefore, the required L_{abs} for a certain ratio of absorption can be reduced with decreasing W_{Si}, so a detector with a narrower Si core can be expected to have a better performance. Meanwhile, the propagation loss of a conventional nanoplasmonic slot waveguide (i.e. one without the silicide layers) also increases with decreasing W_{Si}. The Si-core width should be finally determined by the real application of the nanoplasmonic circuit. Here we just set it 50 nm.

Figure 6.31 depicts the propagation loss (the left axis) and the SPP power absorption in the silicide region (the right axis) as a function of Si_3N_4 width (W_{SiN}) for absorbers with fixed W_{Si} of 50 nm and fixed W_{TaSi_2} of 3 nm. Both the propagation loss and the power absorption in the silicide region increase monotonically with decreasing W_{SiN}. The propagation loss of a Cu–Si_3N_4–Si–Si_3N_4–Cu plasmonic waveguide is also shown as a function of W_{SiN} for comparison. This indicates that the smaller W_{SiN} the better for the proposed detector. W_{SiN} as small as ~1–2 nm is reachable in fabrication. However, the Si_3N_4 layer should be thick enough to prevent diffusion of Cu into the Si core during

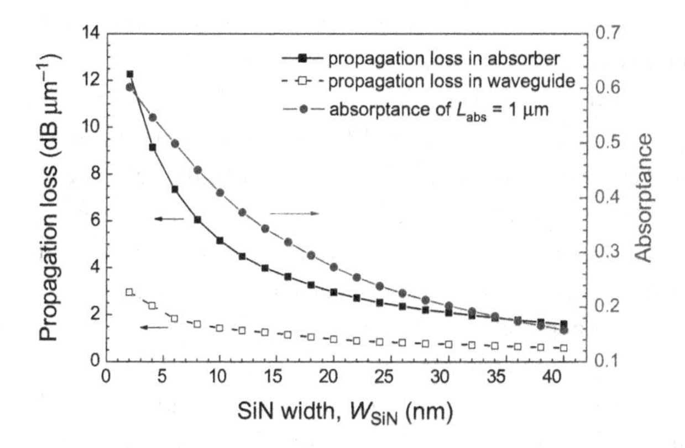

Figure 6.31 The calculated propagation loss (left) and absorption in the silicide region (right) as functions of the Si_3N_4 width (W_{SiN}). The other parameters are set as $W_{TaSi_2} = 3$ nm, $W_{Si} = 50$ nm, and $L_{abs} = 1$ μm. The propagation loss for a corresponding nanoplasmonic slot waveguide without silicide is also shown (as the lower dashed line) for comparison.

subsequent processing. Again, the final insulator thickness should also be determined by the real application of the nanoplasmonic circuits. Here we just set it to 10 nm.

6.4.4 Dark current and speed

The proposed detector actually is two Schottky diodes connected back-to-back. The total dark current is composed of both the electron current and the hole current as follows [39]:

$$I_{dark} = (L_{abs} + L_{con}) \cdot \text{height} \cdot T^2 \cdot \left\{ A_n^{**} \exp\left(-\frac{e\Phi_n}{k_B T}\right) + A_p^{**} \exp\left(-\frac{e\Phi_p}{k_B T}\right) \right\}, \tag{6.10}$$

where T is the absolute temperature, k_B is the Boltzmann constant, A^{**} is the effective Richardson constant (117 A cm^{-2} K^{-2} for electrons in Si and 32 A cm^{-2} K^{-2} for holes in Si). For the abovementioned $TaSi_2$ detector with a responsivity of 0.07 A W^{-1} (i.e. $L_{abs} = 2$ μm, $L_{con} = 0.5$ μm, and height $= 0.34$ μm), I_{dark} is calculated to be ~66 nA at room temperature under a bias of 1 V. The minimum detectable power S_{min} (sensitivity), which is defined as 1 dB above the optical power (in dBm) which can generate photocurrent to the dark current [44], is then calculated to be about −29 dBm. For the detector with the ideal silicide ($L_{abs} = 1$ μm and the other parameters are the same as for the $TaSi_2$ detector), I_{dark} becomes ~39 nA and S_{min} becomes −33 dBm.

The speed of the detector is determined by the time taken by hot carriers to cross the Si core and the RC delay. Assuming that the drift velocity of the carriers is 1×10^7 cm s^{-1} in Si, the transit time is estimated to be ~0.5 ps for $W_{Si} = 50$ nm, which corresponds to a THz cutoff frequency. Therefore, the speed of the proposed detector is mainly limited by the RC delay, which is defined as $f_{max} = 1/(2\pi RC)$. The capacitance (C) of the $TaSi_2$ detector is calculated to be ~1.75 fF using the simple parallel-plate model.

Assuming that the resistivity (ρ) of $TaSi_2$ is ~40 μΩ cm, the piece of $TaSi_2$ film of length 2 μm and thickness 2 nm has a sheet resistance (R) of ~1.3 kΩ. Then, the cutoff frequency is estimated to be ~68 GHz. For the detector with the ideal silicide, assuming the same resistivity of 40 μΩ cm, both R and C are lower due to its short L_{abs}, so the cutoff frequency is then estimated to be ~228 GHz. It should be pointed out that the capacitor of the nanoplasmonic waveguide is ignored in the above simple estimation. The waveguide capacitor can be regarded as a parallel-plate capacitor with multilayer dielectrics of insulator, Si, and insulator, whose capacitance depends on the wavelength (which depends on the real application of the nanoplasmonic circuits), the Si-core width, and the insulator thickness. In the case of a 5-μm waveguide, 50-nm Si-core width, and an Si_3N_4 thickness of 5 nm, the waveguide capacitance is calculated to be ~2.22 fF. Then, the speed of the $TaSi_2$ detector will be reduced to ~30 GHz. Moreover, from the results of Figs. 6.30 and 6.31, we know that the required L_{abs} for a certain ratio of absorption can be reduced with decreasing Si_3N_4 thickness and/or Si-core width. The use of a short L_{abs} can both reduce the dark current and improve the speed of the proposed detector.

6.5 Metallic nanoparticle-based detectors

6.5.1 Device structure

As mentioned above, the low internal quantum efficiency of conventional SBDs results from the fact that the hot carriers excited in the metal silicide layer are isotropically distributed in ***k***-space, and only those whose kinetic energy normal to the silicide/Si interface is larger than Φ_B can be emitted into the Si. The kinetic energy is given by the ratio of the solid angle (Ω) of the escape cone to the solid angle of the ***k***-space sphere (4π), as shown in Fig. 6.32(d). One can increase η_i by reducing Φ_B and by thinning down the silicide layer. Here, an alternative approach, as shown schematically in Figs. 6.32(a) and 6.32(b), is proposed to address this challenge. The detector is an Si p–n junction with electrically floating silicide nanoparticles (NPs) embedded in its space-charge region. The concept of embedded silicide NPs for infrared detection can be traced back to the layered internal photoemission sensor [45, 46]. Figure 6.32(c) depicts the band diagram of such a p–n junction, showing that both electrons and holes photoexcited in the NPs are simultaneously emitted into Si over Φ_B (Φ_n for electrons and Φ_p for holes). Because the photoexcited carriers in the silicide NPs face the silicide/Si interface in all directions, the escape cone in Fig. 6.32(d) extends to the whole ***k***-space sphere [47]. As a result, $P(E)$ approaches unity and η_i approaches $(h\nu - \Phi_B)/(h\nu)$.

6.5.2 LSPR-enhanced absorption

The light absorption can be dramatically enhanced through excitation of localized surface plasmon resonance (LSPR) on the silicide NPs. It is well known that light absorption

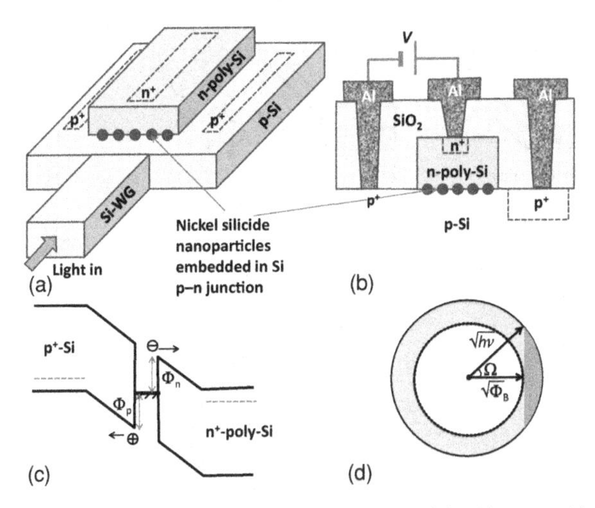

Figure 6.32 (a) A schematic representation of the Si waveguide-integrated photodetector based on embedded metal silicide nanoparticles (NPs). (b) A cross-sectional view of the device. (c) The energy-band diagram of the Si p–n junction with embedded metal NPs under reverse bias. (d) A simplified escape-cone model for Schottky emission shown in momentum space. The shaded cone with angle of Ω represents a thick silicide film on Si whereas the shaded spherical ring represents a very small silicide NP surrounded by Si.

predominates over scattering on a metal NP with diameter well below the skin depth ($\sim$25 nm for common metals in the entire optical region). The absorption cross-section C_{abs} of a metal NP is expressed by $(2\pi/\lambda)\text{Im}[3V(\epsilon_{\text{m}}(\omega) - \epsilon_{\text{d}})/(\epsilon_{\text{m}}(\omega) + 2\epsilon_{\text{d}})]$ according to the point-dipole model, where V is the particle volume, $\epsilon_{\text{m}}(\omega)$ is the complex permittivity of the metal NP, and ϵ_{d} is the permittivity of the embedding medium. The value of C_{abs} can greatly exceed the geometrical cross-section of the metal NP when $\epsilon_{\text{m}}(\omega) = -2\epsilon_{\text{d}}$, known as the LSPR. To verify the above prediction in the case of metal NPs embedded in an Si waveguide, the light absorption in 1-μm-long Si waveguides with continuous silicide films or silicide NPs, as shown schematically in Figs. 6.33(a)–(c), is numerically simulated using a 3D FDTD method. Here, the dimensions of the Si waveguide surrounded by SiO_2 are set to 0.4 μm × 0.5 μm, and the metal is 5-nm-thick NiSi silicide. The NiSi NPs are approximated as nanodisks of height 5 nm and diameter d, periodically arranged on a square grid with a spacing of g. The simulation results are depicted in Figs. 6.33(d) and (e) for TE (with the electric field parallel to the NiSi layer) and TM (with the electric field perpendicular to the NiSi film) polarized light, respectively. The continuous NiSi layer in the middle of the Si waveguide can absorb the TE light more efficiently than can that on the top of the Si waveguide because the light field of the TE mode is maximal in the middle of the waveguide. The absorption of NPs increases with decreasing d. The absorption under conditions of $d = 10$ nm and $g = 5$ nm is even larger than that for the continuous NiSi layer with the same thickness, which could be attributed to the LSPR effect, although the LSPR condition of $\epsilon_{\text{m}}(\omega) = -2\epsilon_{\text{d}}$ is not exactly met ($\epsilon_{\text{m}}(\omega)$

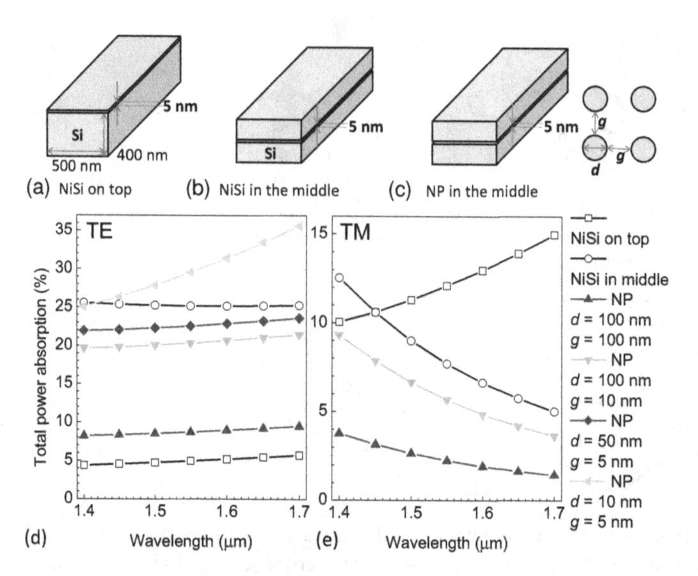

Figure 6.33 (a) The Si waveguide with a 5-nm-thick layer of NiSi placed on top. (b) The Si waveguide with a 5-nm-thick layer of NiSi embedded in the middle. (c) The Si waveguide with NiSi NPs embedded in the middle. (d) The total power absorption in the above three kinds of 1-μm-long waveguide versus wavelength for TE light. (e) The total power absorption for TM light in the three cases. In the case of NiSi film on top of the waveguide, TM light can excite a surface plasmon polariton (SPP) mode, thus being absorbed more effectively than TE light.

of NiSi is about $-20 + 15i$ at 1550 nm, whereas $\epsilon_d = 11.9$). Because $\epsilon_m(\omega)$ of the Ni silicide depends on its phase, being about $8.2 + 60i$ for Ni_3Si, $-0.88 + 29i$ for Ni_2Si, and $-19 + 25i$ for $NiSi_2$ at 1550 nm, and several phases may co-exist in the Ni silicide (relating to the fabrication condition), $\epsilon_m(\omega)$ of the final silicide NPs may be engineered by the fabrication condition or even by adding an additional metal such as Co or Ti to form a ternary silicide. Moreover, the LSPR condition also depends on the silicide NP shape, taking higher-order LSPRs into account. Therefore, more efficient light absorption could be expected once strong LSPR has been excited, which could be achieved by optimizations of fabrication parameters of silicide NPs, thus enabling one to scale down the absorber length to the submicron range. As expected, the TM light cannot excite LSPR. Therefore, our PD is suitable only for TE light detection.

6.5.3 Experimental demonstration

To quickly verify the feasibility of the proposed PD, a proof-of-concept device was fabricated on SOI using an existing mask set designed for conventional germanium PDs. Some of the key fabrication processes are as follows. After boron doping, the Si pattern was defined by a dry etch of Si down to the buried oxide using a layer of SiO_2 as a hard mask. A rectangular window of dimensions 80 μm × 4.4 μm was opened to expose the Si surface. Then, ~1 nm of Ti and ~3 nm of Ni were deposited subsequently, followed by rapid thermal annealing (RTA) at 200 °C for 30 s for silicidation. After removing

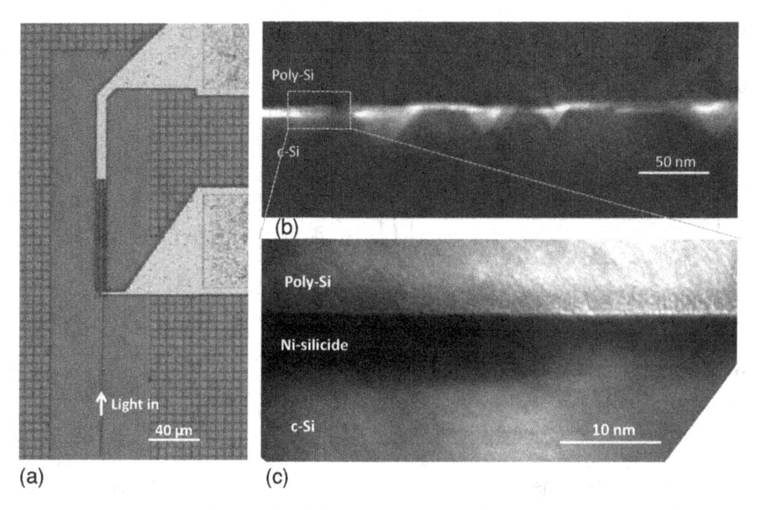

Figure 6.34 (a) Top view of the fabricated device, (b) STEM image, and (c) enlarged XTEM image of the final poly-Si/Ni-silicide/c-Si structure.

the unreacted metal by the use of 90-°C Piranha solution, 200 nm of amorphous Si (a-Si) was deposited, followed by phosphorus implantation, RTA at 700 °C for 5 min, and patterning. During this thermal processing, the Ni silicide agglomerates to form nanoparticles. Meanwhile, the a-Si layer crystallizes to form polycrystalline Si (poly-Si) due to the well-known metal-induced crystallization effect. The p^+-Si and n^+-poly-Si were contacted by Al electrodes to form a p–n diode. Figure 6.34(a) shows the final detector. The scanning transmission electron microscopy (STEM) image shown in Fig. 6.34(b) and the enlarged cross-sectional TEM (XTEM) image shown in Fig. 6.34(c) reveal clearly that the embedded Ni silicide film is discontinuous and Ni diffuses deeply into the c-Si layer at some spots. Meanwhile, a-Si is already crystallized to become poly-Si. However, it should be noted that the above fabrication parameters were chosen on the basis of our experience rather than optimization.

Figure 6.35(a) plots the current–voltage curve of the fabricated PD, which exhibits the typical rectifying property of a p–n diode. The insert shows the photocurrent and dark current on a semi-logarithmic scale. For optical measurement, light from a tunable laser source of power about −2 dBm is coupled into the Si waveguide through a lensed polarization-maintaining fiber. Assuming a coupling loss of ~3 dB, the light power arriving at the detector is estimated to be ~0.32 mW. Figure 6.35(b) plots the responsivity versus wavelength for quasi-TE and TM modes. Unlike for conventional SB PDs, the responsivity of this PD depends both on wavelength and on polarization, and the peak responsivity for TE light reaches ~30 mA W^{-1} at some specific wavelengths. The temporal responses at different biases, measured under the illumination of a pulsed laser, are depicted in Fig. 6.36. They consist of a sharp peak with a full width at high maximum of ~30 ps and a longer-timescale tail. The 3-dB bandwidth extracted from the Fourier transform is ~6 GHz, as shown in the inset.

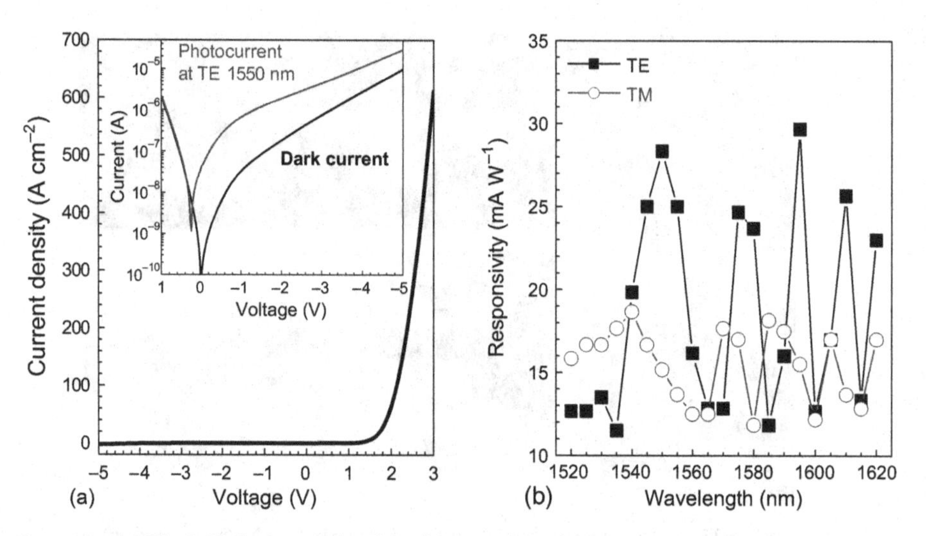

Figure 6.35 (a) The I–V characteristic of the fabricated PD at room temperature. The inset shows the photocurrent and dark current of the PD. The light power arriving at the detector is ~0.32 mW. (b) The responsivity at 5 V reverse bias versus wavelength for quasi-TE and quasi-TM light.

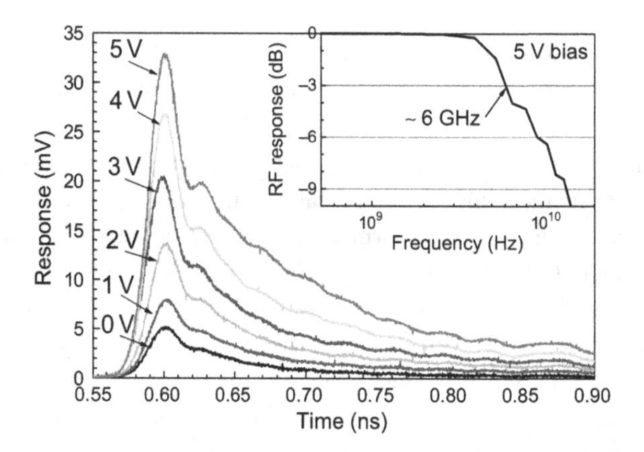

Figure 6.36 The temporal response of the fabricated PD under different reverse biases illuminated by a 1550-nm pulsed laser. The inset shows the Fourier transform, from which the 3-dB bandwidth is estimated to be ~6 GHz.

6.5.4 Issues and solutions

The light propagating along our PD is attenuated by absorption (by the embedded silicide NPs, the poly-Si layer, and the Al electrode) and scattering. The absorption by the Al electrode will not contribute to the photocurrent because the emitted hot carriers are not in the space-charge region. The absorption by the poly-Si may contribute the photocurrent, but it is weak, as is indicated by the fact that the propagation loss of the poly-Si waveguide ($\lesssim$0.01 dB μm^{-1}) is quite small compared with that of the Si waveguide with silicide (either thin film or NPs). The responsivity of our PD is thus mainly limited by the ratio of the light absorbed by the embedded silicide NPs to the overall light attenuation. The silicide NPs in our SB PD are random in location and shape, and their LSPR excitation is wavelength-dependent, making the responsivity also wavelength-dependent. One way to increase the responsivity is to enhance the LSPR effect, which could be achieved by optimizing the fabrication parameters, as mentioned above; the other way is to reduce the unusable light attenuation (especially the absorption of the Al electrode), which could be achieved by optimizing the detector's configuration to move the Al electrodes away from the optical mode. Once LSPRs have been excited on the silicide NPs, they decay into incoherent electron–hole pairs through Landau damping very quickly (i.e. on the femtosecond scale). Thus, the speed of our PD is still limited by the carrier transmitting time (from NPs to the electrodes) and the resistance–capacitance (RC) constant, as for conventional SB PDs.

One disadvantage of the proposed PD is the high dark-current density (e.g. $\sim$2.84 A cm^{-2} at 5 V bias) due to field bunching at the surfaces of NPs. Fortunately, the PD can be scaled down to a submicron scale, as mentioned above, thus the dark current can be reduced to a value comparable to that of conventional PDs. For instance, the dark current will be $\sim$14 nA if the active area is reduced to 0.5 μm^2.

6.6 Conclusions

This chapter describes recently developed ultra-compact plasmonic modulators and detectors for seamless integration into existing Si EPICs. The modulators are based on Cu–insulator–Si–insulator–Cu plasmonic waveguides or Cu–insulator–Si hybrid plasmonic waveguides. The MOS structure of these two kinds of plasmonic waveguides allows a voltage to be applied between the Cu cover and the Si core to induce free-electron accumulation at the insulator/Si interface, which in turn modulates both the real and the imaginary part of the model index of plasmonic waveguides significantly due to the optical mode confinement in this accumulation layer. One can realize EA and MZI modulators on the basis of MISIM plasmonic waveguides, while ring modulators can be realized on the basis of MIS hybrid plasmonic waveguides.

An ultra-thin silicide layer inserted into the MISIM plasmonic waveguide can absorb light significantly. This mechanism, can be employed to realize an ultra-compact Schottky-barrier plasmonic detector. Another detector described in this chapter consists of metallic nanoparticles embedded in an Si p–n junction. The light absorption is dramatically enhanced once the localized surface plasmon resonance has been excited.

For CMOS-compatibility, the devices proposed in this chapter involve only materials that are commonly used in Si CMOS lines. More functional plasmonic devices could be designed if other materials, which are not used in Si CMOS lines but may still be CMOS-compatible, are involved as the active material, such as vanadium dioxide [48], indium tin oxide [49], and graphene [50]. In general, we can say that the capability of tight light confinement and the existence of metal (which can be used as electrodes) in plasmonic waveguides provide more freedom or flexibility to design novel photonic devices than in the case of conventional dielectric waveguides.

References

[1] D. J. Lockwood and L. Pavesi, *Silicon Photonics II: Components and Integration*. Heidelberg: Springer-Verlag, 2011.

[2] G. T. Reed, G. Mashanovich, F. Y. Gardes, and D. J. Thomson, "Silicon optical modulators," *Nature Photonics*, vol. 4, pp. 518–526, 2010.

[3] K. W. Ang, S. Y. Zhu, M. B. Yu, G. Q. Lo, and D. L. Kwong, "High-performance waveguided Ge-on-SOI metal–semiconductor–metal photodetectors with novel silicon–carbon (Si:C) Schottky barrier enhancement layer," *IEEE Photon. Technol. Lett.*, vol. 20, pp. 754–756, 2008.

[4] Q. Xu, D. Fattal, and R. G. Beausoleil, "Silicon microring resonators with 1.5-μm radius," *Opt. Express*, vol. 16, pp. 4306–4315, 2008.

[5] B. Guha, B. B. C. Kyotoku, and M. Lipson, "CMOS-compatible athermal silicon microring resonators," *Opt. Express*, vol. 18, pp. 3487–3493, 2010.

[6] D. K. Gramotnev and S. I. Bozhevolnyi, "Plasmonics beyond the diffraction limit," *Nature Photonics*, vol. 4, pp. 83–91, 2010.

[7] M. I. Stockman, "Nanoplasmonics: Past, present, and glimpse into future," *Opt. Express*, vol. 19, pp. 22029–22106, 2011.

[8] M. Dragoman and D. Dragoman, "Plasmonics: Applications to nanoscale terahertz and optical devices," *Prog. Quant. Electron.*, vol. 32, pp. 1–41, 2008.

[9] E. Ozbay, "Plasmonics: Merging photonics and electronics at nanoscale dimensions," *Science*, vol. 311, pp. 189–193, 2006.

[10] S. I. Bozhevolnyi and J. Jung, "Scaling for gap plasmon based waveguides," *Opt. Express*, vol. 16, pp. 2676–2684, 2008.

[11] R. F. Oulton, V. J. Sorger, D. A. Genov, D. F. P. Pile, and X Zhang, "A hybrid plasmonic waveguide for subwavelength confinement and long-range propagation," *Nature Photonics*, vol. 2, pp. 496–500, 2008.

[12] T. Holmgaard and S. I. Bozhevolnyi, "Theoretical analysis of dielectric loaded surface plasmon-polariton waveguides," *Phys. Rev. B*, vol. 75, pp. 245405–245416, 2007.

[13] A. V. Krasavin and N. I. Zheludev, "Active plasmonics: Controlling signals in Au/Ga waveguide using nanoscale structural transformations," *Appl. Phys. Lett.*, vol. 84, pp. 1416–1418, 2004.

[14] P. Berini and I. D. Leon, "Surface plasmon-polariton amplifiers and lasers," *Nature Photonics*, vol. 6, pp. 16–24, 2012.

[15] K. F. MacDonald and N. I. Zheludev, "Active plasmonics: Current status," *Laser Photo. Rev.*, vol. 6, pp. 562–567, 2010.

[16] P. Neutens, P. V. Dorpe, I. D. Vlaminck, L. Lagae, and G. Borghs, "Electrical detection of confined gap plasmons in metal–insulator–metal waveguides," *Nature Photonics*, vol. 3, pp. 283–286, 2009.

[17] S. Y. Zhu, T. Y. Liow, G. O. Lo, and D. L. Kwong, "Fully complementary metal–oxide–semiconductor compatible nanoplasmonic slot waveguides for silicon electronic photonic integrated circuits," *Appl. Phys. Lett.*, vol. 98, pp. 021107–021109, 2011.

[18] S. Y. Zhu, G. O. Lo, and D. L. Kwong, "Experimental demonstration of vertical Cu–SiO_2–Si hybrid plasmonic waveguide components on an SOI platform," *IEEE Photon. Technol. Lett.*, vol. 24, pp. 1224–1226, 2012.

[19] S. Y. Zhu, T. Y. Liow, G. O. Lo, and D. L. Kwong, "Silicon-based horizontal nanoplasmonic slot waveguides for on-chip integration," *Opt. Express*, vol. 19, pp. 8888–8902, 2011.

[20] S. Y. Zhu, G. O. Lo, and D. L. Kwong, "Components for silicon plasmonic nanocircuits based on horizontal Cu–SiO_2–Si–SiO_2–Cu nanoplasmonic waveguides," *Opt. Express*, vol. 20, pp. 5867–5881, 2012.
[21] S. Y. Zhu, G. O. Lo, and D. L. Kwong, "Performance of ultracompact copper-capped silicon hybrid plasmonic waveguide-ring resonators at telecom wavelengths," *Opt. Express*, vol. 20, pp. 15232–15246, 2012.
[22] S. Y. Zhu, G. O. Lo, and D. L. Kwong, "Nanoplasmonic power splitters based on the horizontal nanoplasmonic slot waveguide," *Appl. Phys. Lett.*, vol. 99, pp. 031112–031114, 2011.
[23] J. Zhang, S. Y. Zhu, H. Zhang *et al.*, "An ultracompact surface plasmon polariton-effect-based polarization rotator," *IEEE Photon. Technol. Lett.*, vol. 23, pp. 1606–1608, 2011.
[24] J. Chee, S. Y. Zhu, G. O. Lo, and D. L. Kwong, "CMOS compatible polarization splitter using hybrid plasmonic waveguide," *Opt. Express*, vol. 20, pp. 25345–25348, 2012.
[25] S. Y. Zhu, G. O. Lo, and D. L. Kwong, "Experimental demonstration of horizontal nanoplasmonic slot waveguide-ring resonators with submicrometer radius," *IEEE Photon. Technol. Lett.*, vol. 23, pp. 1896–1898, 2011.
[26] A. Tardella and J. N. Chazalviel, "Highly accumulated electron layer at a semiconductor/electrolyte interface," *Phys. Rev. B*, vol. 32, pp. 2439–2448, 1985.
[27] R. Soref. R. E. Peale, and W. Buchwald, "Longwave plasmonics on doped silicon and silicides," *Opt. Express*, vol. 16, pp. 6507–6514, 2008.
[28] S. Roberts, "Optical properties of copper," *Phys. Rev.*, vol. 118, pp. 1509–1514, 1960.
[29] S. Y. Zhu, G. O. Lo, and D. L. Kwong, "Electro-absorption modulation in horizontal metal–insulator–silicon–insulator–metal nanoplasmonic slot waveguides," *Appl. Phys. Lett.*, vol. 99, pp. 151114–151116, 2011.
[30] G. T. Reed, *Silicon Photonics: The State of the Art*. New York: John Wiley & Sons, 2008, Chapter 7.
[31] G. Gulsen and M. N. Inci, "Thermal optical properties of TiO_2 films," *Opt. Mater.*, vol. 18, pp. 373–381, 2002.
[32] R. Paily, A. DasGupta, N. DasGupta *et al.*, "Pulsed laser deposition of TiO_2 for MOS gate dielectric," *Appl. Surf. Sci.*, vol. 187, pp. 297–304, 2002.
[33] W. Bogaerts, P. D. Heyn, T. V. Vaerenbergh *et al.*, "Silicon microring resonators," *Laser Photo. Rev.*, vol. 6, pp. 47–73, 2012.
[34] D. Dai, Y. Shi, S. He, L. Wosinski, and L. Thylen, "Silicon hybrid plasmonic submicron-donut resonator with pure dielectric access waveguides," *Opt. Express*, vol. 19, pp. 23671–23682, 2011.
[35] P. Bai, M. X. Gu, X. C. Wei, and E. P. Li, "Electrical detection of plasmonic waves using an ultra-compact structure via a nanocavity," *Opt. Express*, vol. 17, pp. 24349–24357, 2009.
[36] L. Tang, S. E. Kocabas, S. Latif *et al.*, "Nanometer-scale germanium photodetector enhanced by a near-infrared dipole antenna," *Nature Photonics*, vol. 2, pp. 226–229, 2008.
[37] W. A. Cabanski and M. J. Schulz, "Electronic and IR-optical properties of silicide silicon interfaces," *Infrared Phys.*, vol. 32, pp. 29–44, 1991.
[38] C. Scales and P. Berini, "Thin-film Schottky barrier photodetector models," *IEEE J. Quant. Electron.*, vol. 46, pp. 633–643, 2010.
[39] S. Y. Zhu, M. B. Yu, G. O. Lo, and D. L. Kwong, "Near-infrared waveguide-based nickel silicide Schottky-barrier photodetector for optical communications," *Appl. Phys. Lett.*, vol. 92, pp. 081103–081105, 2008.

[40] S. Y. Zhu, G. O. Lo, M. B. Yu, and D. L. Kwong, "Silicide Schottky-barrier phototransistor integrated in silicon channel waveguide for in-line power monitoring," *IEEE Photon. Technol. Lett.*, vol. 21, pp. 185–187, 2009.

[41] A. Noya, M. Takeyama, K. Sasaki, and T. Nakanishi, "First phase nucleation of metal-rich silicide in Ta/Si systems," *J. Appl. Phys.*, vol. 76, pp. 3893–3895, 1994.

[42] J. Pelleg and N. Goldshleger, "Silicide formation in the Ta/Ti/Si system by reaction of codeposited Ta and Ti with Si (100) and Si (111) substrates," *J. Appl. Phys.*, vol. 85, pp. 1531–1539, 1999.

[43] C. Schwarz, U. Scharer, P. Sutter *et al.*, "Application of epitaxial $CoSi_2$/Si/$CoSi_2$ heterostructures to tunable Schottky-barrier detectors," *J. Crystal Growth*, vol. 127, pp. 659–662, 1993.

[44] C. Scales, I. Breukelaar, and P. Berini, "Surface-plasmon Schottky contact detector based on a symmetric metal strip in silicon," *Opt. Lett.*, vol. 35, pp. 529–531, 2010.

[45] R. W. Fathauer, J. M. Iannelli, C. W. Nieh, and S. Hashimoto, "Infrared response from metallic particles embedded in a single-crystal Si matrix: The layered internal photoemission sensor," *Appl. Phys. Lett.*, vol. 57, pp. 1419–1421, 1990.

[46] R. W. Fathauer, S. M. Dejewski, T. George *et al.*, "Infrared photodetectors with tailorable response due to resonant plasmon absorption in epitaxial silicide particles embedded in silicon," *Appl. Phys. Lett.*, vol. 62, pp. 1774–1776, 1993.

[47] F. Raissi, "A possible explanation for high quantum efficiency of PtSi/porous Si Schottky detectors," *IEEE Trans. Electron Devices*, vol. 50, pp. 1134–1137, 2003.

[48] L. A. Sweatlock and K. Diest, "Vanadium dioxide based plasmonic modulators," *Opt. Express*, vol. 20, pp. 8700–8709, 2012.

[49] V. J. Sorger, N. D. Lanzillotti-Kimura, R. M. Ma, and X. Zhang, "Ultra-compact silicon nanophotonic modulator with broadband response," *Nanophotonics*, vol. 1, pp. 17–22, 2012.

[50] Q. Bao and K. P. Loh, "Graphene photonics, plasmonics, and broadband optoelectronic devices," *ACS Nano*, vol. 6, pp. 3677–3692, 2012.

7 Plasmonic biosensing devices and systems

To detect an analyte, surface plasmons whose characteristics are sensitive to the refractive-index variations close to the sensor's surface are excited and measured. Binding of the target analyte onto the sensor's surface will cause changes in refractive index and hence in the measured plasmonic characteristics. Depending on what type of surface plasmon is excited (e.g. surface plasmon polariton (SPP), Fano resonance), which plasmonic characteristic is measured/modulated (e.g. resonance wavelength, transmitted light intensity), and in what manner the bio-functionalization (i.e. binding of the target analyte) is performed, there are many different configurations for plasmonic biosensors, which will be reviewed in this chapter. The ultimate goal is to increase the sensor's sensitivity and the figure of merit. To achieve this goal, one must first understand the physics of the resonances, and then implement a smart structural design. In this chapter, two design methods will be introduced: an N-layer model and a finite-element-method (FEM) model, which are further elaborated by presentation of three biosensor design examples.

7.1 Introduction

A biosensor is a device for detecting an analyte, which typically combines a biological component with a physiochemical detector. For instance, a blood-glucose biosensor uses the enzyme glucose oxidase to break blood glucose down. In doing so it first oxidizes glucose and uses two electrons to reduce the FAD (a component of the enzyme) to $FADH_2$. Then the $FADH_2$ is oxidized by accepting two electrons from the electrode. The resulting current is a measurement of the concentration of glucose.

However, most biosensors work only for a certain type of analyte with some special property. For example, the blood-glucose biosensor requires the analyte (i.e. glucose) to be broken down by the enzyme glucose oxidase. This requirement restricts the sensor's applications. On the other hand, biosensors that are based on the mechanism of surface plasmon resonance (SPR) can be tailored to detect any analyte, as long as a biomolecular-recognition element recognizing the analyte is available. The analyte need not possess any special property.

Over the past two decades, SPR-based biosensor technology has made great progress. Many SPR sensor platforms have been developed, and several commercialized SPR systems are available. They owe their popularity to various advantageous features. Firstly, as mentioned above, the SPR sensor platform can be tailored to detect any analyte. Secondly, SPR biosensing is a label-free technology. One is nonetheless free to add labels to enhance the sensitivity in certain circumstances. Thirdly, the analysis is fast, where a real-time observation is possible. Fourthly, the sensors can work in continuous monitoring mode or one-time analysis mode. All these features make the SPR sensing technology particularly useful in pharmaceutical research, medical diagnostics, environmental monitoring, food safety, security, and defense.

Fundamentally, plasmonic biosensors are optical biosensors that use light to detect and quantify an analyte. For example, the first commercialized SPR biosensor works on the basis of SPPs. A thin layer of gold on a high-refractive-index glass surface can absorb laser light, producing electron waves (i.e. SPPs) on the gold surface. This occurs only at a specific angle and wavelength of incident light, and is highly dependent on the surface of the gold, such that binding of a target analyte to a receptor on the gold surface produces a measurable signal. This is the sensing mechanism of a typical SPR biosensor. Section 7.2 will give more details on the mechanisms. It consists of two parts: Section 7.2.1 describes the resonance condition of a surface plasmon used in plasmonic biosensing; and Section 7.2.2 introduces the parameters to evaluate a plasmonic biosensor, including the sensitivity and the figure of merit.

In addition to the most common SPP-based plasmonic biosensor, there are also other types of SPR biosensors that are based on various types of surface plasmons or plasmonics-related phenomena, e.g. localized surface plasmon resonances (LSPRs), extraordinary transmission (EOT), Fano resonances, and so on. They all have different physics and resonance conditions, leading to different sensor structures. These different sensor structures will be covered in Section 7.3.1. Besides that, the sensing system contains also two other components, namely the modulation part and the bio-functionalization part, which will be discussed in Sections 7.3.2 and 7.3.3, respectively.

Regardless of the structure/modulation/bio-functionalization of plasmonic sensors, the ultimate goal is to improve the performance of the sensor, including the sensitivity, accuracy, precision, repeatability, and lowest detection limit. This can be achieved by combining the efforts from advanced structural design and optimization using computer simulation, controllable nanofabrication, sensitive plasmonic detection, accurate functionalization of the sensing surface to exclusively bind the target analyte, and smart microfluidic system design and fabrication. Here we will particularly focus on one performance characteristics – the sensor sensitivity, which is highly dependent on how the sensor structures are designed and optimized.

Section 7.4 will introduce two design methods: an N-layer model in Section 7.4.1 and an FEM model in Section 7.4.2. To demonstrate how to use these design methods in practical applications, we show in Section 7.5 three examples from our projects in collaboration with experimentalists, hospitals, and medical firms. In brief, Section 7.5.1 shows the case of using the simple N-layer model to design a graphene-on-gold SPR biosensor. The second example, in Section 7.5.2, employs the FEM model to search for the most sensitive nanostructures to detect messenger RNA of moderate size

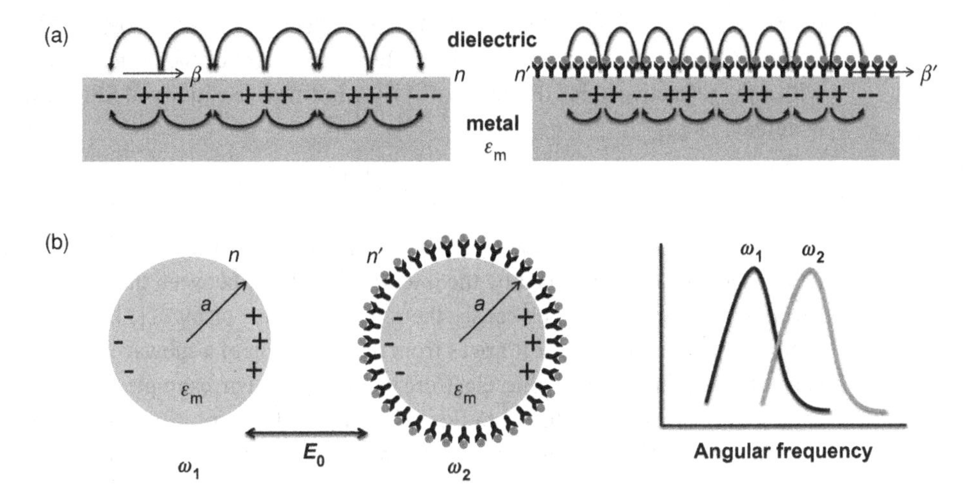

Figure 7.1 The sensing mechanism for (a) an SPP-based and (b) an LSPR-based plasmonic sensor.

(50 nm). The last example, in Section 7.5.3, is to implement the FEM model to design a nanohole-array-based sensing system to detect the nanometer-sized prostate-specific antigen (PSA).

7.2 Plasmonic sensing mechanisms

There are two fundamental excitations of plasmons [1]: SPPs and LSPRs. As shown in Fig. 7.1, SPPs are waves propagating along a metal–dielectric interface, and LSPRs are non-propagating excitations of the conduction electrons of metallic nanostructures coupled to the electromagnetic field of the incident light. The excitation conditions of these two types both depend on the dielectric medium surrounding the metal. This property lays the foundation for plasmonic sensing. When the target analyte is bound onto the metal, the dielectric environment changes, which modifies the plasmon excitation condition. The change of excitation condition is then measured. The resulting output is a measurement of the concentration of the analyte.

7.2.1 Resonance conditions for sensing

For an SPP propagating along the interface of the metal and dielectric in Fig. 7.1(a), the dispersion relation is [1]

$$\beta = k_0 \sqrt{\frac{\epsilon_m \epsilon_d}{\epsilon_m + \epsilon_d}}, \tag{7.1}$$

where β is the propagating constant of the SPP which corresponds to the component of the SPP's wavevector along the propagating direction, $k_0 = \omega/c$ is the wavevector of the propagating wave in vacuum, $\epsilon_m(\omega)$ is the dielectric function of the metal with ω the angular frequency, and $\epsilon_d = n^2$ is the dielectric constant of the dielectric environment with n the refractive index. From Eq. (7.1), the intrinsic sensitivity of SPP-based sensors is defined as [2]

$$S_{\mathrm{SPP}}^{\mathrm{int}} = \frac{\partial \mathrm{Re}\{\beta\}}{\partial n}, \tag{7.2}$$

which is the sensitivity of the SPP's propagating constant to the change in refractive index. This $S_{\mathrm{SPP}}^{\mathrm{int}}$ is influenced by the interaction volume between the SPP wave and the bound molecule layer (which causes the refractive-index change) [2].

On the other hand, an LSPR arises from the scattering of a subwavelength conductive nanostructure in an oscillating electromagnetic field. For example, we assume that a conductive nanoparticle has a radius of a and a dielectric function $\epsilon_m(\omega)$, and the surrounding medium is an isotropic and non-absorbing dielectric with dielectric constant $\epsilon_d = n^2$, as shown in Fig. 7.1(b). The applied oscillating field $\boldsymbol{E}_0$ will induce a dipole moment $\boldsymbol{p}$ inside the nanoparticle:

$$\boldsymbol{p} = \epsilon_0 \epsilon_d \alpha \boldsymbol{E}_0, \tag{7.3}$$

where ϵ_0 is the permittivity of free space and α is the polarizability of the nanoparticle embedded in a particular dielectric medium ϵ_d [1]:

$$\alpha = 4\pi a^3 \frac{\epsilon_m - \epsilon_d}{\epsilon_m + 2\epsilon_d}. \tag{7.4}$$

Equation (7.4) defines the resonance condition of the LSPRs $\mathrm{Re}[\epsilon_m(\omega)] = -2\epsilon_d$, and the intrinsic sensitivity of an LSPR-based sensor is given by

$$S_{\mathrm{LSPR}}^{\mathrm{int}} = \frac{\partial \mathrm{Re}\{\alpha\}}{\partial n}, \tag{7.5}$$

where n is the refractive index of the dielectric medium. As illustrated in Fig. 7.1(b), when an analyte is bound on the nanoparticle, the change of refractive index on the nanoparticle's surface would shift the resonance frequency or wavelength to satisfy the resonance condition $\mathrm{Re}[\epsilon_m(\omega)] = -2\epsilon_d$.

7.2.2 Sensitivity and figure of merit

The intrinsic sensitivity introduced in Section 7.2.1 defines how the plasmon oscillation (characterized by α or β) intrinsically changes with the dielectric environment (n). This sensitivity is independent of the excitation or the modulation of the surface plasmons. To fully quantify the overall sensitivity, an external sensitivity needs to be included:

$$S_{\mathrm{RI}} = \frac{\partial P}{\partial \mathrm{Re}\{\alpha \text{ or } \beta\}} \frac{\partial \mathrm{Re}\{\alpha \text{ or } \beta\}}{\partial n} = S^{\mathrm{ext}} S^{\mathrm{int}}, \tag{7.6}$$

where P is the sensor's optical output, and the external sensitivity $S^{\mathrm{ext}} = \partial P / \partial \mathrm{Re}\{\alpha \text{ or } \beta\}$ defines the sensitivity of the optical output to the variation of the surface plasmon oscillation. This external sensitivity mainly depends on the methods of excitation and modulation.

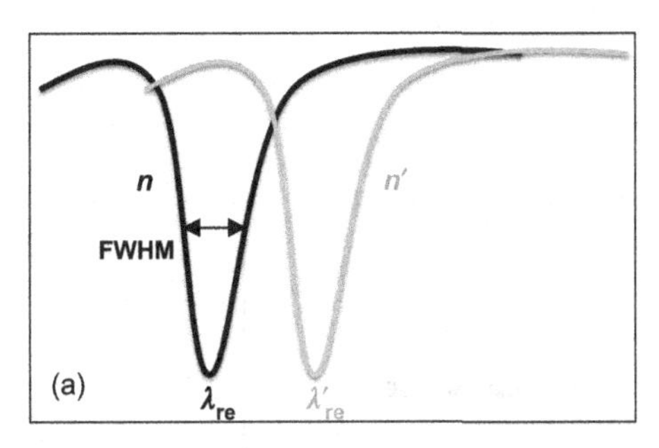

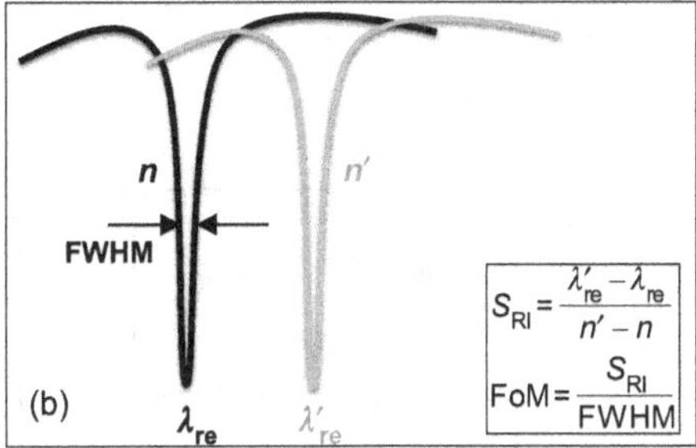

Figure 7.2 The definitions of the sensitivity (S_{RI}) and figure of merit (FoM). The sensor shown in (a) has the same sensitivity as, but a different FoM from, sensor (b).

The overall sensitivity S_{RI} therefore represents the sensitivity of the optical output to the variations of the refractive index of the dielectric environment. For instance, the modulated optical output is the incident wavelength in resonance with the propagating SPP (i.e. λ_{re}). The overall sensitivity is then

$$S_{RI} = \frac{\partial \lambda_{re}}{\partial n} = \frac{\partial \lambda_{re}}{\partial \mathrm{Re}\{\beta\}} \frac{\partial \mathrm{Re}\{\beta\}}{\partial n} = S^{ext} S^{int}. \tag{7.7}$$

In an SPR sensing experiment, however, it is impractical to directly measure the refractive-index change (i.e. ∂n). Instead the sensitivity to the change in measurand (e.g. the analyte concentration c) is more commonly employed, for example,

$$S = \frac{\partial \lambda_{re}}{\partial c} = \frac{\partial \lambda_{re}}{\partial n} \frac{\partial n}{\partial c} = S_{RI} E, \tag{7.8}$$

where the additional term E accounts for the binding efficiency, i.e. how efficiently the presence of analyte at concentration c is converted into a change in the refractive index n. From the design perspective, S_{RI} and E can be dealt with independently. In order to design a highly sensitive SPR sensor structure, the focus should be on optimizing S_{RI}. On the other hand, the binding efficiency E is more relevant when one needs to design a good biomolecular-recognition element to bind the target analyte efficiently.

In addition to the sensitivity, the other term employed to evaluate the performance of a surface plasmon biosensor is the figure of merit (FoM). The FoM is expressed in terms of both the sensitivity and the line width of the resonance peak (i.e. the full width at half maximum (FWHM)),

$$\mathrm{FoM} = \frac{S_{RI}}{\mathrm{FWHM}}. \tag{7.9}$$

In Fig. 7.2, the two sensors have the same sensitivity but different FoMs: that in Fig. 7.2(b) has a greater FoM than does that in Fig. 7.2(a). In general, the sharper the resonance peak, the higher the FoM. Essentially, the target in designing an SPR sensor is to increase both the sensitivity and the FoM.

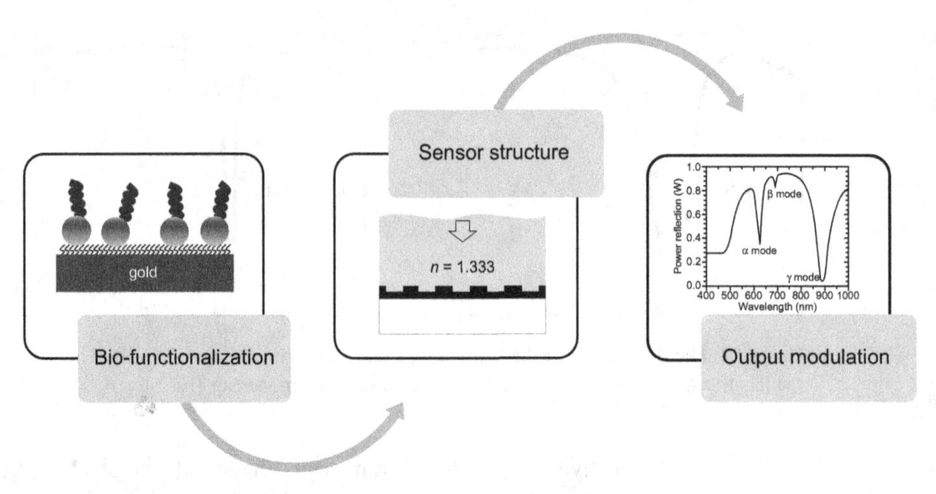

Figure 7.3 An overview of a plasmonic sensing system consisting of three parts: the sensor structure, modulated output, and bio-functionalization.

7.3 Plasmonic biosensing systems

A plasmonic biosensing system consists of three parts, as illustrated in Fig. 7.3. The central part is the sensor's structure where different types of surface plasmons could be excited. The sensor's structure influences both the intrinsic and the external sensitivity. The second part is the output modulation, which mainly affects the external sensitivity. The third part is the bio-functionalization assay formats (i.e. how the target analyte is bound to the sensor), which is the key to determining the binding efficiency.

7.3.1 Sensor structures

Most currently available plasmonic sensors are based on prism-coupled propagating surface plasmons, i.e. SPPs. This is also the most well-studied configuration [2, 3]. Nevertheless there are several other ways to excite the propagating surface plasmons, which will be reviewed first in this section. Following the review of SPP-based plasmonic sensors, the conventional LSPR-based sensors [4–6] will be discussed next. In addition to these two classical surface plasmons, many new plasmonic phenomena have been discovered in recent years and proven to be good for sensing. Two examples are extraordinary transmission (EOT) [7–17] and Fano resonance [18–21].

Propagating surface plasmons

There are at least six different ways to excite a propagating surface plasmon – an SPP: prism coupling, grating coupling, waveguide excitation, charged-particle excitation (e.g. electrons in electron-energy-loss spectra (EELS)), highly focused optical beam excitation, and near-field excitation [1]. Among them, the first two are commonly used in plasmonic sensing.

In the prism-coupling-based configuration shown in Fig. 7.4(a), a light wave passes through a high-refractive-index prism and is totally reflected at the prism/metal

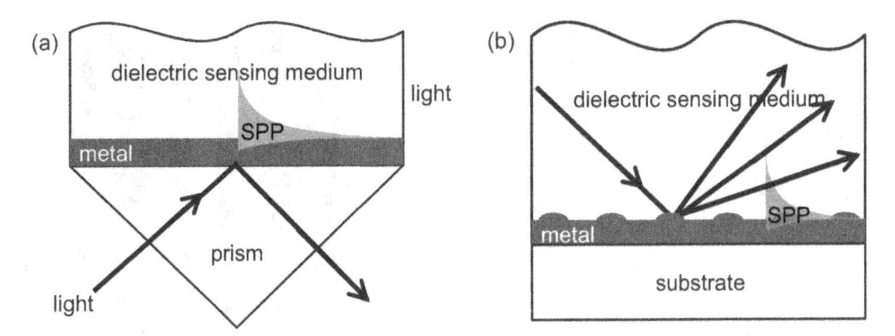

Figure 7.4 Two common structures for SPP-based plasmonic sensors [2]: (a) prism coupling and (b) grating coupling.

interface, generating an evanescent wave penetrating into the thin metal layer (usually 50 nm thick). This evanescent wave propagates along the interface with a propagation constant that can be adjusted to match that of the SPP by controlling the angle of incidence. This prism-coupling method is also referred to as the attenuated-total-reflection (ATR) method. Most currently available commercial plasmonic sensors use similar configurations. Slight variations can be made, for example, by sandwiching the ultra-thin metal film between two layers having identical dielectric constants, to excite long-range propagating surface plasmons, which has been shown to provide a greatly enhanced sensitivity [22].

The SPPs can also be excited by patterning the metal surface with gratings, as shown in Fig. 7.4(b). When light strikes the grating and gets diffracted, the component of the wavevector of the diffracted waves parallel to the interface is increased by an amount that is inversely proportional to the period of the grating and can be adjusted to match that of the SPP. It has been shown that a grating-based surface-plasmon-resonance sensing system can exhibit an extremely high sensitivity to detect a small change of refractive index; this sensitivity could be much higher than that of the prism-based systems [23].

Localized surface plasmons

Besides propagating surface plasmons, efforts have also been made to develop "nano-structured" plasmonic sensors. In this case, changes in the dielectric environment of nano-structured metals, such as nanoparticles, result in measurable shifts of the wavelength and/or magnitude of the LSPR. Various noble-metal nanostructures such as nanoparticles in solution, nanoparticles immobilized on surfaces, and nanoisland films have been developed in this way [4–6]. Another classical example of an LSPR sensor would be an array of zero-mode waveguides consisting of subwavelength holes in a metal film, which provides a simple and highly parallel means for studying single-molecule dynamics at micromolar concentrations with microsecond temporal resolution [24].

Extraordinary transmission

Extraordinary transmission (EOT) was first observed in 1998 [7]. In EOT the presence of tiny holes (with sizes smaller than the wavelength of incident light) in an opaque metal film leads to strongly enhanced transmission of light through the holes at selected

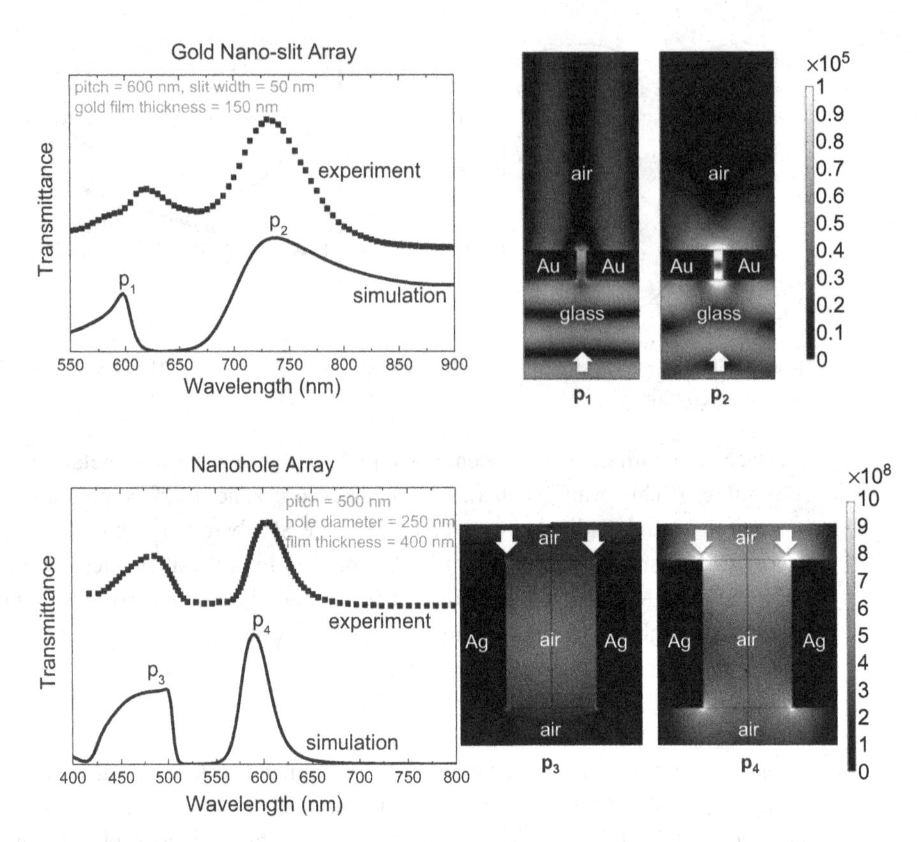

Figure 7.5 The EOT effect of (a) a gold nano-slit array [17] and (b) a silver nanohole array [13]. The measured and simulated transmission spectra are compared, and the nature of each EOT peak is shown by the near-field plots.

wavelengths. This EOT effect is now known to be due to the interaction of the light with resonances in the surface of the metal film. The most studied system for the EOT effect is nano-slit or nanohole arrays patterned in a metal film [7–17]. The EOT peaks could be attributed to SPPs or LSPRs, or sometimes to a waveguide mode or even a more complicated mode.

As shown in Fig. 7.5(a), a nano-slit array (which is similar to a grating-coupling configuration) could excite the SPP (c.f. the p_1 peak), whereby the enhanced plasmonic field originates from the center of the gold surface, decaying gradually into the air. The same nano-slit array supports an LSPR mode (c.f. the p_2 peak) as well, whereby the enhanced field is around the edge of the slit. On the other hand, a nanohole array as shown in Fig. 7.5(b) with two-dimensional confinement in the planar directions supports only the LSPR mode (c.f. the p_4 peak).

There are two advantageous features of EOT sensing. Firstly, it is flexible in terms of the ability to bind the target analyte onto the metal surface, the side wall of the hole, or the substrate surface inside the hole. Secondly, an additional flow-through format is available for a suspended plasmonic nanohole array structure, in contrast to the established flow-over method. The former has a sixfold improvement in response time [25, 26], since it enables rapid transport of reactants to the active surface.

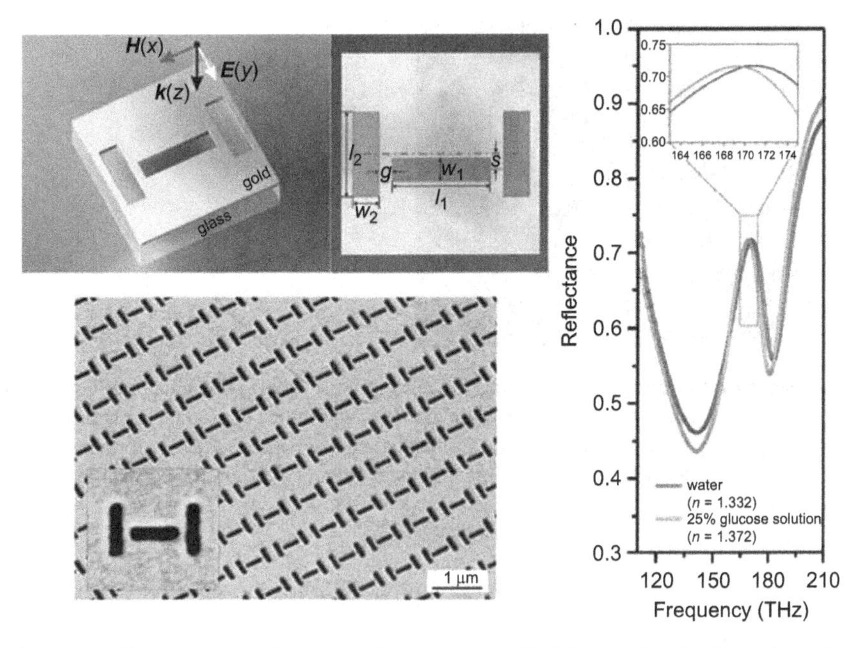

Figure 7.6 Electromagnetically induced transparency (EIT) for sensing. Left panel: a schematic view of the planar metamaterial design and an oblique view of the sample. Right panel: experimental tuning of the reflectance spectrum by changing the liquid which is adjacent to the gold nanostructures from water to 25% aqueous glucose solution. The increase of the refractive index on going from water ($n = 1.332$) to 25% glucose solution ($n = 1.372$) causes a red-shift of the resonance. Adapted from [18]. Reproduced with permission from the American Chemical Society.

Fano resonance

When there are a few nanoparticles arranged in a certain manner, coupling between nanoparticles' optical responses may lead to Fano resonance. Briefly, it is a type of resonant scattering phenomenon that gives rise to an asymmetric line-shape. This line-shape is due to interference between two scattering amplitudes, one due to scattering within a continuum of states (the background process) and the other due to an excitation of a discrete state (the resonant process). Specifically, in a plasmonic system, a Fano resonance arises from the coherent interference of "bright" and "dark" hybridized plasmon modes. Bright plasmon modes are spectrally broadened due to radiative damping, whereas dark plasmon modes do not couple efficiently to light and are therefore not radiatively broadened. In the following, two recent examples within the framework of Fano resonance that are particularly promising for sensing applications are reviewed.

The first example involves the phenomenon of electromagnetically induced transparency (EIT) demonstrated by Harald Giessen's group in 2010 [18]. The sensor consists of an optically bright dipole antenna (spectrally broad) and an optically dark quadrupole antenna (spectrally narrow), which are cutout structures in a thin gold film, as shown in Fig. 7.6. Owing to the close proximity, the two antennas are strongly coupled. It was experimentally demonstrated that this design achieved a sensitivity of 588 nm per refractive-index unit (RIU) and an FoM of 3.8. The theoretically predicted values for

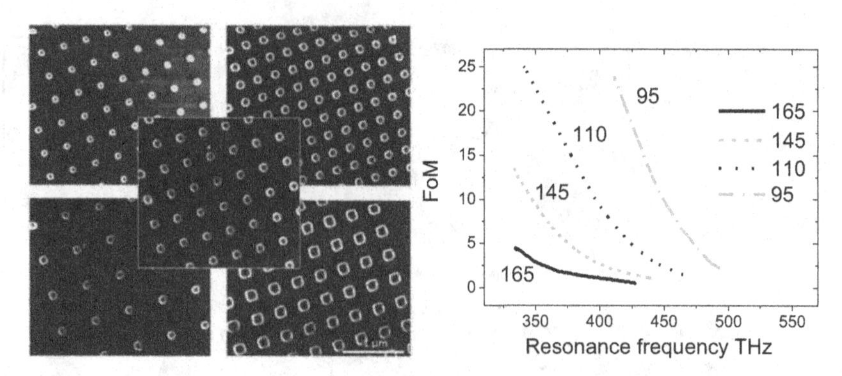

Figure 7.7 Surface lattice resonance for sensing. Left panel: SEM images of gold nanoparticle arrays on quartz with varying particle diameter and lattice constants. Right panel: The FoM for particle arrays as a function of the surface lattice resonance frequency for nanoparticles of diameters 95, 110, 145, and 165 nm. Adapted from [19]. Reproduced with permission from the American Chemical Society.

this design can be as high as 725 nm per RIU for the sensitivity and 7.4 for the FoM [18].

The second example is the achievement of surface lattice resonance by coupling the broad LSPR with a narrow Rayleigh anomaly resonance [19]. The Rayleigh anomaly represents the onset of diffraction in a grating. At the wavelength and angle of incidence corresponding to the Rayleigh anomaly, there is a transition between a diffracted order propagating in the plane of the array and an evanescent diffracted order. At this wavelength and angle, the wavevector of the diffracted order is parallel to the surface of the array, thus enhancing the coupling of the LSPRs. This coupling leads to the appearance of narrow resonances in the transmittance and reflectance spectra of the array [19]. An FoM up to 25 has been experimentally demonstrated by making use of this surface lattice resonance, as shown in Fig. 7.7.

7.3.2 Modulation methods

In general, there are five different plasmonic characteristics to modulate: the resonance angle, resonance wavelength, intensity, phase, and polarization [2]. For angular modulation, the coupling strength is measured at a fixed incident wavelength and multiple angles of incidence, and the angle of incidence yielding the strongest coupling (i.e. the resonance angle) is determined. This is often used in prism-coupling SPP-based sensors. For wavelength modulation, the coupling strength is measured at a fixed angle of incidence and multiple incident wavelengths, and the incident wavelength yielding the strongest coupling (i.e. the resonance wavelength) is determined. Wavelength modulation is the most widely used modulation method, and is often seen in SPP-, LSPR-, EOT-, and Fano-resonance-based plasmonic sensors. One example is shown in Fig. 7.7, where the resonance wavelengths (or frequencies) are modulated. For intensity, phase, or polarization modulation, both the incident wavelength and the angle of incidence

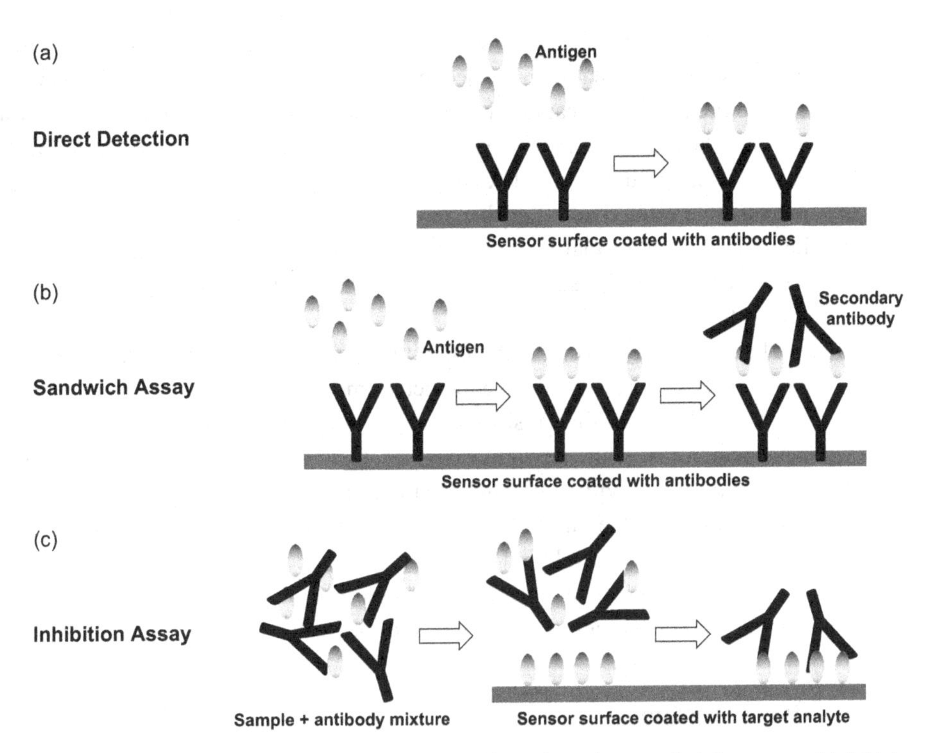

Figure 7.8 The three different assay formats [2]: direct detection, sandwich assay, and inhibition assay.

are fixed, and the change in the intensity, phase, or polarization of the light wave is measured.

7.3.3 Bio-functionalization formats

There are three main bio-functionalization assay formats that can be used to detect the target analytes: direct detection, sandwich assay, and inhibition assay [2]. The choice of assay format depends on the size of the target analyte molecules, the binding characteristics of the biomolecular-recognition elements, and the range of concentrations of analyte to be measured. Here, an example is shown in Fig. 7.8 to illustrate the three bio-functionalization assay formats, where the antigen is the target analyte, and the antibody is the biomolecular-recognition element.

Direct detection is usually preferred when direct binding of the analyte produces a sufficient response. In this format, the analyte in a sample (i.e. the antigen) interacts with a biomolecular-recognition element (i.e. the antibody) immobilized on the sensor surface, as shown in Fig. 7.8(a). The resulting refractive-index change is directly proportional to the concentration of the antigen.

The lowest detection limits of direct biosensors can be improved by using a sandwich assay, which is shown in Fig. 7.8(b). The measurement consists of two steps. Firstly, the sample containing antigens is brought into contact with the sensor, and the antigens are bound to the antibodies on the sensor surface. Secondly, the sensor surface is incubated with a solution containing the secondary antibodies. The secondary antibodies will bind to the previously captured antigens. These additional secondary antibodies would increase the refractive-index change. Moreover, additional labels (e.g. fluorescent dyes, metallic nanoparticles) could be tagged to the secondary antibodies, which would further increase the refractive-index change or even give extra optical output, e.g. fluorescent luminance. This format is preferred when a low concentration of analyte molecules is to be detected or the size of analyte molecules is too small for them to be discerned spectrally.

The third assay format is the inhibition assay, which is particularly useful to detect small analytes (those of relative molecular mass, or molecular weight, $<10\,000$). In the inhibition detection format shown in Fig. 7.8(c), the sensor surface is coated with antigens, in contrast to the previous two formats, in which antibodies are coated. A sample is initially mixed with antibodies, resulting in antigen–antibody pairs and unoccupied antibodies. Then the mixture is brought into contact with the sensor surface. The unoccupied antibodies would bind to the sensor surface. The amount of bound antigens and the initial binding rate are inversely proportional to the antigen concentration. Therefore the antigen concentration can be inferred indirectly.

7.4 Design methods

Among the three components of a plasmonic sensing system, this chapter will mainly focus on the design and optimization of the sensor's structure, in order to achieve the highest possible sensitivity and FoM. In this section, two design methods will be introduced. The first one is an analytical model known as the "N-layer model," in which one calculates the reflection and transmission of light through N layers of materials. It is often used to design a prism-coupled SPP-based plasmonic sensor. When the sensor's structure becomes more complicated, involving nanostructures, such as grating, slits, and holes, a more comprehensive model based on the finite-element method (FEM) is required.

7.4.1 The N-layer model

We consider the situation when light is reflected from an N-layer composite structure, which consists of layer $1, 2, \ldots, N-1, N$ as shown in Fig. 7.9. The thickness of each layer is d_k, where the integer $k = 2, 3, 4, \ldots, N-1$ denotes the layer number. The first and last layers are assumed to be infinitely thick.

The reflection coefficient r (and hence the reflectance $R = |r|^2$) is expressed as [27]

$$r = \frac{(M_{11} + M_{12}q_N)q_1 - (M_{21} + M_{22}q_N)}{(M_{11} + M_{12}q_N)q_1 + (M_{21} + M_{22}q_N)}, \tag{7.10}$$

where the four matrix elements M_{ij} $(i, j = 1, 2)$ are defined as

$$M_{ij} = \left(\prod_{k=2}^{N-1} M_k\right)_{ij}, \quad i, j = 1, 2. \tag{7.11}$$

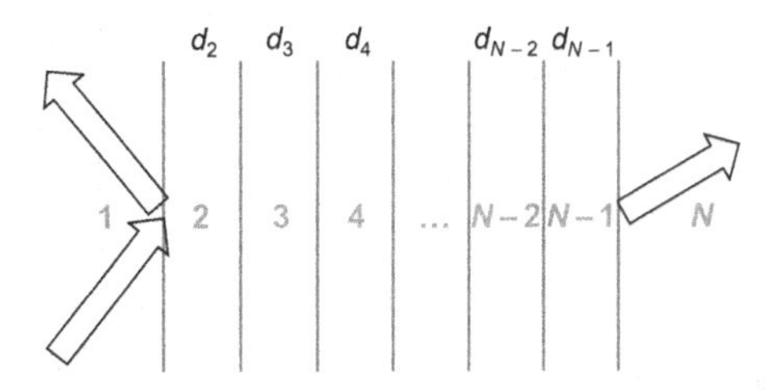

Figure 7.9 The N-layer model for SPR measurement.

Here, M_k is a 2×2 matrix related to the kth (k runs from 2 to $N - 1$) layer's optical property and thickness:

$$M_k = \begin{bmatrix} \cos\beta_k & -\mathrm{i}\sin\beta_k/q_k \\ -\mathrm{i}q_k\sin\beta_k & \cos\beta_k \end{bmatrix},$$

where

$$\beta_k = d_k \frac{2\pi}{\lambda_0}(\epsilon_k - n_1^2 \sin^2\theta_1)^{1/2} \tag{7.12}$$

and

$$q_k = \begin{cases} (\mu_k/\epsilon_k)^{1/2} \dfrac{(\epsilon_k - n_1^2\sin^2\theta_1)^{1/2}}{\epsilon_k^{1/2}} & \text{for a TM wave,} \\ (\epsilon_k/\mu_k)^{1/2} \dfrac{(\epsilon_k - n_1^2\sin^2\theta_1)^{1/2}}{\epsilon_k^{1/2}} & \text{for a TE wave.} \end{cases} \tag{7.13}$$

Here, d_k is the kth layer's thickness, λ_0 is the vacuum wavelength of the incident light, θ_1 is the angle of incidence in the first layer with refractive index n_1, and ϵ_k denotes the kth layer's dielectric permittivity. Through a series of numerical calculations using the above equations, the reflectance from a multilayer structure could be obtained.

7.4.2 The FEM model

The handy N-layer model is useful for layered sensing structures. For complicated sensing structures involving nanostructures, FEM simulation is needed. Figure 7.10 shows a schematic representation of the FEM model. The model solves Maxwell's equations and calculates the light–matter interactions of any geometry.

The model takes two inputs, which are the material property (most commonly the refractive index n or the dielectric permittivity ϵ, where $\epsilon = n^2$), and the exact size of each part of the structure. The whole structure is drawn and divided into small grids. Maxwell's equations are solved for each grid, with the continuity condition being maintained from grid to grid. Simultaneously, a certain boundary condition at all outer boundaries needs to be satisfied as well. Once there is a convergent solution, we are able to know the distributions of the electric field $\boldsymbol{E}$, magnetic field $\boldsymbol{H}$, electric flux density $\boldsymbol{D}$, and magnetic flux density $\boldsymbol{B}$ in the structure. Then post-processing is done

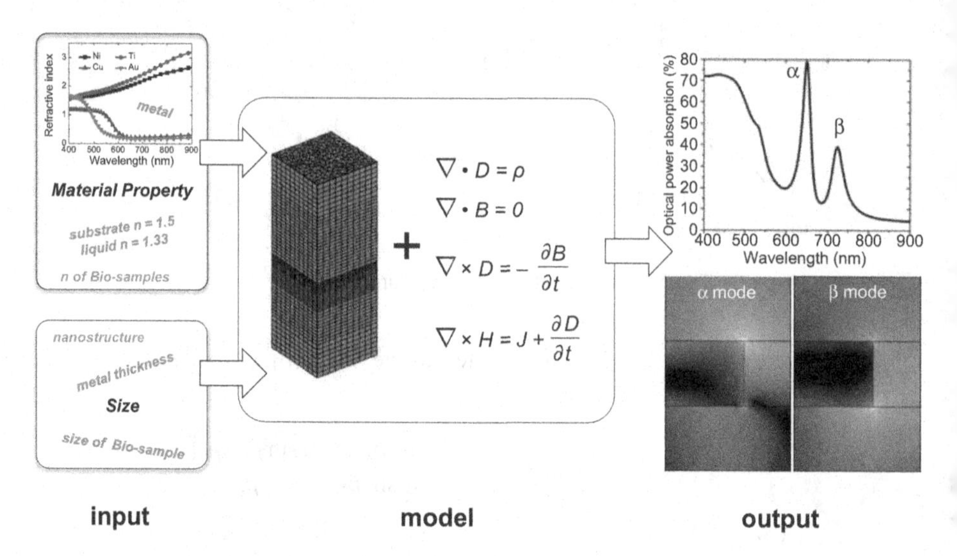

Figure 7.10 A schematic representation of FEM simulation showing the input, the output, and the calculation model.

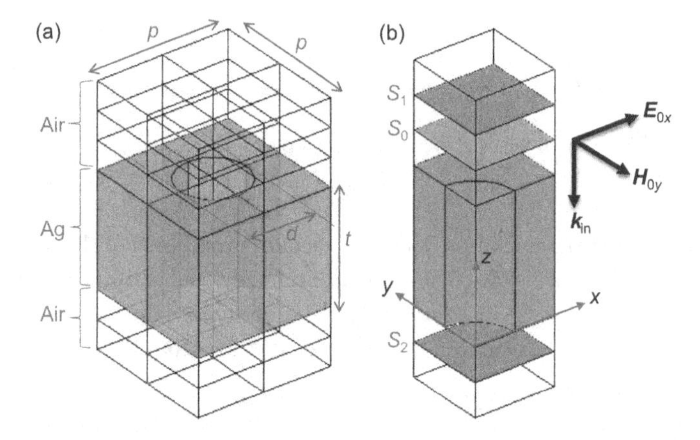

Figure 7.11 (a) The unit cell for simulating light propagating through a two-dimensional cylindrical hole array (repeating in the x and y directions) in a suspended Ag film in air, i.e. air/silver/air, where p, d, and t are the pitch of the array, the diameter of the hole, and the thickness of the Ag film, respectively. (b) A quarter of the unit cell, where S_0 is the surface used to set up the incident light wave, and the surfaces S_1 and S_2 are used to calculate the reflected and transmitted power, respectively. For simplicity, a linearly polarized plane wave traveling in the $-z$ direction with the electric field polarized in the x direction is assumed. Adapted from [28]. Reproduced with permission from OSA The Optical Society.

to extract the needed information, for example, the optical power-absorption spectrum, the near-field distributions, and so on.

Here we will use a nanohole array structure as an example to demonstrate the application of the FEM model [28]. When light illuminates a nanohole array milled in a metallic film, some of the incident optical power is reflected, some is absorbed, and the rest is transmitted. To understand how the plasmonic resonance is generated at the

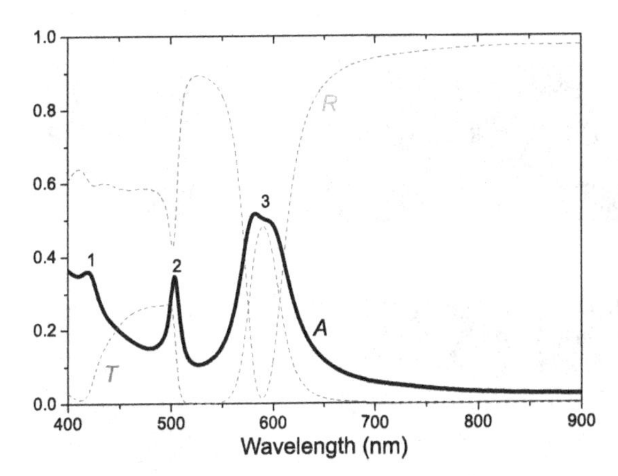

Figure 7.12 Simulated spectra of absorptance A, reflectance R, and transmittance T for a cylindrical hole array ($p = 500$ nm, $d = 250$ nm) fabricated in a 400-nm-thick free-standing Ag film. Adapted from [28]. Reproduced with permission from OSA The Optical Society.

surface of a patterned metal film, we study the nanohole array in a silver film suspended in air. Figure 7.11(a) depicts a unit cell of the nanohole array structure, where the pitch of the hole array is p, the diameter (and the depth) of each hole is d (and t). Owing to the geometrical symmetry, only a quarter of the unit cell needs to be simulated, as shown in Fig. 7.11(b).

In the simulation, the three-dimensional Maxwell equations are solved using the FEM. The material property of silver $\epsilon_{\text{silver}}(\lambda_0)$ is taken from Palik's handbook [29]. A normally incident plane-wave source ($\lambda_0 = 400$–900 nm) is set up at the surface S_0. For simplicity, a linearly polarized plane wave traveling in the $-z$ direction with the electric field polarized in the x direction is assumed. As the incident wave strikes the structure, it will be absorbed, reflected, and transmitted. The absorbed power P_{abs} is computed through the volume integration of the resistive heating (i.e. $\int_V \boldsymbol{J} \cdot \boldsymbol{E}\, \mathrm{d}V$) in the silver film, with $\boldsymbol{J}$ the current density and $\boldsymbol{E}$ the electric field. The reflected power P_{ref} and the transmitted power P_{tra} are calculated through the surface integration of the power flow (i.e. $\oint_S (\boldsymbol{E} \times \boldsymbol{H}) \cdot \mathbf{n}\, \mathrm{d}S$) over the surfaces S_1 and S_2, respectively. Here $\boldsymbol{E} \times \boldsymbol{H}$ represents the Poynting vector, and $\boldsymbol{n}$ is the unit normal to the surface S.

The absorptance A, reflectance R, and transmittance T are the ratios of the absorbed power, the reflected power, and the transmitted power to the incident power, respectively. The simulated spectra of absorptance, reflectance, and transmittance are plotted in Fig. 7.12, where the pitch of the hole array is 500 nm and the diameter (the depth) of each hole is 250 nm (400 nm). The absorptance spectrum of the nanohole array (A) exhibits three resonance peaks: peak 1 at 420 nm, peak 2 at 504 nm, and peak 3 at 582 nm.

To explore the physics behind this, the total electric field $|\boldsymbol{E}| = (E_x^2 + E_y^2 + E_z^2)^{1/2}$ distributions (V m^{-1}) in the x–y planes and in the z–x plane are shown in Fig. 7.13. The two x–y planes are cut at the front ($z = t$) and back ($z = 0$) silver/air interfaces, respectively, whereas the z–x plane is cut through the center of the nanohole ($y = 0$). From the plot, strong electric fields are observed around the rim of the nanohole for two absorption peaks: peaks 1 and 3, which show a distinct feature of the SPR. However,

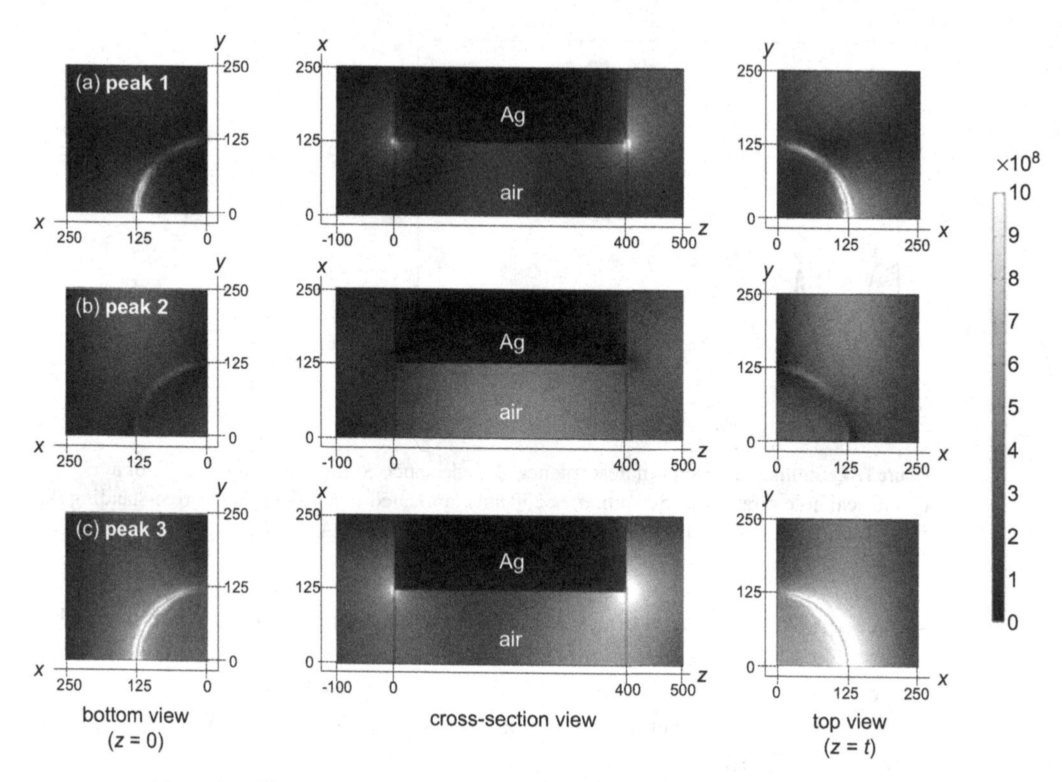

Figure 7.13 The electric-field distribution (V m^{-1}) from the bottom view (x–y plane, $z = 0$), the cross-section view (z–x plane, $y = 0$), and the top view (x–y plane, $z = t$), at three wavelengths: (a) 420 nm, (b) 504 nm, and (c) 582 nm, which correspond to the absorptance peaks 1, 2 and 3 shown in Fig. 7.12(b). Adapted from [28]. Reproduced with permission from OSA The Optical Society.

peak 2 does not show similar characteristics, which suggests that some other mechanism such as one involving the waveguide modes is responsible.

In the example (air/silver/air structure) presented above, we have demonstrated that the absorptance A, reflectance R, and transmittance T can be calculated, and the electromagnetic field distributions can also be plotted to show vividly what the plasmonic modes look like. In most sensing experiments, it is more convenient to measure the reflectance R. The resonant excitation of the surface plasmons is commonly observed as a dip in the reflectance at a specific wavelength. The dip in fact results from the absorption of optical energy in the metal, but the optical power absorption is difficult to measure. On the other hand, our FEM model can directly calculate the optical energy absorption, which does show a peak at the same resonance wavelengths. Therefore we can use either the reflectance R or the absorptance A, or even the transmittance T, to find the resonance wavelengths in our FEM model.

The surface plasmons are extremely sensitive to changes in the refractive index near the metal surface within the range of the surface plasmon field. Such a change may result in a shift in the resonance wavelength of the incident light or a change in the

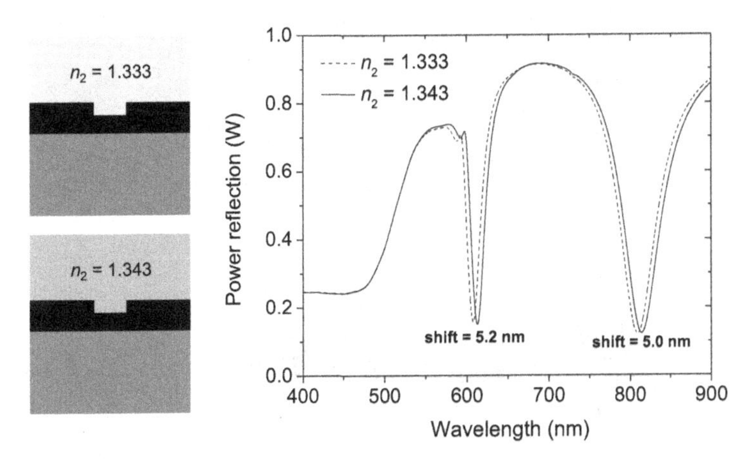

Figure 7.14 An example to show how to evaluate the sensitivity. Two independent simulations are performed with two different refractive indexes, 1.333 and 1.343. For the change in refractive index of 0.01, the resonance wavelength shift defines the sensitivity of a plasmon mode.

intensity of the reflected light. To theoretically predict the sensitivity to the refractive-index change, we will need to run a pair of simulations (with two different dielectric environment) to see this shift. Then the sensitivity is calculated by dividing the shift of the resonance wavelength by the refractive-index change, with the answer in units of nm per refractive-index unit (RIU).

Figure 7.14 shows an example of how the sensitivity can be evaluated, where simulations with $n_2 = 1.333$ and $n_2 = 1.343$ are performed separately. We plot their power reflection spectra on a single figure and can easily see the difference between the two calculations. From Fig. 7.14, we observe two surface plasmon modes in each simulation. The first mode around a resonance wavelength of 600 nm will shift by 5.2 nm and the second mode around 800 nm will shift by 5.0 nm. This corresponds to a sensitivity of around 500–520 nm per RIU. By applying the same approach, we are able to model and evaluate the sensitivity of any proposed sensing structure.

7.5 Plasmonic biosensor design examples

To demonstrate the applications of the two design methods introduced in Section 7.4, here we show three examples. The first one is based on the N-layer model, and the other two make use of the FEM model.

7.5.1 Graphene-based biosensor design

An SPR-based graphene biosensor in which a graphene sheet is coated above a gold thin film has been presented [30]. This graphene-on-gold structure was successfully fabricated in 2010 [31], confirming the feasibility of the proposed configuration. The proposed biosensor uses the attenuated-total-reflection (ATR) method to detect the refractive-index change near the sensor surface, which is due to the binding of

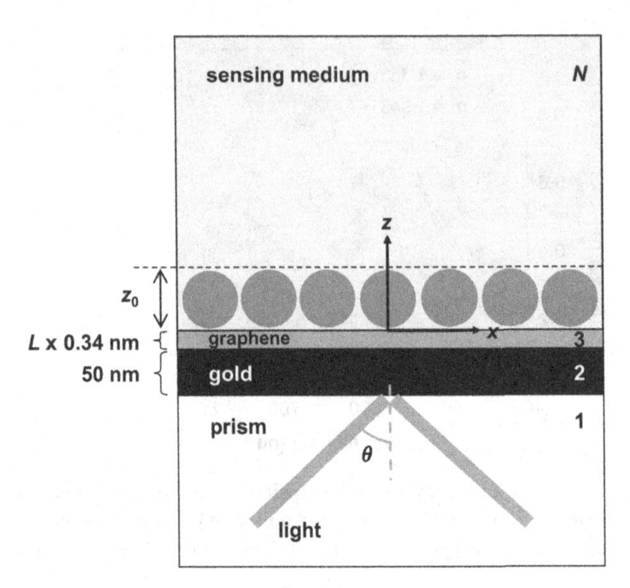

Figure 7.15 The N-layer model for the SPR biosensor: prism–Au (50 nm)–graphene ($L \times$ 0.34 nm)–sensing medium, where L is the number of graphene layers, and $z_0 = 100\,\mathrm{nm}$ is the thickness of the biomolecule layer. Adapted from [30]. Reproduced with permission from OSA The Optical Society.

biomolecules. Our calculations show that the proposed graphene-on-gold SPR biosensor (with L graphene layers) is $(1 + 0.025L)\Gamma$ (where $\Gamma > 1$) times more sensitive than the conventional gold thin-film SPR biosensor. The sensitivity improvement is due to the more efficient binding of biomolecules on graphene (represented by the factor Γ) as well as the optical property of graphene.

Figure 7.15 shows the configuration of the graphene-on-gold SPR biosensor, where the gold surface is covered with an SF10 glass prism. A light wave (of vacuum wavelength λ_0) passes through the prism and is totally reflected at the prism/metal interface, generating an evanescent wave. This evanescent wave can penetrate the thin gold layer (50 nm thick) and propagate along the x direction with a propagation constant of $k_x = n_{\text{prism}}(2\pi/\lambda_0)\sin\theta$, where n_{prism} is the refractive index of the prism, and θ is the angle of incidence. The propagation constant k_x can be adjusted to match that of the SPP by controlling the angle of incidence θ. The interaction of the light wave and the SPP can alter the characteristics of the light wave, such as the totally reflected intensity R. The plot of R versus the angle of incidence θ is called the SPR curve. To generate such a curve, we use the generalized N-layer model described in Section 7.4.1. The total reflection of the system for a TM wave is then

$$R = \left| \frac{(M_{11} + M_{12}q_N)q_1 - (M_{21} + M_{22}q_N)}{(M_{11} + M_{12}q_N)q_1 + (M_{21} + M_{22}q_N)} \right|^2 . \tag{7.14}$$

Here, the first layer is the prism with refractive index $n_1 = n_{\text{prism}}$. The kth layer (k runs from 2 to $N - 1$) has a thickness of d_k and a local dielectric function $\epsilon(\lambda_0)$ or refractive index $n(\lambda_0)$. In this study, we focus on the following system using a He–Ne laser to provide the incident light ($\lambda_0 = 633$ nm): layer one – prism ($n_1 = 1.723$); layer two – Au ($d_2 = 50$ nm, $n_2 = 0.1726 + 3.4218\text{i}$); layer three – graphene ($d_3 = L \times 0.34$ nm, $n_3 = 3 + 1.149\,106\text{i}$); and layer four – water ($n_4 = 1.33$), where L is the number of graphene layers. The biomolecular recognition element (i.e. graphene) will bind the biomolecules (e.g. ssDNA of size $z_0 = 100$ nm) present in the water, which produces an increment of the refractive index Δn on the graphene surface ($0 < z < z_0$).

Following the definition in Eq. (7.8), the sensitivity in our case is defined as the ratio of the change in the sensor output P, e.g. the angle of incidence, to the change in number of moles of biomolecules, M:

$$S^L = \frac{\Delta P^L}{\Delta M} = \frac{\Delta P^L}{\Delta n}\frac{\Delta n}{\Delta M} = S^L_{\text{RI}}E, \tag{7.15}$$

for different numbers of graphene layers L, where Δn is the change in refractive index. The overall sensitivity S^L consists of two components: (i) the sensitivity to refractive-index change $S^L_{\text{RI}} = \Delta P^L/\Delta n$; and (ii) the binding efficiency $E = \Delta n/\Delta M$, which represents the binding efficiency of the target biomolecule on the biomolecular-recognition element. Compared with the gold surface, graphene binds biomolecules more strongly and stably, due to the π-stacking interactions [32, 33] between graphene's hexagonal cells and the carbon-based ring structures in biomolecules. Therefore the enhanced binding efficiency using graphene, which is described by $E_{\text{graphene}} = \Gamma E_{\text{conventional}}$ (where $\Gamma > 1$), is one of the important mechanisms to increase the overall sensitivity. Note that the exact value of Γ requires experimental measurement.

In addition to the increased E, the coating of the graphene sheet can also increase the sensitivity to refractive-index change S^L_{RI}. Using the N-layer model, we calculate and show in Fig. 7.16(a) the SPR curves for the conventional biosensor ($L = 0$) (thin lines) and the monolayer graphene biosensor ($L = 1$) (thick lines) before (dashed lines) and after (solid lines) the binding of biomolecules, assuming that the same refractive-index change of $\Delta n = 0.005$ applies. For each SPR curve, under the resonance condition, the excitation of an SPP is recognized as a minimum in the totally reflected intensity R (i.e. the ATR minimum). The angle of incidence at the ATR minimum is called the SPR angle. We observe that the binding of biomolecules will shift the SPR curve toward a larger SPR angle. For example, there is an SPR angle shift of $\Delta P^0 = 0.26$ and $S^0_{\text{RI}} = 52$ for the conventional SPR biosensor, and a shift of $\Delta P^1 = 0.266$ and $S^1_{\text{RI}} = 53.2$ for the monolayer graphene biosensor as indicated in Fig. 7.16(a). With reference to the conventional SPR biosensor, of sensitivity S^0_{RI}, we plot the sensitivity enhancement $\Delta S^L_{\text{RI}} = S^L_{\text{RI}} - S^0_{\text{RI}}$ as a function of the number of graphene layers L in Fig. 7.16(b). It is found that the use of more graphene layers is able to provide an increased sensitivity. By fitting the curve in Fig. 7.16(b), we obtain a sensitivity enhancement of $1 + 0.025L$, i.e., $\Delta S^L_{\text{RI}}/S^0_{\text{RI}}$ can be as high as 25% for $L = 10$.

This improvement in sensitivity results from the optical property of graphene. At $\lambda_0 = 633$ nm, the refractive index of graphene is $n_{\text{graphene}} = 3 + 1.149\,106\text{i}$ [34], and the dielectric function of graphene is $\epsilon_{\text{graphene}} = 7.68 + 6.89\text{i}$. This means that graphene is a dielectric material at 633 nm. Coating of the gold surface with a dielectric film will

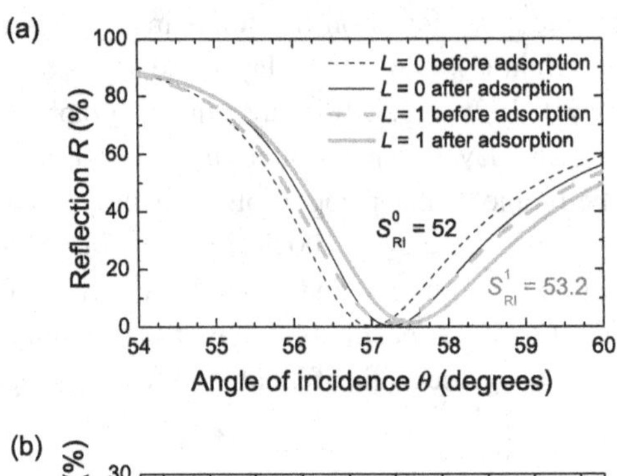

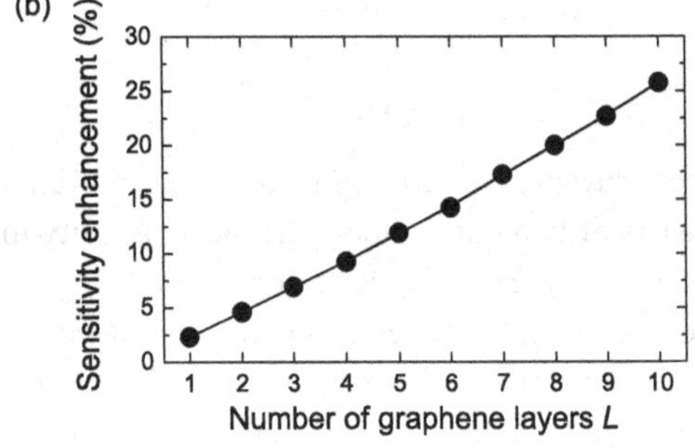

Figure 7.16 (a) The SPR curves for the conventional biosensor ($L = 0$) (black thin lines) and the monolayer graphene biosensor ($L = 1$) (gray thick lines) for He–Ne laser light ($\lambda_0 = 633$ nm): prism (1.723)–Au (50 nm, 0.1726 + 3.4218i)–graphene ($L \times 0.34$ nm, 3 + 1.149 106i)–water (1.33) before (dashed lines) and after (solid lines) the binding of biomolecules, assuming the same refractive index change $\Delta_n = 0.005$. (b) The sensitivity enhancement $\Delta S_{\mathrm{RI}}^{L}/S_{\mathrm{RI}}^{0}$ as a function of the number of graphene layers L. Adapted from [30]. Reproduced with permission from OSA The Optical Society.

change the power flow in the different layers and modify the field of the SPP. In such a way, the position of the SPR angle and the depth and width of the ATR minimum will change with increasing thickness of graphene, eventually leading to an increased S_{RI}. However, it is worth mentioning that adding more graphene layers will broaden the SPR curves, which may cause difficulties in SPR measurement and also degrade the FoM.

In summary, compared with a conventional gold thin-film SPR biosensor, the addition of a few graphene layers is able to increase the sensitivity by a factor of $(1 + 0.025L)\Gamma$ (where $\Gamma > 1$). The increased sensitivity results from two properties of graphene: (a) graphene binds biomolecules with carbon-based ring structures strongly and stably, so graphene can be used as the biomolecular-recognition element to enhance the binding efficiency E by a factor of Γ; and (b) graphene's optical property modifies the SPR curves and increases the sensitivity to refractive-index change S_{RI} by 25% for $L = 10$. Thus the overall sensitivity S will be increased fivefold if graphene binds four times ($\Gamma = 4$) as many biomolecules; likewise, if $\Gamma = 2$, S will be increased by a factor of 2.5. Since fabrication of graphene-on-Au (111) layers has already been successfully demonstrated using a transfer-printing technique, this design configuration is expected to be implemented easily [30].

7.5.2 Messenger RNA detection

All organisms use messenger RNA (mRNA) to carry genetic information for protein synthesis, including cancer cell proteins such as survivin proteins and PLK1 proteins. One of the cancer therapies, called "RNA interference (RNAi)," specifically destroys the mRNA of the cancer cell protein and thus controls the protein expression. To monitor the therapy or post-surgical recurrence, we need to detect the concentration of the targeted mRNA after the RNA interference. Ultimately, we want to develop a cancer diagnostic system that is based on plasmonic sensing, in order to monitor the concentration of the mRNA of a certain cancer cell protein (such as PLK1) with high accuracy, high speed, and low cost.

The most important part in a plasmonic sensor is the metal, where free electrons oscillate collectively and form the surface plasmons. Therefore patterning the metal film into different nanostructures will dramatically influence the surface plasmon modes, and affect the sensitivity to the environment's refractive index. In this section, we will take four patterns as examples and compare their performances.

Figure 7.17 shows the four metallic nanostructures and the corresponding spectra of optical power reflection: (a) a grating, (b) a nano-slit array, (c) a nanohole array, and (d) a nanoisland array. Initially, with a water sensing medium (refractive index 1.333), there are a few fundamental resonances for each structure. For instance, the nano-slit array has the α, β, and γ modes as shown in Fig. 7.17(b). After the 50-nm-sized PLK1 mRNA molecules (with an assumed refractive index of 1.343) are bound to the metallic structures, all the resonances are red-shifted. For example, the α, β, and γ modes in the nano-slit array have red-shifted by 1.6 nm, 0.2 nm, and 0.8 nm, respectively. Among the four structures, the most shifted resonances are observed in the slit array and the grating structure, while the nanoisland array shows the least shift.

In general, the nanohole and nanoisland arrays are two-dimensional, so the oscillation of free electrons at the metal–dielectric boundary is restricted in the two planar directions [6]. Localized surface plasmon modes are favorable under these restrictions, with very limited penetrations of electromagnetic field into the dielectric sensing medium. On the other hand, the nano-slit array and grating structures are one-dimensional. Propagating surface plasmonic modes can be sustained at the centers of metal strips. These propagating modes will have much greater field penetration depth. In addition, they originate from the centers of the metal strips (not the edges), where most of the biomolecules are bound. As a result, the one-dimensional nano-slit array and the grating structure have larger resonance shifts than do the two-dimensional nanohole and nanoisland arrays in detecting the PLK1 mRNA (of moderate size, 50 nm).

Nevertheless, there are still some differences between the slit array and the grating structures. The slit array has an additional transmission scheme (light illuminates from the substrate side). But a grating only has the conventional reflection scheme (light illuminates from the analyte side) in which the scattering of the aqueous medium causes noise. In addition, the slit array allows the analytes to flow through the metal film to improve the binding efficiency [26], whereas this cannot occur in the case of a grating. Moreover, with a slit array one may use thinner metal film in the transmission scheme

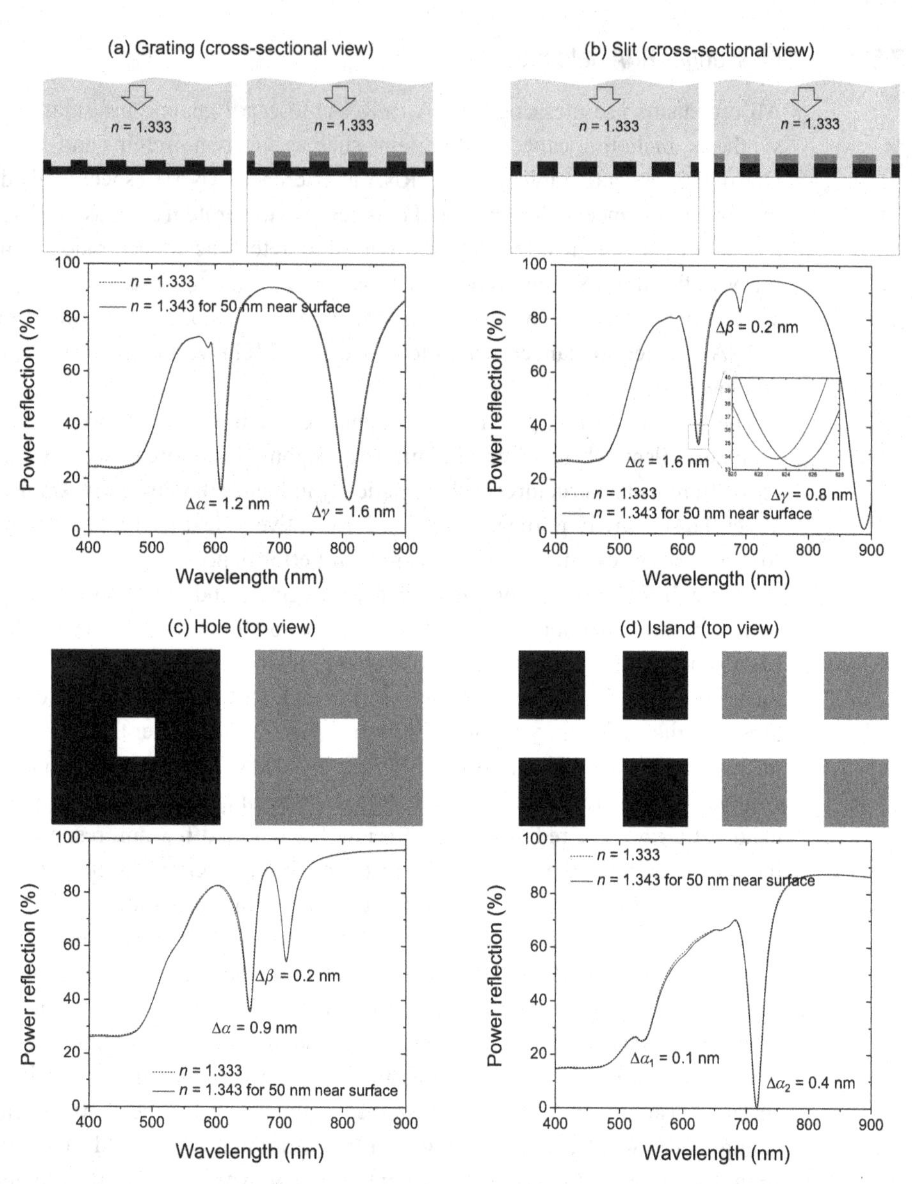

Figure 7.17 The simulated resonance shift upon binding 50-nm-sized PLK1 mRNA molecules for four structures: (a) a grating, (b) a nano-slit array, (c) a nanohole array, and (d) a nanoisland array. In all four structures, the metal film is 100 nm thick and the array's pitch is 400 nm. The grating is 25 nm wide and 50 nm deep; the slit is 25 nm wide; the nanohole is of dimensions 100 nm × 100 nm; and the nanoisland is of dimensions 300 nm × 300 nm. For each structure, the resonance shift is calculated by assuming a local refractive index of 1.343 (50 nm above the metal surface), on top of a background refractive index of water (1.333).

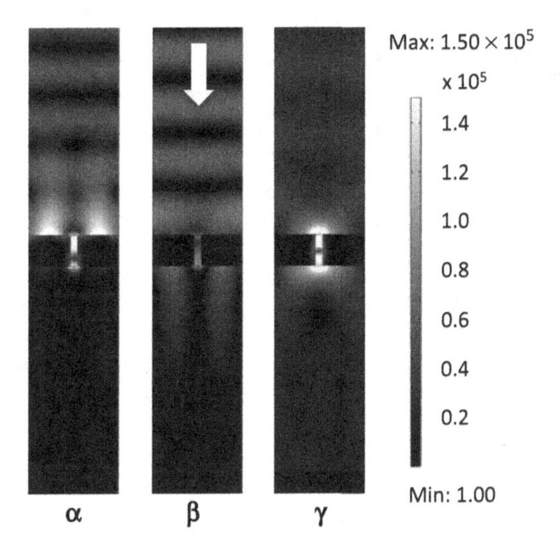

Figure 7.18 The simulated near field distributions for the α, β, and γ modes in the slit array (cross-sectional view). Light is incident from the analyte side (reflection scheme).

to save on the cost of materials [35]. Hence our study will focus on the slit array structure.

For the slit array structure, the near-field distributions were studied in order to understand why the α mode has the largest shift. Figure 7.18 compares the near-field distributions of the α, β, and γ modes in a cross-sectional view for the slit array. From the field distributions, α is a propagating mode excited at the water/metal interface; β is a propagating mode excited at the substrate/metal interface; and γ is a localized mode excited around the slits' edges and corners. The refractive-index change (due to binding of RNA molecules) usually occurs on the water side of the metal. Since the α mode originates from the water–metal surface, the largest overlap is between the electromagnetic fields of the α mode and the PLK1 mRNA molecules. As a result, the α mode is the most sensitive mode to the refractive-index change.

Furthermore, the effects of the size and refractive index of the analytes (which could be any mRNA molecules rather than PLK1 mRNA) on the resonance shift (in nm) were investigated with the slit array structure. Figure 7.19 shows the resonance shift as a function of the size and refractive index of the analytes. The shift varies linearly with the refractive index, that is, the sensitivity (defined as the resonance shift divided by the refractive-index change) is constant. On the other hand, the shift varies nonlinearly with the analyte's size. This is due to the electromagnetic fields of the plasmonic mode not being able to cover the whole analyte if the analyte is too large [36]. This indicates that, although the sensitivity increases when the size of the molecule increases, the rate of incrementation will slow down.

Finally, the effects of the dimensions of the nano-slit array structure on the resonance shift were studied. We find that both the plasmonic modes and the resonance shift are sensitive to changes of pitch, width, and thickness of the slit array structures. In

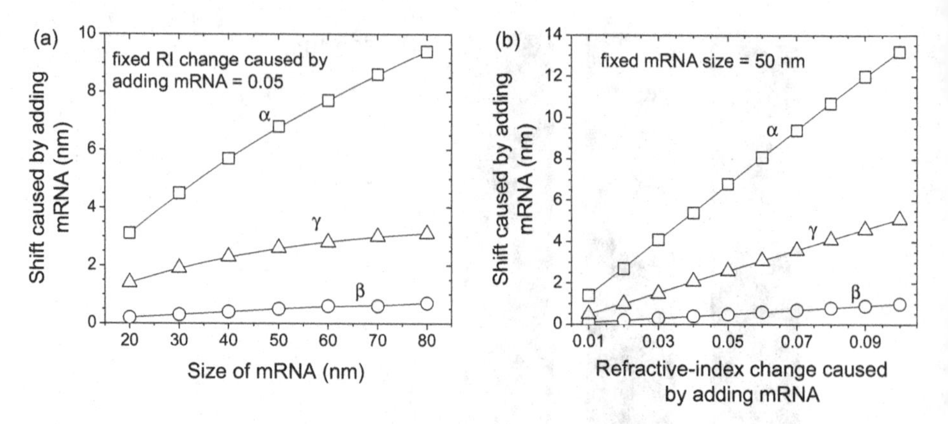

Figure 7.19 The effects of the analyte size (a) and the analyte-induced refractive-index change (b) on the shift of resonance.

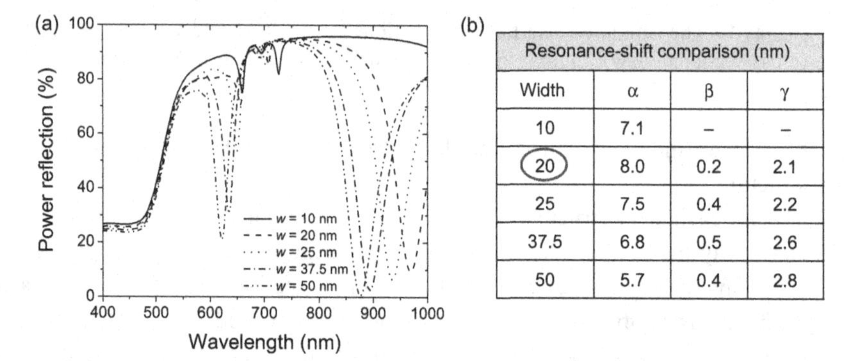

Resonance-shift comparison (nm)			
Width	α	β	γ
10	7.1	–	–
20	8.0	0.2	2.1
25	7.5	0.4	2.2
37.5	6.8	0.5	2.6
50	5.7	0.4	2.8

Figure 7.20 The effects of the width on the plasmonic modes (a) of, and the resonance shift (b) in, the slit array structure, where the slit pitch is 450 nm and the thickness of the metal film is 150 nm.

Fig. 7.20(a) we show how the power-reflection spectrum changes with the width of the slit array, where the slit pitch is 450 nm and the thickness of the metal film is 150 nm. All three plasmonic modes are blue-shifted when the slit width increases. For each slit width, these modes will be red-shifted upon binding the analytes (assuming a refractive-index change of 0.05). The shifts of each mode (in nm) for all the widths are shown in Fig. 7.20(b). The resonance shift is maximized with the 20-nm-wide slits.

In summary, we have shown an example of designing an appropriate plasmonic sensor for mRNA detection. Four structures, namely the grating, nano-slit array, nanoisland array, and nanohole array have been studied with the comprehensive FEM modeling method. The studies show that the nano-slit array and grating structures have much higher sensitivities than those of the nanoisland and nanohole arrays for detecting 50-nm-sized PLK1 mRNA. This is attributed to the propagating surface plasmon mode supported by the one-dimensional nano-slit array or grating structures, which possesses a larger field penetration depth to cover the entire PLK1 mRNA molecules. These

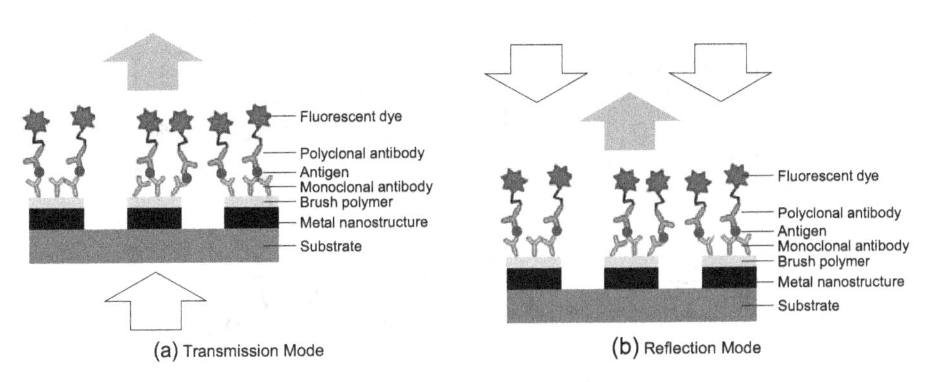

Figure 7.21 Highly sensitive LSPR detection with fluorescent dyes in (a) transmission and (b) reflection mode.

structures can be further optimized to detect 50-nm-sized PLK1 mRNA or other types of mRNA.

7.5.3 Point-of-care clinical screening of PSA

The third example is to demonstrate the feasibility of LSPR technology for use in clinical point-of-care (POC) as well as highly sensitive and highly selective laboratory-based biomedical device platforms for the detection of prostate-specific antigen (PSA), which is widely accepted as the biomarker for prostate cancer. The focus here is to enhance the detection sensitivity of existing LSPR technology, such as adding fluorescence tags.

This approach is illustrated in Fig. 7.21, where monoclonal PSA antibodies are specifically bound on gold nanostructures by synthesizing a non-fouling oligo ethylene glycol methacrylate (OEGMA) brush polymer on the gold nanostructures. Then these monoclonal antibodies will bind the PSA antigens in the sample. Next, the PSA antigens will capture the secondary polyclonal PSA antibodies (which are labeled with fluorescent dyes). These fluorescent dyes are excited by the plasmonic signal generated by the gold nanostructures on the substrate, whose intensity is proportional to the PSA concentration in the sample.

The detection should be conducted under dark-field conditions in order to improve the signal-to-noise ratio. The detector can be a photomultiplier tube (PMT), a photodiode, or a charge-coupled-device (CCD) camera, and the light source can be white light, a bright light-emitting diode (LED), or a laser. In general, there are two detection modes: transmission and reflection. In transmission mode, as illustrated in Fig. 7.21(a), the light source is illuminated from the substrate side and the fluorescent signal is collected at the analyte side. In the reflection mode, illustrated in Fig. 7.21(b), a dark-field lens can be used to illuminate the light from the side and collect the fluorescence at the center.

In this way, the transmission mode provides a larger space (at the analyte side) for adding optical filters to eliminate the unwanted scattered light and increase the signal-to-noise ratio. However, the surface plasmon excitation is usually stronger at the side of the light incidence, so the fluorescent molecules could be more strongly excited in the

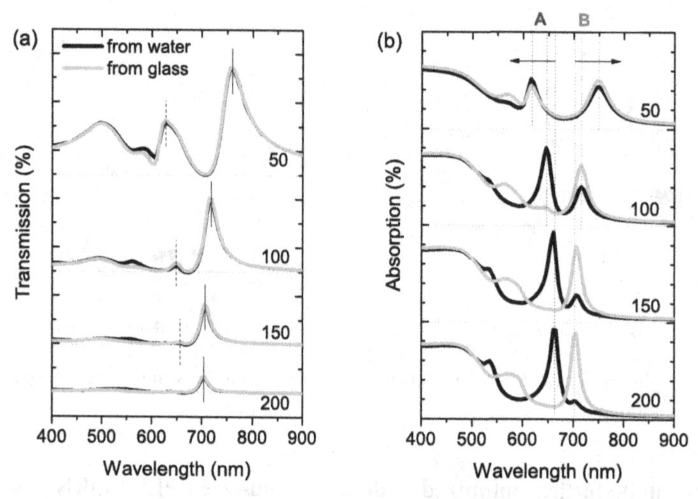

Figure 7.22 Spectra of (a) transmission and (b) absorption for nanohole arrays (of pitch 400 nm and diameter 150 nm) milled in gold films of thicknesses 50, 100, 150, and 200 nm, for which the light illuminates the structures from the water (black) or glass (gray) side. Adapted from [35]. Reproduced with permission from the IEEE.

reflection mode. Given these pros and cons, it is difficult to decide which mode should be adopted in the engineering design of a sensor. Additional information acquired from the simulation is required [35].

The present study is based on a water–gold–glass structure, where a nanohole array is patterned in a gold film that is deposited onto a glass substrate and the array is assumed in the water environment. The square-lattice nanohole array is fully characterized by three parameters: the array pitch, the hole diameter, and the thickness of the gold film. In this study, our focus is to determine which side of the structure should be illuminated, and the main influencing parameter is the thickness of the gold film. The other two parameters (pitch and diameter) tune the spectral location of the resonance and adjust the optical power distributions in the planes perpendicular to the direction in which light is propagating [28]. Assuming the fluorescent molecules' excitation wavelength is 647 nm, we have optimized our nanohole array structure to have a pitch of 400 nm and a hole diameter of 150 nm. Then various thicknesses of gold film will be examined carefully both in reflection and in transmission mode.

Figure 7.22 shows the transmission and absorption spectra for nanohole arrays milled in gold films of thicknesses 50, 100, 150, and 200 nm. In the calculations, two illumination conditions are considered: from the water (black) and from the glass (gray) side. It is clear from Fig. 7.22(a) that the transmission spectrum varies with the thickness of gold film both for reflection and for transmission mode. As the film thickness is reduced from 200 nm to 50 nm gradually, the transmission peak around 700 nm undergoes a red-shift and becomes more prominent, and the other transmission peak around 650 nm gradually appears and experiences a blue-shift. More importantly, we observe that the transmission spectrum is invariant regardless of which side of the structure is

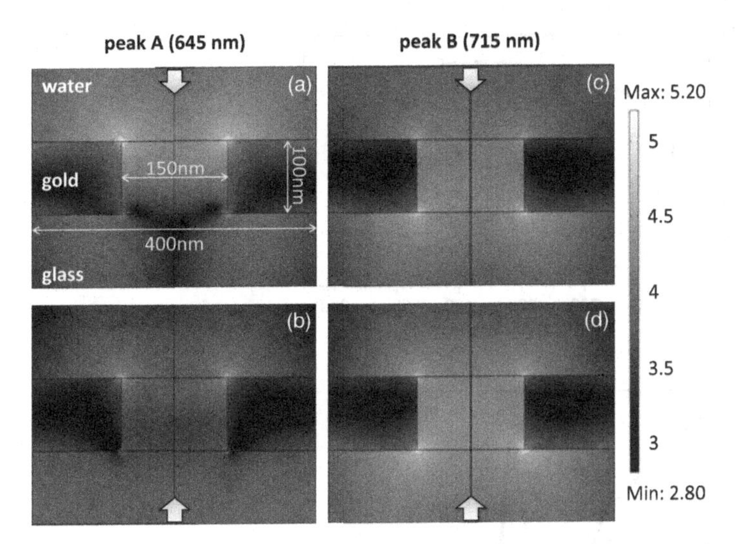

Figure 7.23 Cross-sectional views of the optical field distributions ($V\,m^{-1}$) (on a logarithmic scale) for two cases in which the light illuminates the sample from the water, (a) and (c), or glass, (b) and (d), side at two peak wavelengths, labeled as peak A, (a) and (b), and peak B, (c) and (d). The hole array under analysis is composed of many 150-nm-diameter and 100-nm-deep holes arranged with pitch 400 nm. Adapted from [35]. Reproduced with permission from the IEEE.

illuminated. This finding is consistent with the experimental observations in the pioneering paper by T. W. Ebbesen [7], and agrees with the classical Helmholtz reciprocity principle [37].

On the other hand, the absorption spectrum shown in Fig. 7.22(b) gives us different but more interesting information. The illumination condition does influence the absorption spectrum. There are two major absorption peaks: peaks A are at about 610–660 nm and peaks B at about 700–750 nm. For a greater film thickness such as 200 nm, peak A is dominant when illuminating from the water side, whereas peak B is dominant when illuminating from the glass side. However, for a smaller film thickness such as 50 nm, both peak A and peak B are present.

To better understand peaks A and B in the absorption spectrum, we pick the 100-nm-thick hole array and show its near-field distribution in Fig. 7.23 (cross-sectional view). Among the four plots, (a) and (c) represent the case in which the sample is illuminated from the water side, and (b) and (d) represent the other case. Moreover, (a) and (b) show the optical fields at peak A (resonance wavelength 645 nm), and (c) and (d) are for peak B (715 nm). We find that the plots corresponding to peak A always show evidence of the SPR at the water/gold interface. The signal in (a) is stronger than that in (b), which means that illuminating from the water side leads to stronger SPR at the water/gold interface. On the other hand, plots (c) and (d), corresponding to peak B, exhibit SPR primarily at the glass/gold interface and secondarily at the water/gold interface, because the signal at the glass/gold interface is stronger. This is true no matter which side is illuminated. However, the signal in (d) is slightly stronger than that in (c).

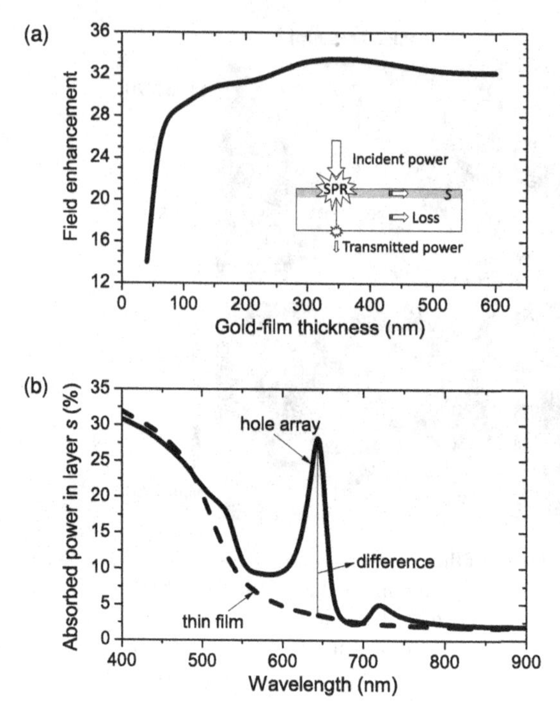

Figure 7.24 Illumination from the water side. (a) The peak-A optical field enhancement γ at the rim of holes of the water/gold interface is shown as a function of the gold-film thickness. The inset shows the optical power flow when light is illuminating from the water side. (b) The absorbed optical power in layer S (i.e. the 10-nm-thick gold surface layer near the water/gold interface), which is represented by the hatched area in the inset of (a), for a 100-nm-thick nanohole array (solid line) and a 100-nm-thick thin film (dashed line). Adapted from [35]. Reproduced with permission from the IEEE.

Knowing that peak A correlates to SPR at the water/gold interface and peak B correlates to SPR at the glass/gold interface, the present study focuses on the former because the analyte (attached with the fluorescent molecules) is bound at the water/gold interface. Besides, in the measurement, we need to optimize the optical fields around the fluorescent labels in order to excite them more efficiently. Hence, in the following studies, we will focus on the optical field enhancement $\gamma = E/E_{0x}$ at the resonance wavelength of peak A, where E_{0x} is the electric-field component of the incident light wave.

Figure 7.24(a) shows for the reflection mode a plot of the optical field enhancement γ as a function of the gold-film thickness, where γ is evaluated at the rim of the holes near the water/gold interface. It is found that γ increases with the film thickness rapidly before it saturates around a value of 32 with a variation of 3.5% when a critical film thickness of 150 nm is reached.

In the reflection mode, the incident optical power first excites the SPR at the water/gold interface, next passes through the bulk metal film with some optical loss (which is related to the imaginary part of the metal's refractive index), and then induces some weak response at the glass/gold interface before it finally transmits out. To get more information on the SPR at the water/gold interface, we specifically calculate the absorbed optical power in the 10-nm-thick gold surface layer near the water/gold interface, i.e. the layer S (hatched area) in the inset of Fig. 7.24(a). The absorbed optical power in layer S consists of the power used to excite the water/gold SPR and the optical loss in that 10-nm-thick layer. To demonstrate this, in Fig. 7.24(b), we plot the absorbed power in layer S for a 100-nm-thick nanohole array and a 100-nm-thick thin film. We know that the surface-layer power absorption for the thin-film structure is purely due to the optical loss. Thus the difference in surface-layer power absorption between the hole array and the thin film is the optical power to excite the water/gold SPR, assuming that the loss in their surface layers is the same.

We did the same analysis for other thicknesses (not shown) and find that the trend of layer-S power absorption is the same as that of the total power absorption, which is shown in Fig. 7.22(b) (black curves). As the gold film becomes thinner, the resonance peak A continues to decrease in magnitude, while the optical loss in the 10-nm-thick surface layer is independent of the film thickness. This suggests that less optical power is used to excite the water/gold SPR as the film thickness decreases. Consequently, a relatively thick film ($>$150 nm) is required in the reflection mode.

On the other hand, if the biosensor is operated in the transmission mode, as shown in Fig. 7.25(a), we observe that the optimized film thickness to maximize γ is much thinner, namely about 60 nm. In this case, the light is incident on the glass side and passes through the gold film to the water side; only part of the optical power (i.e. that which has reached layer S) is used to excite the water/gold SPR, as illustrated in the inset of Fig. 7.25(a). Intuitively it is expected that an extremely thin film that is even optically transparent is desirable. Our Fig. 7.22(b) (gray lines) clearly demonstrates this point, since peak A's magnitude increases continuously with decreasing film thickness.

However, this is true only when the film thickness is reduced to 60 nm. Beyond 60 nm, the value magnitude of the optical field enhancement starts to drop, as shown in Fig. 7.25(a). To find out why this happens, we plot the S-layer power absorption for the 60-nm-thick and 30-nm-thick hole arrays (solid lines) and thin films (dashed lines) in Fig. 7.25(b). We find that the 30-nm-thick thin film allows more optical power to reach the 10-nm-thick gold surface layer, as the absorbed power in layer S is substantially increased. This is because 30 nm is very close to the skin depth of gold in the visible spectrum. However, most of the absorbed optical power is not contributing to the resonant excitation of SPR, as we see that the difference in surface-layer power absorption between a hole array and a thin film is reduced as the film thickness decreases from 60 nm to 30 nm. In short, if one is illuminating from the glass side, a much thinner (but thicker than the skin depth) gold film must be used. From another perspective, using a thinner gold film is good for saving on material costs. Comparing thicknesses of 150 nm for reflection mode and 60 nm for transmission mode, 60% less gold film is used.

Besides knowing the optimized film thickness for each mode, the magnitude of the optical field enhancement γ is also important, because the excitation efficiency of the fluorescent molecules is proportional to the plasmonic optical power (i.e. the square of the optical field). As shown in Figs. 7.24(a) and 7.25(a), γ is saturated at a value of about 32 when illuminating from the water side and maximized at about 22 when illuminating from the glass side. In terms of the plasmonic optical power (which is proportional to the

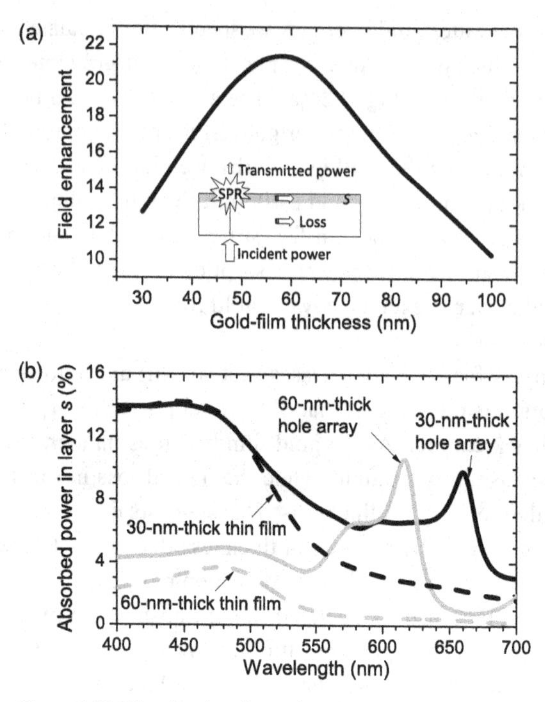

Figure 7.25 Illumination from the glass side. (a) The peak-A optical field enhancement γ at the rim of holes of the water/gold interface is shown as a function of the gold-film thickness. The inset shows the optical power flow when light is illuminating from the glass side. (b) The absorbed optical power in the layer S (i.e. the 10-nm-thick gold surface layer near the water/gold interface), which is represented by the hatched area in the inset of (a), for 30- and 60-nm-thick nanohole arrays (solid lines) and 30- and 60-nm-thick thin films (dashed lines). Adapted from [35]. Reproduced with permission from the IEEE.

number of photons per second), the reflection mode is twice as large as the transmission mode, which could generate much stronger fluorescence signals. Fundamentally, this is because in reflection mode the incoming optical power is directly used to excite the water/gold SPR, but in transmission mode the incoming optical power first suffers the loss and then excites the SPR.

In summary, we have presented the design procedure of choosing an appropriate detection mode (reflection or transmission) for PSA screening, through an in-depth study on exciting the SPR of a water–gold (holes)–glass structure illuminated from either the water side (reflection mode) or the glass side (transmission mode). Our studies show that both modes can excite the resonance at the water/gold interface, which is used in surface plasmon fluorescence spectroscopy. This resonance correlates different absorption spectra and optical field distributions when changing the illumination direction or the gold-film thickness, due to the changed optical power flow and surface-layer power absorption. In the reflection mode, a thicker gold film (>150 nm) is preferred, whereas a thinner film (60 nm) is necessary in the transmission mode, in order to optimize the sensing performance. The reflection mode is superior in providing larger optical field enhancements (32 compared with 22), i.e. in terms of the plasmonic optical power to generate the fluorescence signal, it is twice as large as the transmission mode. However, the transmission mode has more space to add optical filters to increase the signal-to-noise ratio in real engineering designs, and it also uses much thinner gold films (60% less material), so the material costs are much lower. These findings could assist people in determining the favorable illumination direction and gold-film thickness in the design of metallic-nanostructure-based plasmonic sensing systems.

References

[1] S. A. Maier, *Plasmonics: Fundamentals and Applications*. Berlin: Springer, 2007.

[2] J. Homola, "Present and future of surface plasmon resonance biosensors," *Anal. Bioanal. Chem.*, vol. 377, pp. 528–539, 2003.

[3] I. Abdulhalim, M. Zourob, and A. Lakhtakia, "Surface plasmon resonance for biosensing: A mini-review," *Electromagnetics*, vol. 28, pp. 214–242, 2008.

[4] J. N. Anker, W. P. Hall, O. Lyanders *et al.*, "Biosensing with plasmonic nanosensors," *Nature Mater.*, vol. 7, pp. 442–453, 2008.

[5] M. E. Stewart, C. R. Anderton, L. B. Thompson *et al.*, "Nanostructured plasmonic sensors," *Chem. Rev.*, vol. 108, pp. 494–521, 2008.

[6] B. Sepúlveda, P. C. Angelomé, L. M. Lechuga, and L. M. Liz-Marzán, "LSPR-based nanobiosensors," *Nano Today*, vol. 4, pp. 244–251, 2009.

[7] T. W. Ebbesen, H. J. Lezec, H. F. Ghaemi, T. Thio, and P. A. Wolff, "Extraordinary optical transmission through sub-wavelength hole arrays," *Nature*, vol. 391, pp. 667–669, 1998.

[8] H. F. Ghaemi, T. Thio, D. E. Grupp, T. W. Ebbesen, and H. J. Lezec, "Surface plasmons enhance optical transmission through subwavelength holes," *Phys. Rev. B*, vol. 58, pp. 6779–6782, 1998.

[9] A. Degiron, H. J. Lezec, W. L. Barnes, and T. W. Ebbesen, "Effects of hole depth on enhanced light transmission through subwavelength hole arrays," *Appl. Phys. Lett.*, vol. 81, pp. 4327–4329, 2002.

[10] W. L. Barnes, W. A. Murray, J. Dintinger, E. Devaux, and T. W. Ebbesen, "Surface plasmon polaritons and their role in the enhanced transmission of light through periodic arrays of subwavelength holes in a metal film," *Phys. Rev. Lett.*, vol. 92, 107401, 2004.

[11] J. Prikulis, P. Hanarp, L. Olofsson, D. Sutherland, and M. Kall, "Optical spectroscopy of nanometric holes in thin gold films," *Nano Lett.*, vol. 4, pp. 1003–1007, 2004.

[12] H. J. Lezec and T. Thio, "Diffracted evanescent wave model for enhanced and suppressed optical transmission through subwavelength hole arrays," *Opt. Express*, vol. 12, pp. 3629–3651, 2004.

[13] A. Degiron and T. W. Ebbesen, "The role of localized surface plasmon modes in the enhanced transmission of periodic subwavelength apertures," *J. Opt. A: Pure Appl. Opt.*, vol. 7, pp. S90–S96, 2005.

[14] C. Genet and T. W. Ebbesen, "Light in tiny holes," *Nature*, vol. 445, pp. 39–46, 2007.

[15] F. J. Garcia-Vidal, L. Martin-Moreno, T. W. Ebbesen, and L. Kuipers, "Light passing through subwavelength apertures," *Rev. Mod. Phys.*, vol. 82, pp. 729–787, 2010.

[16] T. Sannomiya, O. Scholder, K. Jefimovs, C. Hafner, and A. B. Dahlin, "Investigation of plasmon resonances in metal films with nanohole arrays for biosensing applications," *Small*, vol. 7, pp. 1653–1663, 2011.

[17] K.-L. Lee, C.-W. Lee, W.-S. Wang, and P.-K. Wei, "Sensitive biosensor array using surface plasmon resonance on metallic nanoslits," *J. Biomed. Opt.*, vol. 12, 044023, 2007.

[18] N. Liu, T. Weiss, M. Mesch *et al.*, "Planar metamaterial analogue of electromagnetically induced transparency for plasmonic sensing," *Nano Lett.*, vol. 10, pp. 1103–1107, 2010.

[19] P. Offermans, M. C. Schaafsma, S. R. K. Rodriguez *et al.*, "Universal scaling of the figure of merit of plasmonic sensors," *ACS Nano*, vol. 5, pp. 5151–5157, 2011.

[20] Z. Fang, J. Cai, Z. Yan *et al.*, "Removing a wedge from a metallic nanodisk reveals a Fano resonance," *Nano Lett.*, vol. 11, pp. 4475–4479, 2011.

[21] W.-S. Chang, J. B. Lassiter, P. Swanglap *et al.*, "A plasmonic Fano switch," *Nano Lett.*, vol. 12, pp. 4977–4982, 2012.

[22] Y. Wang, A. Brunsen, U. Jonas, J. Dostlek, and W. Knoll, "Prostate specific antigen biosensor based on long range surface plasmon-enhanced fluorescence spectroscopy and dextran hydrogel binding matrix," *Anal. Chem.*, vol. 81, pp. 9625–9632, 2009.

[23] X. F. Li and S. F. Yu, "Extremely high sensitive plasmonic refractive index sensors based on metallic grating," *Plasmonics*, vol. 5, pp. 389–394, 2010.

[24] M. J. Levene, J. Korlach, S. W. Turner *et al.*, "Zero-mode waveguides for single-molecule analysis at high concentrations," *Science*, vol. 299, pp. 682–686, 2003.

[25] F. Eftekhari, C. Escobedo, J. Ferreira *et al.*, "Nanoholes as nanochannels: Flow-through plasmonic sensing," *Anal. Chem.*, vol. 81, pp. 4308–4311, 2009.

[26] A. A. Yanik, M. Huang, A. Artar, T. Y. Chang, and H. Altug, "Integrated nanoplasmonic–nanofluidic biosensors with targeted delivery of analytes," *Appl. Phys. Lett.*, vol. 96, 021101, 2010.

[27] M. Yamamoto, "Surface plasmon resonance (SPR) theory: Tutorial," *Rev. Polarography*, vol. 48, pp. 209–237, 2002.

[28] L. Wu, P. Bai, and E. P. Li, "Designing surface plasmon resonance of subwavelength hole arrays by studying absorption," *J. Opt. Soc. Am. B*, vol. 29, pp. 521–528, 2012.

[29] E. D. Palik, Ed., *Handbook of Optical Constants of Solids*. San Diego, CA: Academic Press, 1991.

[30] L. Wu, H. S. Chu, W. S. Koh, and E. P. Li, "Highly sensitive graphene biosensors based on surface plasmon resonance," *Opt. Express*, vol. 18, pp. 14395–14400, 2010.

[31] B. Song, D. Li, W. P. Qi *et al.*, "Graphene on Au(111): A highly conductive material with excellent adsorption properties for high-resolution bio/nanodetection and identification," *ChemPhysChem*, vol. 11, pp. 585–589, 2010.

[32] G. B. McGaughey, M. Gagné, and A. K. Rappé, "π-Stacking interactions alive and well in proteins," *J. Biol. Chem.*, vol. 273, pp. 15458–15463, 1998.

[33] Z. Tang, H. Wu, J. R. Cort *et al.*, "Constraint of DNA on functionalized graphene improves its biostability and specificity," *Small*, vol. 6, pp. 1205–1209, 2010.

[34] M. Bruna and S. Borinia, "Optical constants of graphene layers in the visible range," *Appl. Phys. Lett.*, vol. 94, 031901, 2009.

[35] L. Wu, P. Bai, X. Zhou, and E. P. Li, "Reflection and transmission modes in nanohole-array-based plasmonic sensors," *IEEE Photon. J.*, vol. 3, pp. 441–449, 2012.

[36] A. A. Yanik, M. Huang, O. Kamohara *et al.*, "An optofluidic nanoplasmonic biosensor for direct detection of live viruses from biological media," *Nano Lett.*, vol. 10, pp. 4962–4969, 2010.

[37] B. Hapke, *Theory of Reflectance and Emittance Spectroscopy*. Cambridge: Cambridge University Press, 1993.

Index

Printed in the United States
by Baker & Taylor Publisher Services